T0245457

CAMBRIDGE LIBRARY COLLECTION

Books of enduring scholarly value

Life Sciences

Until the nineteenth century, the various subjects now known as the life sciences were regarded either as arcane studies which had little impact on ordinary daily life, or as a genteel hobby for the leisured classes. The increasing academic rigour and systematisation brought to the study of botany, zoology and other disciplines, and their adoption in university curricula, are reflected in the books reissued in this series.

My Life

Alfred Russel Wallace (1823–1913) was a British naturalist, explorer, geographer and biologist, best remembered as the co-discoverer, with Darwin, of natural selection. His extensive fieldwork and advocacy of the theory of evolution led to him being considered one of the nineteenth century's foremost biologists. He was later moved by a variety of personal experiences to examine the concept of spirituality, but his exploration into the potential for compatibility between spiritualism and natural selection alienated him from the scientific community. He was also a social activist, highly critical of unjust social and economic systems in nineteenth-century Britain, and one of the first prominent scientists to express concern over the environmental impact of human activity. This autobiography was first published in 1905. Volume 1 covers his childhood, his early social activism, and his expeditions to the Amazon and the Malay archipelago, which established his reputation.

Cambridge University Press has long been a pioneer in the reissuing of out-of-print titles from its own backlist, producing digital reprints of books that are still sought after by scholars and students but could not be reprinted economically using traditional technology. The Cambridge Library Collection extends this activity to a wider range of books which are still of importance to researchers and professionals, either for the source material they contain, or as landmarks in the history of their academic discipline.

Drawing from the world-renowned collections in the Cambridge University Library, and guided by the advice of experts in each subject area, Cambridge University Press is using state-of-the-art scanning machines in its own Printing House to capture the content of each book selected for inclusion. The files are processed to give a consistently clear, crisp image, and the books finished to the high quality standard for which the Press is recognised around the world. The latest print-on-demand technology ensures that the books will remain available indefinitely, and that orders for single or multiple copies can quickly be supplied.

The Cambridge Library Collection will bring back to life books of enduring scholarly value (including out-of-copyright works originally issued by other publishers) across a wide range of disciplines in the humanities and social sciences and in science and technology.

My Life

A Record of Events and Opinions

VOLUME 1

ALFRED RUSSEL WALLACE

CAMBRIDGE
UNIVERSITY PRESS

CAMBRIDGE UNIVERSITY PRESS

Cambridge, New York, Melbourne, Madrid, Cape Town,
Singapore, São Paolo, Delhi, Tokyo, Mexico City

Published in the United States of America by Cambridge University Press, New York

www.cambridge.org
Information on this title: www.cambridge.org/9781108029582

© in this compilation Cambridge University Press 2011

This edition first published 1905
This digitally printed version 2011

ISBN 978-1-108-02958-2 Paperback

MY LIFE

Alfred R Wallace

MY LIFE

A RECORD OF EVENTS AND OPINIONS

BY

ALFRED RUSSEL WALLACE

AUTHOR OF

"MAN'S PLACE IN THE UNIVERSE," "THE MALAY ARCHIPELAGO," "DARWINISM,"
"GEOGRAPHICAL DISTRIBUTION OF ANIMALS," "NATURAL
SELECTION AND TROPICAL NATURE," ETC.

*WITH FACSIMILE LETTERS, ILLUSTRATIONS
AND PORTRAITS*

TWO VOLUMES

VOLUME I

LONDON: CHAPMAN & HALL, LD.

1905

PREFACE

THE present volumes would not have been written
had not the representatives of my English and
American publishers assured me that they would
probably interest a large number of readers.

I had indeed promised to write some account of
my early life for the information of my son and
daughter, but this would have been of very limited
scope, and would probably not have been printed.

Having never kept a diary, except when abroad,
nor preserved any of the earlier letters of my friends,
I at first thought that I had no materials for any
full record of my life and experiences. But when I
set to work in earnest to get together whatever
scattered memoranda I could find, the numerous
letters I possessed from men of considerable eminence, dating from my return home in 1862, together
with a few of my own returned to me by some of
my correspondents, I began to see that I had a fair
amount of material, though I was very doubtful how
far it would interest any considerable number of
readers.

As several of my friends have assured me that a
true record of a life, especially if sufficiently full as to

illustrate development of character so far as that is due to environment, would be extremely interesting, I have kept this in mind, perhaps unduly, though I am not at all sure that my own conclusions on this point are correct.

It is difficult to write such a record as mine (extending to the memories of nearly eighty years) without subjecting one's self to the charge of diffuseness or egotism, and I cannot hope to escape this altogether. But as my experiences have been certainly varied, if not exciting, I trust that the frequent change of scene and of occupation, together with the diversity of my interests and of the persons with whom I have been associated, may render this story of my life less tedious than might have been anticipated.

My thanks are due to those friends who have assisted me with facts or illustrations, and especially to Mrs. Arthur Waugh, who has been so kind as to make the very full Index to my book.

OLD ORCHARD, BROADSTONE,
September, 1905.

CONTENTS

CHAPTER I

PAGE

MY RELATIVES AND ANCESTORS I

CHAPTER II

USK : MY EARLIEST MEMORIES 20

CHAPTER III

HERTFORD : THE HOME OF MY BOYHOOD 30

CHAPTER IV

HERTFORD : MY SCHOOL LIFE 46

CHAPTER V

HERTFORD : MY HOME LIFE 63

CHAPTER VI

LONDON WORKERS, SECULARISTS AND OWENITES . . . 79

CHAPTER VII

BEDFORDSHIRE : SURVEYING 106

CHAPTER VIII

BEDFORDSHIRE : TURVEY 118

CHAPTER IX

BEDFORDSHIRE : SILSOE AND LEIGHTON BUZZARD . . . 129

CONTENTS

CHAPTER X

PAGE

KINGTON AND RADNORSHIRE 140

CHAPTER XI

BRECKNOCKSHIRE 160

CHAPTER XII

SHROPSHIRE AND JACK MYTTON 170

CHAPTER XIII

GLAMORGANSHIRE : NEATH 178

CHAPTER XIV

FIRST LITERARY EFFORTS 199

CHAPTER XV

REMARKS ON MY CHARACTER AT TWENTY-ONE 223

CHAPTER XVI

LONDON AND LEICESTER 229

CHAPTER XVII

RESIDENCE AT NEATH 241

CHAPTER XVIII

THE JOURNEY TO THE AMAZON 264

CHAPTER XIX

" IN MEMORIAM " 289

CHAPTER XX

IN LONDON, AND VOYAGE TO SINGAPORE 302

CHAPTER XXI

THE MALAY ARCHIPELAGO—SINGAPORE, MALACCA, BORNEO . 337

CONTENTS

CHAPTER XXII

PAGE

CELEBES, THE MOLUCCAS, NEW GUINEA, TIMOR, JAVA, AND SUMATRA 356

CHAPTER XXIII

LIFE IN LONDON, 1862–1871 — SCIENTIFIC AND LITERARY WORK 385

CHAPTER XXIV

HOME LIFE—MY FRIENDS AND ACQUAINTANCES—SIR CHARLES LYELL 409

ILLUSTRATIONS

ALFRED RUSSEL WALLACE. 1902 . . . . . *Frontispiece*

MY FATHER. AGE 35 . . . . . . *To face p.* 16
   (*From a miniature*)

MY MOTHER. AGE 18 . . . . „ 16
   (*From a miniature*)

MY BIRTH-PLACE. KENSINGTON COTTAGE, USK . „ 20

THE GRAMMAR SCHOOL, HERTFORD . . . . „ 49
   (*From an engraving in Turner's " History."* 1830)

LLANBISTER, RADNORSHIRE . . . . . . „ 150
   (*Pencil sketch by W. G. Wallace.* 1840)

"A LONELY CHAPEL" . . . . „ 150
   (*Pencil sketch by W. G. Wallace.* 1840)

THE BEACONS (Looking south) . . . . . „ 165
   (*From a photograph by Mr. Symonds Neale*)

PLAN OF SUMMIT OF BEACONS (Looking north) . . „ 165

SECTION THROUGH SUMMITS OF BEACONS . . . „ 165

OUR ECCENTRIC NEIGHBOUR AT DEVYNOCK . . „ 166
   (*From a sketch by W. G. Wallace*)

"MAEN LLIA," UPPER VALE OF NEATH . . . „ 167
   (*From a photograph by Miss Neale*)

"WHITTERN" . . . . „ 170
   (*An outdoor sketch by W. G. Wallace.* 1842)

SAMUEL OSGOOD . . . . „ 188
   (*From a sketch by W. G. Wallace.* 1843)

IN DERBYSHIRE . . . . „ 238
   (*From pencil sketch by A. R. Wallace.* 1844)

A VILLAGE NEAR LEICESTER . . . . . . „ 238
   (*Pencil drawing by A. R. Wallace.* 1844)

FREE LIBRARY, NEATH . . . . „ 246
   (*Designed by A. R. Wallace.* 1847)

YSGWD GLADYS, VALE OF NEATH . . . . . „ 251

PORTH-YR-OGOF, VALE OF NEATH . . . . „ 250

"Maen Madoc," Upper Vale of Neath . . . *To face p.* 249
    (*From three photographs by Miss Neale*)

Latin Inscription on "Maen Madoc" . . .   ,,   251

Alfred R. Wallace. 1848 . . . . . .   ,,   264
    (*From a daguerrotype*)

Fishes of the Rio Negro :
    (*From drawings by A. R. Wallace*)

    1. Cynodon Scombroides. Fam. Characinidæ   ,,   285
      (One-fourth natural size)

    2. Xiphostoma Lateristriga. Fam. Chara-
        cinidæ . . . .   ,,   285
      (One-third natural size)

    3. Pimelodus Holomelas. Fam. Siluridæ .   ,,   286
      (One-third natural size)

    4. Plecostomus Guacari. Fam. Loricariidæ .   ,,   286
      (One-third natural size)

    5. Pterophyllum Scalara. Fam. Cichlidæ .   ,,   286
      (One-third natural size)

    6. Cichlosoma Severum. Fam. Cichlidæ. .   ,,   286
      (One-third natural size)

H. E. Wallace. Age 8 . . . . . .   ,,   289
    (*From a pencil sketch by Miss Townsend*)

Herbert Edward Wallace. Age 20 . . .   ,,   289
    (*From a silhouette*)

The Rio Negro . . . .   ,,   320
    (*From observations made in the years* 1851 *and*
      1852 *by Alfred R. Wallace*)

Enlarged Map of the River Uaupés . . .   ,,   320

Alfred R. Wallace. 1853 . . . . . .   ,,   324
    (*From a photograph by Mr. Sims*)

Native House, Wokan, Aru Islands . . .   ,,   357
    (Where I lived two weeks in March, 1859)

My Faithful Malay Boy—Ali. 1855–1862 . .   ,,   383

Map of the Malay Archipelago . . . .   ,,   384

Alfred R. Wallace. 1869 . . . . . .   ,,   385
    (*From a photograph by Mr. Sims*)

ERRATA.

Vol. I., page 107, line 11, *for* " Earl Cowper " *read* " Earl de Grey."
Vol. I., page 320, line 18, *for* " Juambari " *read* " Inambari."

" My Life."

MY LIFE

A RECORD OF EVENTS AND OPINIONS

CHAPTER I

MY RELATIVES AND ANCESTORS

OUR family had but few relations, and I myself never saw a grandfather or grandmother, nor a true uncle, and but one aunt —my mother's only sister. The only cousins we ever had, so far as I know, were that sister's family of eight or nine, all but two of whom emigrated to South Australia in 1838. Of the two who remained in England, the daughter had married Mr. Burningham, and had only one child, a daughter, who has never married. The son, the Rev. Percy Wilson, had a family, none of whom, however, I have ever met, though I have recently had a visit from a son of another cousin, Algernon, with whom I had a considerable correspondence.

My father was practically an only son, an elder boy dying when three months old ; and as his father died when he was a boy of twelve, and his mother when he was an infant, he had not much opportunity of hearing about the family history. I myself left home before I was fourteen, and only rarely visited my parents for short holidays, except once during my recovery from a dangerous illness, so that I also had little opportunity of learning anything of our ancestors on the paternal side, more especially as my father seldom spoke of his youth, and I as a boy felt no interest in his genealogy. Neither did my eldest brother William—with whom I lived

till I was of age—ever speak on the subject. The little I
have gleaned was from my sister Fanny and from a recent
examination of tombstones and parish registers, and especially
from an old Prayer-book (1723) which belonged to my grandfather Wallace, who had registered in it the dates of the births
and baptisms of his two sons, while my father had continued
the register to include his own family of nine children, of whom
I am the only survivor.

My paternal grandfather was married at Hanworth,
Middlesex, in 1765, and the parish register describes him as
William Wallace, of Hanworth, bachelor, and his wife as
Elizabeth Dilke, of Laleham, widow. Both are buried in
Laleham churchyard, where I presume the former Mrs. Dilke
had some family burial rights, as my grandfather's brother,
George Wallace, is also buried there. The register at Hanworth contains no record of my father's birth, but the church
itself shows that quite a small colony of Wallaces lived at
Hanworth. On a long stone in the floor of the chancel is
the name of JAMES WALLACE, ESQ., who died February 7,
1778, aged 87 years. He was therefore thirty-five years older
than my grandfather, and may have been his uncle. Then
follows ADMIRAL SIR JAMES WALLACE, who died on March
6, 1803, aged 69 years; and FRANCES SLEIGH, daughter of
the above JAMES WALLACE, ESQ., who died December 12,
1820, aged 69 years.

Also, on a small stone in the floor of the nave, just outside
the chancel, we find MARY WALLACE, who died December
5, 1812, aged 39 years. She may, therefore, not improbably
have been a daughter, or perhaps niece, of the admiral.

Here, then, we have four Wallaces buried in the same
church as that in which my grandfather was married, and of
which place he was a resident at the time. As Hanworth is
a very small place, the total population of the parish being
only 750 in 1840, it is hardly probable that my grandfather
and the others met there accidentally. I conclude, therefore,
that James Wallace was probably an uncle or cousin, and
that all were in some way related. As there is no record of

my father's birth at Hanworth, it is probable that his parents had left the place and gone to live either at Laleham or in London.

How or why my grandfather came to live at Hanworth (probably with his brother George, who is also buried at Laleham), I can only conjecture from the following facts. Baron Vere of Hanworth is one of the titles of the Dukes of St. Albans since 1750, when Vere Beauclerc, third son of the first Duke, was created Baron, and his son became fifth Duke of St. Albans in 1787. It is to be presumed that the village and a good deal of the land was at that time the property of this family, though they appear to have parted with it not long afterwards, as a Mr. Perkins owned the park and re-built the church in 1812. The St. Albans family have a tomb in the church. Now, my father's name was Thomas Vere Wallace, and it therefore seems probable that his father was a tenant of the first Baron Vere, and in his will he is styled "Victualler." He probably kept the inn on the estate.

The only further scrap of information as to my father's family is derived from a remark he once made in my hearing, that his uncles at Stirling (I think he said) were very tall men. I myself was six feet when I was sixteen, and my eldest brother William was an inch taller, while my brother John and sister Fanny were both rather tall. My father and mother, however, were under rather than over middle height, and the remark about his tall uncles was to account for this abnormal height by showing that it was in the family. As all the Wallaces of Scotland are held to be various branches of the one family of the hero Sir William Wallace, we have always considered ourselves to be descended from that famous stock; and this view is supported by the fact that our family crest was said to be an ostrich's head with a horseshoe in its mouth, and this crest belongs, according to Burke's "Peerage," to Craigie-Wallace, one of the branches of the patriot's family.

Of my mother's family I have somewhat fuller details, though not going any further back. Her father was John

Greenell, of Hertford, who died there in 1824 at the age of
79. He had two daughters, Martha, who married Thomas
Wilson, Esq., a solicitor, and agent for the Portman estate,
and Mary Anne, my mother. Their mother died when the
two girls were two and three years old. Mr. Greenell married
a second time, and his widow lived till 1828, so that my elder
brothers and sisters may have known her, but she was only
their step-grandmother. Mr. Greenell had died four years
earlier. Although he lived to such a comparatively recent
period, I have not been able to ascertain what was his business. His father, however, my mother's grandfather, who
died in 1797, aged 80, was for many years an alderman, and
twice Mayor of Hertford (in 1773 and 1779), as stated in the
records of the borough. He was buried in St. Andrew's
churchyard.

There is also in the same churchyard a family tomb, in
which my father and my sister Eliza are buried, but which
belonged to a brother of my mother's grandfather, William
Greenell, as shown by the following inscription :—

" Under this tomb with his beloved wife are deposited the remains of
WILLIAM GREENELL,
A native of this parish, who resided 56 years in St. Marylebone,
In the County of Middlesex,
Where he acquired an ample fortune,
With universal esteem and unblemished reputation.
He died the 17th day of January, 1791, aged 71."

There is also an inscription to his wife Ann, who died a
year earlier, and is described as the "wife and faithful friend
of William Greenell, of Great Portland Street, Marylebone."
As the tomb was not used for any other interment till my
sister's death in 1832, it seems likely that William Greenell
had no family, or that if he had they had all removed to
other parts of England.

My mother's mother was a Miss Hudson, whose cousin I
remember as owner of the Town-mill in Hertford, and his
daughters were my sisters' playfellows and friends, but this
family is now extinct so far as the town is concerned.
A sister of my grandfather Greenell married Mr. John

Roberts, whose son lived many years at Epsom, and this family is also extinct by the death of an only son in early manhood, and of an only daughter at an advanced age in 1890.

Through the kindness of Mr. J. B. Wohlmann, late headmaster of the Grammar School, I learn that in the parish registers of births, deaths, and marriages in Hertford, and also in Chauncey's "History and Antiquities of Hertfordshire" and in Clutterbuck's "History of Herts," there are considerable numbers of Greenells (the name being variously spelt, as Grinell, Greenhill, etc.), going back continuously to 1579. I possess an old seal with a coat-of-arms which belonged to my grandfather, and was believed to be those of the Greenell family—a cross on a shield with seven balls on the cross, and a leopard's head for a crest. The balls indicate the name, "Grenaille" being French for shot; and the family were not improbably French refugees after the massacre of St. Bartholomew in 1572.

My mother had several large oil-paintings of the Greenell ancestors which came to her from her sister, Mrs. Wilson, when the Wilsons went to South Australia. Being inconveniently large for our small houses and our frequent removals, they were given to the Miss Roberts above mentioned, who had a large house at Epsom, and on her death they passed with the house to some relatives of her mother, who had no kinship whatever with the Greenells. One of these portraits was that of the great-uncle William Greenell, of Marylebone, who was an architect, and is represented with the design of some public building which, we were told, he had had the honour of himself showing to the king, George the Second or Third. He is shown as a young man, and I was said to resemble him, not only in features, but in a slight peculiarity in one eyebrow, which was indicated on the portrait. I wished to obtain a photograph of this portrait a few years ago, but the present owner refused to allow it to be copied, having, I fancy, some exaggerated idea of its value as a work of art.

Other friends or relatives of the Greenell family were named Russell and Pugh, and are buried at Hertford. A large gentleman's mourning ring in memory of Richard Russell, Esq., was given me by Miss Roberts, as, I presume, the person after whom I was given my second name, though probably from an error in the register mine is always spelt with one *l*, and this peculiarity was impressed upon me in my childhood. Another ring is from Miss Pugh, a friend of my mother's, and, I believe, one of the Russell family. We also possess a very beautiful pastel miniature of Mrs. Frances Hodges, who was a Miss Russell, and who died in 1809, and is buried at All Saints, Hertford ; but the precise relationship, if any, of the Russells to the Greenells I have not been able to ascertain.

One other point may be here mentioned. There seems to have been some connection by marriage between the Wallace and Greenell families before my father's marriage, as shown by the fact that his elder brother, who died in infancy, was named William Greenell Wallace, and it seems not unlikely that his mother, Mrs. Dilke, had been a Miss Greenell before her first marriage.

I will now say a few words about my father's early life, and the various family troubles which, though apparently very disadvantageous to his children, may yet have been on the whole, as is so often the case, benefits in disguise.

My father, Thomas Vere Wallace, was twelve years old when his father died, but his stepmother lived twenty-one years after her husband, and I think it not improbable that she may have resided in Marylebone near William Greenell the architect, and that my father went to school there. The only thing I remember his telling us about his school was that his master dressed in the old fashion, and that he had a best suit entirely of yellow velvet.

When my father left school he was articled to a firm of solicitors—Messrs. Ewington and Chilcot, Bond Court, Walbrook, I think, as I find this name in an old note-book of my

father's—and in 1792, when he had just come of age, he was
duly sworn in as an Attorney-at-Law of the Court of King's
Bench. He is described in the deed of admittance as of
Lamb's Conduit Street, where he probably lodged while pur-
suing his legal studies, it being near the Inns of Court and at
the same time almost in the country. He seems, however,
never to have practised law, since he came into property
which gave him an income of about £500 a year. This I
heard from my sister Fanny.

From this time till he married, fifteen years later, he
appears to have lived quite idly, so far as being without any
systematic occupation, often going to Bath in the season,
where he used to tell us he had met the celebrated Beau
Brummell and other characters of the early years of the
nineteenth century. An old note-book shows that he was
fond of collecting epitaphs from the churchyards of the
various places he visited ; among which are Brighton, Lowestoft, Bognor, Ryegate, Godalming, Sevenoaks, Chichester,
etc. Most of these are commonplace reflections on the
uncertainty of life or equally commonplace declarations of
faith in the orthodox heaven, but here and there are more
original efforts. This is one at Chichester on Henry Case,
aged 28—

> " Here lies a brave soldier whom all must applaud,
> Much hardship he suffer'd at home and abroad,
> But the hardest Engagement he ever was in
> Was the Battle of Self in the Conquest of Sin."

In the following, at Woodford, Essex, the village poet
has been severely practical :—

> "ON WILLIAM MEARS, PLUMBER.
>
> " Farewell, old friend, for thou art gone
> To realms above, an honest Man.
> A plumber, painter, glazier, was your trade,
> And in sodering pipes none could you exceed.
> In Water-work you took great delight
> And had power to force it to any Height,
> But in Water-closets great was your skill,
> For each branch was subordinate to your will.

> But now your Glass is run—your work is done,
> And we scarcely can find such another man.
> Now mourn ye all, and your great loss deplore,
> For this useful man is gone for evermore."

The following seems to be a heartfelt and worthy tribute to a good man—Mr. Mark Sanderson, of Chepstow, aged 66 :—

> " Loving, belov'd, in all relations true,
> Exposed to follies but subdued by few,
> Reader, reflect, and copy if you can
> The social virtues of this honest man."

One more I will give, as it is at least original, from a tombstone at Lowestoft, Suffolk—

> " In memory of
> CHARLES WARD,
> Who died May, 1770,
> Aged 60.
> A dutiful Son, a loving Brother, and an affectionate Husband.

This Stone is not erected by Susan his wife. She erected a Stone to John Salter her second Husband, forgetting the affection of Charles Ward her first Husband."

In some other old MSS. and note-books are a number of quotations in prose and verse, mostly from well-known writers or not of any interest, but among them are a few that seem worth preserving.

The following epitaph by a Dominican friar on Pope Clement the Fourth is remarkable for the ingenuity of the verse, which is equally good when the words and sense are inverted :—

> " Laus tua, non tua fraus, virtus non copia rerum
> Scandere te fecit, hoc decus eximium,
> Pauperibus tua das, nunquam stat janua clausa,
> Fundere res quæris, nec tua multiplicas,
> Conditio tua sit stabilis ! non tempore parvo
> Vivere te faciat, hic Deus omnipotens."

(The same reversed.)

> " Omnipotens Deus hic faciat te vivere parvo
> Tempore ! non stabilis sit tua conditio !
> Multiplicas tua nec quæris res fundere clausa

> Janua stat, nunquam das tua pauperibus,
> Eximium decus hoc fecit te scandere rerum
> Copia, non virtus, fraus tua non tua Laus."

My friend, Mr. Comerford Casey, has kindly given me the following elegant translation of the above :—

> " Not by intrigue but merit, not by wealth
> But worth you rose.   This is your title, this,
> That you bestowed your goods on those in need.
> Your hospitable door was never closed :
> More eager ever to alleviate
> The wants of others than to gather gain.
> May your prosperity be lasting, Pope !
> May God all-powerful grant you length of days ! "

> (*The same read backwards.*)

> " May God omnipotent remove you soon
> From earth !   May your prosperity be short !
> You grasp at gain and shun expense : your door,
> Inhospitable Pope, stands ever shut.
> Naught to the poor you give : your power is due
> To wealth not worth : by intrigue you have risen."

In faded ink and very old handwriting, probably my grandfather's, is the following charade, the answer to which is not given, but it is worth preserving for its style :—

> " My first's the proud but hapless Child of danger,
> Parent of highest honours and of woe ;
> Too long my second to the brave a stranger
> Heaps useless laurels on the soldier's brow.
> My whole by dext'rous artifice contrives
> To gain the prize by which he stands accurst,
> And plung'd in infamy when most he thrives,
> He gains my second whilst he gives my first."

I myself believe the answer to be " cut-purse "—a Shakespearean word in common use in the eighteenth century, and applying to all the terms of the charade with great accuracy. But few of my friends think this solution good enough.

The following is in my father's writing, and as it is comparatively easy, I leave the answer to my young reader's ingenuity :—

"A RIDDLE.

"O Doctor, Doctor, tell me can you cure
   Or say what 'tis I ail? I'm feverish sure!
Sometimes I'm very hot, and sometimes warm,
Sometimes again I'm cool, yet feel no harm.
Part bird, part beast, and vegetable part,
Cut, slash'd, and wounded yet I feel no smart.
I have a skin, which though but thin and slender,
Yet proves to me a powerful defender.
When stript of that, so desperate is my case,
I'm oft devoured in half an hour's space."

One more enigma in my father's writing is interesting because founded on a custom common in my youth, but which has now wholly passed away.

"Kitty, a fair but frozen maid,
   Kindled a flame I still deplore,
The hood-wink'd Boy was called in aid
   So fatal to my suit before.
Tell me, ye fair, this urchin's name
   Who still mankind annoys;
Cupid and he are not the same,
Though each can raise or quench a flame,
   And both are hood-wink'd boys."

My sister told me (and from what followed it was pretty certainly the case) that while he remained a bachelor my father lived up to his income or very nearly so; and from what we know of his after life this did not imply any extravagance or luxurious habits, but simply that he enjoyed himself in London and the country, living at the best inns or boarding-houses, and taking part in the amusements of the period, as a fairly well-to-do, middle-class gentleman.

After his marriage in 1807 he lived in Marylebone, and his ordinary household expenses, of course, increased; and as by 1810 he had two children and the prospect of a large family, he appears to have felt the necessity of increasing his income. Having neglected the law so long, and probably having a distaste for it, he apparently thought it quite hopeless to begin to practise as a solicitor, and being entirely devoid of business habits, allowed himself to be persuaded into undertaking one of the most risky of literary speculations, the starting a new illustrated magazine, devoted apparently to art, antiquities, and general literature. A few numbers were issued, and I remember, as a boy, seeing an elaborate engraving of the Portland Vase, which was one of the illustrations ; and in those days before photography, when all had to be done by skilled artists and engravers, such illustrations were ruinously expensive for a periodical brought out by a totally unknown man. Another of these illustrations is now before me, and well shows the costly nature of the work. It is on large paper, 11½ by 8½ inches to the outer line of the engraving, the margins having been cut off. It is headed " Gallery of Antiquities, British Museum, Pl. I.," and contains forty distinct copper-plate engravings of parts of friezes, vases, busts, and full-length figures, of Greek or Roman art, all drawn to scale, and exquisitely engraved in the best style of the period. The plate is stated at the foot to be " Published for the Proprietor, May 1st, 1811," four years after my father's marriage. It shows that the work must have been of large quarto size, in no way of a popular character, and too costly to have any chance of commercial success. After a very few numbers were issued the whole thing came to grief, partly, it was said, by the defalcations of a manager or bookkeeper, who appropriated the money advanced by my father to pay for work and materials, and partly, no doubt, from the affair being in the hands of persons without the necessary business experience and literary capacity to make it a success.

A few old letters are in my possession, from a Mr. E. A. Rendall to my father, written in 1812 and 1813, relating to the affair. They are dated from Bloomsbury Square, and are exceedingly long and verbose, so that it is hardly possible to extract anything definite from them. They refer chiefly to the mode of winding up the business, and urging that the engraved plates, etc., may be useful in a new undertaking. He proposes, in fact, to commence another magazine with a different name, which he says will cost only sixty guineas a number, and can be published at half a crown. He refers

to the *General Chronicle* as if that were the title of the
recently defunct magazine, and he admits that my father
may rightly consider himself an ill-used man, though wholly
denying that he, Mr. Rendall, had any part in bringing about
his misfortunes.

The result was that my father had to bear almost the
whole loss, and this considerably reduced his already too
scanty income.   Whether he made any other or what efforts
to earn money I do not know, but he continued to live in
Marylebone till 1816, a daughter Emma having been born
there in that year ; but soon after he appears to have removed to St. George's, Southwark, in which parish my
brother John was born in 1818.   Shortly afterwards his affairs
must have been getting worse, and he determined to move
with his family of six children to some place where living
was as cheap as possible ; and, probably from having introductions to some residents there, fixed upon Usk, in Monmouth-
shire, where a sufficiently roomy cottage with a large garden
was obtained, and where I was born on January 8, 1823.   In
such a remote district rents were no doubt very low and
provisions of all kinds very cheap—probably not much more
than half London prices.   Here, so far as I remember, only
one servant was kept, and my father did most of the garden
work himself, and provided the family with all the vegetables
and most of the fruit which was consumed.   Poultry, meat,
fish, and all kinds of dairy produce were especially cheap ;
my father taught the children himself ; the country around
was picturesque and the situation healthy ; and, notwithstanding his reverse of fortune, I am inclined to think that this
was, perhaps, the happiest portion of my father's life.

In the year 1828 my mother's mother-in-law, Mrs. Rebecca
Greenell, died at Hertford, and I presume it was in consequence of this event that the family left Usk in that year,
and lived at Hertford for the next nine or ten years, re-
moving to Hoddesdon in 1837 or 1838, where my father died
in 1843.   These last fifteen years of his life were a period of
great trouble and anxiety, his affairs becoming more and

more involved, till at last the family became almost wholly
dependent on my mother's small marriage settlement of less
than a hundred a year, supplemented by his taking a few
pupils and by a small salary which he received as librarian to
a subscription library. While at Hoddesdon my sister Fanny
got up a small boarding-school for young ladies in a roomy,
old-fashioned house with a large garden, where my father
passed the last few years of his life in comparative freedom
from worry about money matters, because these had reached
such a pitch that nothing worse was to be expected.

During the latter part of the time we lived at Hertford
his troubles were great. He appears to have allowed a
solicitor and friend whom he trusted to realize what remained
of his property and invest it in ground-rents which would
bring in a larger income, and at the same time be perfectly
secure. For a few years the income from this property was
duly paid him, then it was partially and afterwards wholly
stopped. It appeared that the solicitor was himself engaged
in a large building speculation in London, which was certain
to be ultimately of great value, but which he had not capital
enough to complete. He therefore had to raise money, and
did so by using funds entrusted to him for other purposes,
among them my father's small capital, in the absolute belief
that it was quite as safe an investment as the ground-rents
in which it was supposed to be invested. But, unfortunately,
other creditors pressed upon him, and he was obliged to
sacrifice the whole of the building estate at almost a nominal
price. Out of the wreck of the solicitor's fortune my father
obtained a small portion of the money due, with promises to
pay all at some future time; and I recollect his having
frequently to go to London by coach to interview the
solicitor, and try to get some security for future payment.
Among the property thus lost were some legacies from my
mother's relations to her children, and the whole affair got
into the hands of the lawyers, from whom small amounts
were periodically received which helped to provide us with
bare necessaries.

As a result of this series of misfortunes the children who

reached their majority had little or nothing to start with in earning their own living, except a very ordinary education, and a more or less efficient training. The oldest son, William, was first articled to a firm of surveyors at Kington, Herefordshire, probably during the time we resided at Usk. He then spent a year or two in the office of an architect at Hertford, and finally a year in London with a large builder named Martin then engaged in the erection of King's College, in order to become familiar with the practical details of building. He may be said, therefore, to have had a really good professional education. At first he got into general land-surveying work, which was at that time rather abundant, owing to the surveys and valuations required for carrying out the Commutation of Tithes Act of 1836, and also for the enclosures of commons which were then very frequent. During the time I was with him we were largely engaged in this kind of work in various parts of England and Wales, as will be seen later on ; but the payment for such work was by no means liberal, and owing to the frequent periods of idleness between one job and another, it was about as much as my brother could do to earn our living and travelling expenses.

About the time I went to live with my brother my sister Fanny entered a French school at Lille to learn the language and to teach English, and I think she was a year there. On her return she started the school at Hoddesdon, but after my father's death in 1843, she obtained a position as a teacher in Columbia College, Georgia, U.S.A., then just established under the Bishop of Georgia ; and she only returned after my brother William's death in 1846, when the surviving members of the family in England were reunited, and lived together for two years in a cottage near Neath, in Glamorganshire.

My brother John, at the age of fourteen or fifteen, was apprenticed, first to Mr. Martin and then to Mr. Webster, a London builder living in Albany Street, Regent's Park, where he became a thorough joiner and carpenter. He afterwards worked for a time for Cubitt and other large builders ; then, when he came to live with me at Neath, he learnt surveying and a little architecture. When I went to the Amazon, he took a small dairy-farm at too high a rent, and not making this pay, in 1849 he emigrated to California at the height of the first rush for gold, joined several mining camps, and was moderately successful. About five years later he came home, married Miss Webster, and returning to California, settled for some years at Columbia, a small mining town in Tuolumne County. He afterwards removed to Stockton, where he practised as surveyor and water engineer till his death in 1895.

My younger brother, Herbert, was first placed with a trunk maker in Regent Street, but not liking this business, afterwards came to Neath and entered the pattern-shops of the Neath Ironworks. After his brother John went to California he came out to me at Para, and after a year spent on the Amazon as far as Barra on the Rio Negro, he returned to Para on his way home, where he caught yellow fever, and died in a few days at the early age of twenty-two. He was the only member of our family who had a considerable gift of poesy, and was probably more fitted for a literary career than for any mechanical or professional occupation.

It will thus be seen that we were all of us very much thrown on our own resources to make our way in life ; and as we all, I think, inherited from my father a certain amount of constitutional inactivity or laziness, the necessity for work that our circumstances entailed was certainly beneficial in developing whatever powers were latent in us ; and this is what I implied when I remarked that our father's loss of his property was perhaps a blessing in disguise.

Of the five daughters, the first-born died when five months old; the next, Eliza, died of consumption at Hertford, aged twenty-two. Two others, Mary Anne and Emma, died at Usk at the ages of eight and six respectively ; while Frances married Mr. Thomas Sims, a photographer, and died in London, aged eighty-one.

On the whole, both the Wallaces and the Greenells seem to have been rather long-lived families when they reached manhood or womanhood. The five ancestral Wallaces of whom I have records had an average age of seventy years, while the five Greenells had an average of seventy-six. Of our own family, my brother John reached seventy-seven, and my sister Fanny eighty-one. My brother William owed his death to a railway journey by night in winter, from London to South Wales in the miserable accommodation then afforded to third-class passengers, which, increased by a damp bed at Bristol, brought on severe congestion of the lungs, from which he never recovered.

I will now give a short account of my father's appearance and character. In a miniature of himself, painted just before his marriage, when he was thirty-five years old, he is represented in a blue coat with gilt buttons, a white waistcoat, a thick white neck-cloth coming up to the chin and showing no collar, and a frilled shirt-front. This was probably his wedding-coat, and his usual costume, indicating the transition from the richly coloured semi-court dress of the earlier Georgian period to the plain black of our own day. He is shown as having a ruddy complexion, blue eyes, and carefully dressed and curled hair, which I think must have been powdered, or else in the transition from light brown to pure white. As I remember him from the age of fifty-five onwards, his hair was rather thin and quite white, and he was always clean-shaved as in the miniature. He continued to wear the frilled shirt and thick white neckties, but never wore any outer clothing but black, of the cut we now term a dress-suit, but the coat double-breasted, and the whole rather loose fitting. He also wore large shoes and black cloth gaiters out-of-doors. This dress he never altered, having at first one new suit a year, but latterly I think only one every second or third year; but he always had one for Sundays and visiting, which was kept in perfect order. The second was for everyday wear; and when gardening or doing any other work likely to be injurious to his clothes, I think I remember him wearing a thin home-made holland jacket and a gardener's apron.

In figure he was somewhat below the middle height. He

MY FATHER.   AGE 35.
(*From a miniature.*)

[*To face p.* 16, VOL. I.

MY MOTHER. AGE 18.

(*From a miniature.*)

[*To face p.* 16, VOL. I.

was fairly active and fond of gardening and other country occupations, such as brewing beer and making grape or elder wine whenever he had the opportunity; and during some years at Hertford he rented a garden about half a mile away, in order to grow vegetables and have some wholesome exercise. He had had some injury to one of his ankles which often continued to trouble him, and gave him a slight lameness, and in consequence of this he never took very long walks. He was rather precise and regular in his habits, quiet and rather dignified in manners, and somewhat of what is termed a gentleman of the old school. Of course, he always wore a top-hat—a beaver hat as it was then called, before silk hats were invented—the only other headgear being sometimes a straw hat for use in the garden in summer.

In character he was quiet and even-tempered, very religious in the orthodox Church-of-England way, and with such a reliance on Providence as almost to amount to fatalism. He was fond of reading, and through reading clubs or lending libraries we usually had some of the best books of travel or biography in the house. Some of these my father would read to us in the evening, and when Bowdler's edition of Shakespeare came out he obtained it, and often read a play to the assembled family. In this way I made my first acquaintance with Lear and Cordelia, with Malvolio and Sir Andrew Aguecheek, with the thrilling drama of the *Merchant of Venice*, with Hamlet, with Lady Macbeth, and other masterpieces. At one time my father wrote a good deal, and we were told it was a history of Hertford, or at other times some religious work; but they never got finished, and I do not think they would ever have been worth publishing, his character not leading him to do any such work with sufficient thoroughness. He dabbled a little in antiquities and in heraldry, but did nothing systematic, and though he had fair mental ability he possessed no special talent, either literary, artistic, or scientific. He sketched a little, but with a very weak and uncertain touch, and among his few scrap- and note-books that have been preserved, there is hardly

anything original except one or two short poems in the
usual didactic style of the period, but of no special merit.
I will, however, give here the only two of these that my
mother had preserved, and which are, no doubt, the best
products of his pen.   They were evidently both written
at Usk.

"USK BRIDGE—A SIMILE.

" As on this archéd pile I lately strolled
  And viewed the tide that deep beneath it roll'd,
  Eastward impetuous rushed the foamy wave,
  Each quick ingulph'd—as mortals in the grave ;
  All noisy, harsh, impetuous, was the roar,
  Like the world's bustle—and as quickly o'er.
  For when a few short steps I westward made
  The river here a different scene displayed,
  Its noisy roar seemed now a distant hum,
  Calm was the surface—and the stream was dumb,
  Silent though swift its course—and such I cried
  The life of man !   In youth swoll'n high with pride,
  The passions raging, noisy, foaming, bold,
  Like the rough stream a constant tumult hold.
  But when his steps turn towards the setting sun
  And more than half his wayward course is run,
  By age, and haply by religion's aid,
  His pride subdued, his passions too allay'd,
  With quiet pace—yet swiftly gliding, he
  Rolls to the ocean of Eternity ! "

ON THE CUSTOM OBSERVED IN WALES OF DRESSING THE GRAVES WITH FLOWERS ON PALM SUNDAY.

" The sounding bell from yon white turret calls
  The villagers within those sacred walls,
  And o'er the solemn precincts of the dead,
  Where lifts the church its grey time-honoured head,
  That place of rest where parents, children, sleep,
  Where heaves the turf in many a mould'ring heap
  Affection's hand hath gaily decked the ground
  And spring's sweet gifts profusely scatter'd round.
  Pleas'd memory still delights to linger here
  And many a cheek is moistened with a tear.
  The wife, the child, the parent, and the friend
  In soft regret by these sweet trophies bend.
  Nor let the selfish sneer, the proud upbraid,
  The tribute thus by love, by duty paid,

In nature's purest sentiments its source,
Here nature speaks with a resistless force.
What though these flow'rets speedily decay
Yet they our love, our tenderest thoughts display,
Of friends departed a memorial sweet
With which their relics thus we fondly greet,
' Our minds revisit those we loved when here,
Tho' lost to sight, to memory still they're dear.' "

In consequence of this custom the Sunday before Easter was called in Wales "Flowering Sunday," and was looked forward to by most families as an event of special interest, and by children as quite a festival. It is always a pretty sight when even a grave here and there is nicely adorned with fresh flowers, but when a whole churchyard is so decorated, at least as regards all but the oldest tombs, it becomes really beautiful. The long procession during the morning of women and children carrying baskets of flowers, and coming in from various directions, often from many miles distant, adds greatly to the interest of the scene. This custom seems to be one of the expressions of the idealism and poetry characteristic of most Celtic peoples.

CHAPTER II

USK : MY EARLIEST MEMORIES

My earliest recollections are of myself as a little boy in short
frocks and with bare arms and legs, playing with my brother
and sisters, or sitting in my mother's lap or on a footstool
listening to stories, of which some fairy-tales, especially
" Jack the Giant-Killer," " Little Red Riding Hood," and
" Jack and the Beanstalk," seem to live in my memory ; and of
a more realistic kind, " Sandford and Merton," which perhaps
impressed me even more deeply than any. I clearly remember the little house and the room we chiefly occupied, with a
French window opening to the garden, a steep wooded bank
on the right, the road, river, and distant low hills to the left.
The house itself was built close under this bank, which
was quite rocky in places, and a little back yard between the
kitchen and a steep bit of rock has always been clearly
pictured before me as being the scene of my earliest attempt
to try an experiment, and its complete failure. "Æsop's
Fables " were often read to me, and that of the fox which
was thirsty and found a pitcher with a little water in the
bottom but with the opening too small for its mouth to reach
it, and of the way in which it made the water rise to the top
by dropping pebbles into it, puzzled me greatly. It seemed
quite like magic. So one day, finding a jar or bucket standing
in the yard, I determined to try and see this wonderful thing.
I first with a mug poured some water in till it was about an
inch or two deep, and then collected all the small stones I
could find and put into the water, but I could not see that
the water rose up as I thought it ought to have done. Then

20

MY BIRTH-PLACE. KENSINGTON COTTAGE, USK.

[*To face p.* 20, VOL. I.

I got my little spade and scraped up stones off the gravel path, and with it, of course, some of the soft gravel, but instead of the water rising, it merely turned to mud; and the more I put in the muddier it became, while there seemed to be even less water than before. At last I became tired and gave it up, and concluded that the story could not be true; and I am afraid this rather made me disbelieve in experiments out of story-books.

The river in front of our house was the Usk, a fine stream on which we often saw men fishing in coracles, the ancient form of boat made of strong wicker-work, somewhat the shape of the deeper half of a cockle-shell, and covered with bullock's hide. Each coracle held one man, and it could be easily carried to and from the river on the owner's back. In those days of scanty population and abundant fish the river was not preserved, and a number of men got their living, or part of it, by supplying the towns with salmon and trout in their season. It is very interesting that this extremely ancient boat, which has been in use from pre-Roman times, and perhaps even from the Neolithic Age, should continue to be used on several of the Welsh rivers down to the present day. There is probably no other type of vessel now in existence which has remained unchanged for so long a period.

But the chief attraction of the river to us children was the opportunity it afforded us for catching small fish, especially lampreys. A short distance from our house, towards the little village of Llanbadock, the rocky bank came close to the road, and a stone quarry had been opened to obtain stone, both for building and road-mending purposes. Here, occasionally, the rock was blasted, and sometimes we had the fearful delight of watching the explosion from a safe distance, and seeing a cloud of the smaller stones shoot up into the air. At some earlier period very large charges of powder must have been used, hurling great slabs of rock across the road into the river, where they lay, forming convenient piers and standing-places on its margin. Some of these slabs were eight or ten feet long and nearly as wide; and it was these that formed our favourite fishing-stations, where we sometimes found shoals of small lampreys, which could be scooped up in basins or old saucepans, and were then fried for our dinner or supper, to our great enjoyment.   I think what we caught must have been the young fish, as my recollection of them is that they were like little eels, and not more than six or eight inches long, whereas the full-grown lampreys are from a foot and a half to nearly three feet long.

The lamprey was a favourite dish with our ancestors, and is still considered a luxury in some districts, while in others it is rejected as disagreeable, and the living fish is thought to be even poisonous.   This is, no doubt, partly owing to its wriggling, snake-like motions, and its curious sucking mouth, by which it sticks on the hand and frightens people so much that they throw it away instantly.   But the Rev. J. G. Wood, in his very interesting "Natural History," tells us that he has caught thousands of them with his bare hands, and has often had six or seven at once sticking to his hand without causing the slightest pain or leaving the least mark.   The quantity of these fish is so great in some rivers that they would supply a large amount of wholesome food were there not such a prejudice against them.   Since this period of my early child-hood I do not think I have ever eaten or even seen a lamprey.

At this time I must have been about four years old, as we left Usk when I was about five, or less.   My brother John was four and a half years older, and I expect was the leader in most of our games and explorations.   My two sisters were five and seven years older than John, so that they would have been about thirteen and fifteen, which would appear to me quite grown up; and this makes me think that my recollections must go back to the time when I was just over three, as I quite distinctly remember two, if not three, besides myself, standing on the flat stones and catching lampreys.

There is also another incident in which I remember that my brother and at least one, if not two, of my sisters took part.   Among the books read to us was "Sandford and Merton," the only part of which that I distinctly remember is when the two boys got lost in a wood after dark, and while Merton could do nothing but cry at the idea of having to pass the night without supper or bed, the resourceful Sandford comforted him by promising that he should have both, and set him to gather sticks for a fire, which he lit with a tinder-box and match from his pocket. Then, when a large fire had been made, he produced some potatoes which he had picked up in a field on the way, and which he then roasted beautifully in the embers, and even produced from another pocket a pinch of salt in a screw of paper, so that the two boys had a very good supper. Then, collecting fern and dead leaves for a bed, and I think making a coverlet by taking off their two jackets, which made them quite comfortable while lying as close together as possible, they enjoyed a good night's sleep till daybreak, when they easily found their way home.

This seemed so delightful that one day John provided himself with the matchbox, salt, and potatoes, and having climbed up the steep bank behind our house, as we often did, and passed over a field or two to the woods beyond, to my great delight a fire was made, and we also feasted on potatoes with salt, as Sandford and Merton had done. Of course we did not complete the imitation of the story by sleeping in the wood, which would have been too bold and dangerous an undertaking for our sisters to join in, even if my brother and I had wished to do so.

Another vivid memory of these early years consists of occasional visits to Usk Castle. Some friends of our family lived in the house to which the ruins of the castle were attached, and we children were occasionally invited to tea, when a chief part of our entertainment was to ascend the old keep by the spiral stair, and walk round the top, which had a low parapet on the outer side, while on the inner we looked down to the bottom of the tower, which descended below the ground-level into an excavation said to have been the dungeon. The top of the walls was about three feet thick, and it was thus quite safe to walk round close to the parapet, though there was no protection on the inner edge but the few herbs and bushes that grew upon it. For many years this small fragment of a mediæval castle served to illustrate for me the stories of knights and giants and prisoners immured in dark and dismal dungeons. In our friend's pretty grounds, where we often had tea, there was a summer-house with a table formed of a brick-built drum, with a circular slate slab on the top, and this peculiar construction seemed to us so appropriate that we named it the little castle, and it still remains a vivid memory.

Our house was less than a quarter of a mile from the old bridge of three arches over the river Usk, by which we reached the town, which was and is entirely confined to the east side of the river, while we lived on the west. The walk there was a very pleasant one, with the clear, swift-flowing river on one side and the narrow fields and wooded steep bank on the other; while from the bridge itself there was a very beautiful view up the river-valley, of the mountains near Abergavenny, ten miles off, the conical sugar-loaf in the centre, the flat-topped mass of the Blorenge on the left, and the rocky ridge of the Skirrid to the right. These names were so constantly mentioned that they became quite familiar to me, as the beginning of the unknown land of Wales, which I also heard mentioned occasionally.

My eldest brother William was about eighteen when I was four, and was articled to Messrs. Sayce, a firm of land surveyors and estate agents at Kington, in Herefordshire. I have an indistinct recollection of his visiting us occasionally, and of his being looked up to as very clever, and as actually bringing out a little monthly magazine of literature, science, and local events, of which he brought copies to show us. I particularly remember one day his pointing out to the family that the reflection of some hills in the river opposite us was sometimes visible and sometimes not, though on both occasions in equally calm and clear weather. He explained the cause of this in the magazine, illustrated by diagrams, as being due to changes of a few inches in the height of the water, but this, of course, I did not understand at the time.

I may here mention a psychological peculiarity, no doubt

common to a considerable proportion of children of the same age, that, during the whole period of my residence at Usk, I have no clear recollection, and can form no distinct mental image, of either my father or mother, brothers or sisters. I simply recollect that they existed, but my recollection is only a blurred image, and does not extend to any peculiarities of feature, form, or even of dress or habits. It is only at a considerably later period that I begin to recollect them as distinct and well-marked individuals whose form and features could not be mistaken—as, in fact, being *my* father and mother, my brothers and sisters ; and the house and surroundings in which I can thus first recollect, and in some degree visualize them, enable me to say that I must have been then at least eight years old.

What makes this deficiency the more curious is that, during the very same period at which I cannot recall the personal appearance of the individuals with whom my life was most closely associated, I *can* recall all the main features and many of the details of my outdoor, and, to a less degree, of my indoor, surroundings. The form and colour of the house, the road, the river close below it, the bridge with the cottage near its foot, the narrow fields between us and the bridge, the steep wooded bank at the back, the stone quarry and the very shape and position of the flat slabs on which we stood fishing, the cottages a little further on the road, the little church of Llanbadock and the stone stile into the churchyard, the fishermen and their coracles, the ruined castle, its winding stair and the delightful walk round its top— all come before me as I recall these earlier days with a distinctness strangely contrasted with the vague shadowy figures of the human beings who were my constant associates in all these scenes. In the house I recollect the arrangement of the rooms, the French window to the garden, and the blue-papered room in which I slept, but of the people always with me in those rooms, and even of the daily routine of our life, I remember nothing at all.

I cannot find any clear explanation of these facts in modern psychology, whereas they all become intelligible from the phrenological point of view. The shape of my head shows that I have *form* and *individuality* but moderately developed, while *locality, ideality, colour,* and *comparison* are decidedly stronger. Deficiency in the first two caused me to take little notice of the characteristic form and features of the separate individualities which were most familiar to me, and from that very cause attracted less close attention ; while the greater activity of the latter group gave interest and attractiveness to the ever-changing combinations in outdoor scenery, while the varied opportunities for the exercise of the physical activities, and the delight in the endless variety of nature which are so strong in early childhood, impressed these outdoor scenes and interests upon my memory. And throughout life the same limitations of observation and memory have been manifest. In a new locality it takes me a considerable time before I learn to recognize my various new acquaintances individually ; and looking back on the varied scenes amid which I have lived at home and abroad, while numerous objects, localities, and events are recalled with some distinctness, the people I met, or, with few exceptions, those with whom I became fairly well acquainted, seem but blurred and indistinct images.

In the year 1883, when for the first time since my childhood I revisited, with my wife and two children, the scenes of my infancy, I obtained a striking proof of the accuracy of my memory of those scenes and objects. Although the town of Usk had grown considerably on the north side towards the railway, yet, to my surprise and delight, I found that no change whatever had occurred on our side of the river, where, between the bridge and Llanbadock, not a new house had been built, and our cottage and garden, the path up to the front door, and the steep woody bank behind it, remained exactly as pictured in my memory. Even the quarry appeared to have been very little enlarged, and the great flat stones were still in the river exactly as when I had stood upon them with my brother and sisters sixty years before. The one change I noted here was that the well-remembered stone stile into the village churchyard had been replaced by a wooden one. We also visited the ruined castle, ascended the winding stair, and walked round the top wall, and everything seemed to me exactly as I knew it of old, and neither smaller nor larger than my memory had so long pictured it. The view of the Abergavenny mountains pleased and interested me as in childhood, and the clear-flowing Usk seemed just as broad and as pleasant to the eye as my memory had always pictured it.

There is one other fact connected with my mental nature which may be worth noticing here. This is an often-repeated dream, which occurred at this period of my life, and, so far as I can recall, then only. I seemed first to hear a distant beating or flapping sound, as of some creature with huge wings ; the sound came nearer and nearer, till at last a deep thud was heard and the flapping ceased. I then seemed to feel that the creature was clinging with its wings outspread against the wall of the house just outside my window, and I waited in a kind of fearful expectation that it would come inside. I usually awoke then, and all being still, went to sleep again.

I think I can trace the origin of this dream. At a very early period of these recollections I was shown on the outside of a house, at or near Usk, a hatchment or funeral escutcheon —the coat-of-arms on a black lozenge-shaped ground often put up on the house of a deceased person of rank or of ancient lineage. At the time I only saw an unmeaning jumble of strange dragon-like forms surrounded with black, and I was told that it was there because somebody was dead ; and when this curious dream came I at once associated it with the hatchment, and directly I heard the distant flapping of wings, I used to say to myself (in my dream), " The hatchment is coming ; I hope it will not get in." So far as I can remember, this was the only dream—at all events, the only vivid and impressive one—I had while living at Usk, and it came so often, and so exactly in the same form, as to become quite familiar to me. It was, in fact, the form my childish nightmare took at that period, and though I was always afraid of it, it was not nearly so distressing as many of the nightmares I have had since.

I may here add another illustration of how vividly these scenes of my childhood remain in my memory. My father was very fond of Cowper's poems, and often used to read them aloud to us children. Two of these especially impressed themselves on my memory. That about the three kittens and the viper, ending with the lines—

> " With outstretched hoe I slew him at the door,
> And taught him never to come there no more ! "

was perhaps the favourite, and whenever I heard it or read it in after years, the picture always in my mind was of the door-step of the Usk cottage with the kittens and the viper in the attitudes so picturesquely described. The other one was the fable of the sheep, who, on hearing some unaccustomed noise, rushed away to the edge of a pit, and debated whether it would be wise to jump into it to escape the unknown danger, but were persuaded by a wise old bell-wether that this would be foolish, he being represented as saying—

> " What ! jump into the pit your lives to save,
> To save your lives leap into the grave ! "

And as almost the only sheep I had seen close at hand were in the little narrow field between our house and the bridge, I always associated the scene with that field, although there was no pit of any kind in it. So, in after years, when I became fascinated by the poems of Hood, the beautiful and pathetic verses beginning—

> " I remember, I remember,
> The house where I was born,
> The little window where the sun
> Came peeping in each morn ;
> He never came a wink too soon,
> Nor brought too long a day,
> But now I often wish the night
> Had borne my breath away,"

always brought to my mind the memory of the little blue-papered room at Usk, which faced somewhat east of south, and into which, therefore, the sun did " come peeping in each morn "—at least, during a large portion of the year.

So far as I can remember or have heard, I had no illness of any kind at Usk, which was no doubt due to the free outdoor life we lived there, spending a great part of the day in the large garden or by the riverside, or in the fields and woods around us. As will be seen later on, this immunity ceased as soon as we went to live *in* a town. I remember only one childish accident. The cook was taking away a frying-pan with a good deal of boiling fat in it, which for some reason I wanted to see, and, stretching out my arm over it, I suppose to show that I wanted it lowered down, my fore arm went into the fat and was badly scalded. I mention this only for the purpose of calling attention to the fact that, although I vividly remember the incident, I cannot recall that I suffered the least pain, though I was told afterwards that it was really a severe burn. This, and other facts of a similar kind, make me think that young children suffer far less pain than adults from the same injuries. And this is quite in accordance with the purpose for which pain exists, which is to guard the body against injuries dangerous to life, and giving us the impulse to escape rapidly from any danger. But as infants cannot escape from fatal dangers, and do not even know what things are dangerous and what not, only very slight sensations of pain are at first required, and such only are therefore developed, and these increase in intensity just in proportion as command over the muscles giving the power of rapid automatic movements become possible. The sensation of pain does not, probably, reach its maximum till the whole organism is fully developed in the adult individual. This is rather a comforting conclusion in view of the sufferings of so many infants needlessly massacred through the terrible defects of our vicious social system.

I may add here a note as to my personal appearance at this age. I was exceedingly fair, and my long hair was of a very light flaxen tint, so that I was generally spoken of among the Welsh-speaking country people as the "little Saxon."

CHAPTER III

HERTFORD : THE HOME OF MY BOYHOOD

MY recollections of our leaving Usk and of the journey to
London are very faint, only one incident of it being clearly
visualized—the crossing the Severn at the Old Passage in
an open ferry-boat. This is so very clear to me, possibly
because it was the first time I had ever been in a boat. I
remember sitting with my mother and sisters on a seat at
one side of the boat, which seemed to me about as wide as a
small room, of its leaning over so that we were close to the
water, and especially of the great boom of the mainsail, when
our course was changed, requiring us all to stoop our heads for
it to swing over us. It was a little awful to me, and I think
we were all glad when it was over and we were safe on land
again. We must have travelled all day by coach from Usk
to the Severn, then on to Bristol, then from Bristol to London.
I think we must have started very early in the morning
and have reached London late in the evening, as I do not
remember staying a night on the way, and the stage then
travelled at an average speed of ten miles an hour over good
roads and in the summer time. The monotony of the journey
probably tired me so that it left no impression ; but besides
the ferry-boat the only other incident I can clearly recall is our
sleeping at an old inn in London, and our breakfast there the
next morning. I rather think the inn was the Green Man, or
some such name, in Holborn, and the one thing that lives in
my memory is that in the morning my mother ordered coffee
for breakfast, and said to the waiter, "Mind and make it
good." The result of which injunction was that it was nearly

black, and so strong that none of the party could drink it, till boiling water was brought for us to dilute it with. I, of course, had only milk and water, with perhaps a few drops of coffee as a special luxury.

Of the next few months of my life I have also but slight recollections, confined to a few isolated facts or incidents. On leaving the inn we went to my aunt's at Dulwich. Mrs. Wilson was my mother's only sister, who had married a solicitor, who, besides having a good practice, was agent for Lord Portman's London property. I remember being much impressed with the large house, and especially with the beautiful grounds, with lawns, trees, and shrubs such as I had never seen before. There were here also a family of cousins, some about my own age, and the few days we stayed were very bright and enjoyable.

I rather think that my father, and perhaps my brother also, had left Usk a few days before us to make arrangements for the family at Hertford, and I think that I was taken to a children's school at Ongar, in Essex, kept by two ladies—the Misses Marsh. I think it was at this place, because my father had an old friend there, a Mr. Dyer, a clergyman. There were a number of little boys and girls here about my own age or younger, and what I chiefly remember is playing with them in the playground, garden, and house. The playground was a gravel yard on one side of the house, and there we occasionally found what I here first heard called " thunderbolts "—worn specimens of belemnites—fossils of the chalk formation. We all believed that they fell down during thunderstorms. One rather exciting incident alone stands out clear in my memories of this place. There was a garden sloping down to a small pond in the centre, with rather steep banks and surrounded by shrubs and flower-beds. This was cut off from the house and yard by a low iron fence with a gate which was usually kept locked, and we were not allowed to play in it. But one day the gardener had left it open, and we all went in, and began pulling and pushing an old-fashioned stone roller. After a little while, as we were pushing it along a path which went down to the pond, it suddenly began to go quickly down hill, and as we could not stop it, and were afraid of being pulled into the water, we had to let go, and the roller rushed on, splashed into the pond, and disappeared. We were rather frightened, and were, of course, lectured on the narrow escape we had had from drowning ourselves. This is really all I recollect of my first experience of a boarding-school.

My next recollections are of the town of Hertford, where we lived for eight or nine years, and where I had the whole of my school education. We had a small house, the first of a row of four at the beginning of St. Andrew's Street, and I must have been a little more than six years old when I first remember myself in this house, which had a very narrow yard at the back, and a dwarf wall, perhaps five feet high, between us and the adjoining house. The very first incident I remember, which happened, I think, on the morning after my arrival, was of a boy about my own age looking over this wall, who at once inquired, "Hullo! who are you?" I told him that I had just come, and what my name was, and we at once made friends. The stand of a water-butt enabled me to get up and sit upon the wall, and by means of some similar convenience he could do the same, and we were thus able to sit side by side and talk, or get over the wall and play together when we liked. Thus began the friendship of George Silk and Alfred Wallace, which, with long intervals of absence at various periods, has continued to this day.

The way in which we were brought together throughout our boyhood is very curious. While at Hertford I lived altogether in five different houses, and in three of these the Silk family lived next door to us, which involved not only each family having to move about the same time, but also that two houses adjoining each other should on each occasion have been vacant together, and that they should have been of the size required by each, which after the first was not the same, the Silk family being much the largest. When we moved to our second house, George's grandmother had an old house opposite to us, and we were thus again brought

together. Besides this, for the greater part of the time we
were schoolfellows at the Hertford Grammar School ; and it is
certainly a curious coincidence that this the earliest acquaintance of my childhood, my playmate and schoolfellow, should
be the only one of all my schoolfellows who were also friends,
that I have ever seen again or that, so far as I know, are
now alive.

The old town of Hertford, in which I passed the most
impressionable years of my life, and where I first obtained a
rudimentary acquaintance with my fellow-creatures and with
nature, is, perhaps, on the whole, one of the most pleasantly
situated county towns in England, although as a boy I did
not know this, and did not appreciate the many advantages I
enjoyed. Among its most delightful features are numerous
rivers and streams in the immediately surrounding country,
affording pleasant walks through flowery meads, many picturesque old mills, and a great variety of landscape. The
river Lea, coming from the south-west, passes through the
middle of the town, where the old town mill was situated in
an open space called the Wash, which was no doubt liable to
be flooded in early times. The miller was reputed to be one
of the richest men in the town, yet we often saw him standing at the mill doors in his dusty miller's clothes as we passed
on our way from school. He was a cousin of my mother's
by marriage, and we children sometimes went to tea at his
house, and then, as a great treat, were shown all over the
mill with all its strange wheels and whirling millstones, its
queer little pockets, on moving leather belts, carrying the
wheat up to the stones in a continual stream, the ever-rattling
sieves and cloths which sifted out the bran and pollard, and
the weird peep into the dark cavern where the great dripping
water-wheel went on its perpetual round. Where the river
passed under the bridge close by, we could clamber up and
look over the parapet into the deep, clear water rushing over
a dam, and also see where the stream that turned the wheel
passed swiftly under a low arch, and this was a sight that
never palled upon us, so that almost every fine day, as we

passed this way home from school, we gave a few moments
to gazing into this dark, deep water, almost always in shadow
owing to high buildings on both sides of it, but affording a
pleasant peep to fields and gardens beyond.

After passing under the bridge, the river flowed on among
houses and workshops, and was again dammed up to supply
another mill about half a mile away, and to form the river
Lea navigation.   There was also, in my time, a small lateral
stream carried off to pump water to the top of a wooden
water-tower to supply part of the town, so that about half a
mile from the middle of the town there were four distinct
streams side by side, though not parallel, which I remember
used to puzzle me very much as to their origin.   In addition
to these there was another quite distinct river, the Beane,
which came from the north-west till it was only a furlong
from the Lea at the town bridge, when it turned back to the
north-east, and entered that river half a mile lower down,
enclosing between the two streams the fine open space of
about thirty acres called Hartham, which was sufficiently
elevated to be always dry, and which was at once a common
grazing field and general cricket and playground, the turf
being very smooth and good, and seldom requiring to be
rolled.   The county cricket matches were played here, and it
was considered to be a first-rate ground.

Here, too, in the river Beane, which had a gentle stream
with alternate deep holes and sandy shallows, suitable for
boys of all ages, was our favourite bathing-place, where,
not long after our coming to Hertford, I was very nearly
being drowned.   It was at a place called Willowhole, where
those who could swim a little would jump in, and in a few
strokes in any direction reach shallower water.   I and my
brother John and several schoolfellows were going to bathe,
and I, who had undressed first, was standing on the brink,
when one of my companions gave me a sudden push from
behind, and I tumbled in and went under water immediately.
Coming to the surface half dazed, I splashed about and went
under again, when my brother, who was four and a half years
older, jumped in and pulled me out.   I do not think I had

actually lost consciousness, but I had swallowed a good deal
of water, and I lay on the grass for some time before I got
strength to dress, and by the time I got home I was quite
well. It was, I think, the first year, if not the first time, I
had ever bathed, and if my brother had not been there it is
quite possible that I might have been drowned. This gave
me such a fright that though I often bathed here afterwards,
I always went in where the water was shallow, and did not
learn to swim, however little, till several years later.

Few small towns (it had then less than six thousand inhabitants) have a more agreeable public playground than
Hartham, with the level valley of the Lea stretching away to
Ware on the east, the town itself just over the river on the
south, while on the north, just across the river Beane, was a
steep slope covered with scattered fir trees, and called the
Warren, at the foot of which was a footpath leading to the
picturesque little village and old church of Bengeo. This
path along the Warren was a favourite walk of mine either
alone or with a playmate, where we could scramble up the
bank, climb up some of the old trees, or sit comfortably upon
one or two old stumpy yews, which had such twisted branches
and stiff spreading foliage as to form delightful seats. This
place was very little frequented, and our wanderings in it
were never interfered with.

In the other direction the river Beane, as already stated,
flows down a picturesque valley from the north, but I do
not remember walking much beyond Bengeo. A little way
beyond Hartham, toward Ware, another small stream, the
Rib, came from the north, with a mill-stream along the west
side of Ware Park, but this also was quite unexplored by us.
Just out of the town, to the south-west, the river Mimram
joined the Lea. This came through the village of Herting-
fordbury, about a mile off, and then through the fine park of
Panshanger, about two miles long and containing about a
thousand acres. This park was open to the public, and we
occasionally went there to visit the great oak tree which was,
I believe, one of the finest grown large oaks in the kingdom.
It was one of the sights of the district.

About three-quarters of a mile from the centre of the town, going along West Street, was a mill called Horn's Mill, which was a great attraction to me. It was an old-fashioned mill for grinding linseed, expressing the oil, and making oil-cake. The mill stood close by the roadside, and there were small low windows always open, through which we could look in at the fascinating processes as long as we liked. First, there were two great vertical millstones of very smooth red granite, which shone beautifully from the oil of the ground seeds. These were fixed on each side of a massive vertical wooden axis on a central iron axle, revolving slowly and silently, and crushing the linseed into a fine oily meal. A curved fender or scoop continually swept the meal back under the rollers with an excentric motion, which was itself altogether new to us, and very fascinating; and, combined with the two-fold motion of the huge revolving stones, and their beautiful glossy surfaces, had an irresistible attraction for us which never palled.

But this was only one part of this delightful kind of peep-show. A little way off an equally novel and still more complex operation was always going on, accompanied by strange noises always dear to the young. Looking in at other windows we saw numbers of workmen engaged in strange operations amid strange machinery, with its hum and whirl and reverberating noises. Close before us were long erections like shop counters, but not quite so high. Immediately above these, at a height of perhaps ten or twelve feet, a long cylindrical beam was continually revolving with fixed beams on each side of it, both higher up and lower down. At regular intervals along the counter were great upright wooden stampers shod with iron at the bottom. When not in action these were supported so that they were about two feet above the counter, and just below them was a square hole. As we looked on a man would take a small canvas sack about two feet long, fill it quite full of linseed meal from a large box by his side, place this bag in a strong cover of a kind of floorcloth with flaps going over the top and down each side. The sack of meal thus prepared would be then dropped into the hole, which it entered easily. Then a thin board of hard wood, tapered to

the lower edge, was pushed down on one side of it, and outside this again another wedge-shaped piece was inserted.
The top of this was now just under the iron cap of the heavy
pile or rammer, and on pulling a rope, this was freed and
dropped on the top of the wedge, which it forced halfway
down.   In a few seconds it was raised up again, and fell upon
the wedge, driving it in a good deal further, and the third
blow would send it down level with the top of the counter.
Then when the rammer rose up, another rope was pulled, and
it remained suspended ; a turn of a handle enabled the first
wedge to be drawn out and a much thicker one inserted, when,
after two or three blows, this became so hard to drive that the
rammer falling upon it made a dull sound and rebounded a
little ; and as the process went on the blows became sharper,
and the pile would rebound two or three times like a billiard
ball rebounding again and again from a stone floor, but in
more rapid succession.   This went on for hours, and when
the process was finished, the meal in the sack had become so
highly compressed that when taken out it was found to be
converted into a compact oilcake.   In this mill there were,
I think, three or four counters parallel to each other, and on
each, perhaps, six or eight stamps, and when all these were
at work together, but rebounding at different rates and with
different intensities of sound, the whole effect was very strange,
and the din and reverberation almost deafening, but still at
times somewhat musical.   During this squeezing process the
oil ran off below through suitable apertures, but was never
seen by us.   I believe these old stamping-mills are now all
replaced by hydraulic presses, which get more oil out and
leave the cake harder, but the process would be almost silent
and far less picturesque.

A very interesting and beautiful object connected with
the water-supply of the neighbourhood was the New River
Head or Chadwell Spring, the source of the original New
River brought to London by Sir Hugh Myddleton.   It is
about two-thirds of the distance from Hertford to Ware, and
is situated in a level meadow not far from the high-road, and

about a quarter of a mile from the main river. As I knew it, it was a circular pond nearly a hundred feet in diameter, filled with the most crystal clear water, and very deep in the centre, where the springs were continually bubbling upward, keeping up a good stream which supplied a considerable part of the water in the New River. But its chief beauty was, that the centre was filled with great flocculent masses of green confervæ, while the water in the centre appeared to have a blue tint, producing exquisite shades of blue and green in ever-varying gradations, which were exceedingly beautiful. In fact only once have I seen another spring which equalled it in beauty, in the little island of Semau, near Timor, and that was by no means equal in colour-effects, but only in the depth and purity of the water and the fine rock-basins that contained it. I am informed that now this beautiful Chadwell Spring has been entirely destroyed by the boring of deep wells in the neighbourhood, which have drawn off the springs that supplied it, and that it is now little more than a mud-hole, the whole New River supply being drawn from the river Lea or pumped up from deep wells near Ware. Thus does our morbid civilization destroy the most beautiful works of nature. This spring was, I believe, unequalled in the whole kingdom for simple beauty.

While the country to the north and west of the town was characterized by its numerous streams, mills, and rich meadows, that to the east and south was much higher and drier, rising gradually in low undulations to about four hundred feet and upwards at from four to five miles away. This district was all gravelly with a chalk subsoil, the chalk in many places coming up to the surface, while in others it was only reached at a depth of ten or twenty feet. In the total absence of any instruction in nature-knowledge at that period, my impression, and that of most other boys, no doubt, was, that in some way chalk was the natural and universal substance of which the earth consisted, the only question being how deep you must go to reach it. All this country was thickly dotted with woods and coppices, with numbers of parks and old manor houses ; and as there were abundance of lanes and footpaths, it offered greater attractions to us boys than the more culti- vated districts to the north and west. Walking along the London Road, in about a mile and a half we reached Hertford Heath at a height of three hundred feet above the sea, and half a mile further was Haileybury College, then a training college for the East India Company, now a public school. All round here the country was woody and picturesque ; but our favourite walk, and that of the Grammar School boys, on fine half-holidays in summer, was to what we called the racing- field, a spot about two miles and a half south of the town. As this walk is typical of many of the best features of this part of the town's surroundings, it may be briefly described.

From the south-west corner of All Saints' Churchyard was a broad pathway bounded by hedges, called Queen's-bench Walk, near the top of which was a seat, whence there was a nice view over the town, and the story was that the seat had been put there for Queen Elizabeth, who admired the view. This led into a lane, and further on to an open footpath across a field to Dunkirk's Farm. In this field, about fifty yards to the left, was a spring of pure water carefully bricked round, and as springs were not by any means common, we seldom went this way without running down to it to take a drink of water and admire its purity and upward bubbling out of the earth. At Dunkirk's Farm we crossed the end of Morgan's Walk, a fine straight avenue of lofty elms (I think) about three-quarters of a mile long, terminating in a rather large house—Brickenden Bury. In after years, when I became acquainted with Hood as a serious writer, the scene of that wonderful poem which begins with the verse—

> "'Twas in a shady Avenue,
> Where lofty Elms abound—
> And from a tree
> There came to me
> A sad and solemn sound,
> That sometimes murmur'd overhead
> And sometimes underground "—

was always associated with this Morgan's Walk of my boyhood, an association partly due to the fact that sometimes a woodman was at work felling trees not far off, and this recalled another verse—

> " The Woodman's heart is in his work,
>   His axe is sharp and good :
> With sturdy arm and steady aim
>   He smites the gaping wood ;
>   From distant rocks
>   His lusty knocks
> Re-echo many a rood."

Leaving the avenue we crossed a large field, descending into a lane in a hollow, whence a little further on a path led us along the outside of Bayfordbury Park, the old oak palings of which were well covered with lichen and ivy. Following this path about a mile further by hedges and little brooks and small woods, we came out into a sloping grass field of irregular shape and almost entirely surrounded by woods, while little streamlets, usually with high banks on one side and low banks or gravel heaps on the other, offered the most enticing places for jumping and for playing the exciting game of follow-my-leader. This we called the racing-field ; why I never heard, as it was certainly not suited for horse-racing, though admirably adapted for boyish games and sports. When the boarders of the Grammar School came here, usually accompanied by some of the day-scholars and in charge of one of the masters, or ushers as we then called them, this was the end of our walk, and we were all free to amuse ourselves as we liked till the hour fixed for our return. We then broke up into parties. Some lay down on the grass to rest or to read, some wandered into the woods bird-nesting, some played leap-frog or other games. Here again in after years when I read " The Dream of Eugene Aram," I always associated it with our games in the racing-field, although the place described was totally unlike it—

> " Like sportive deer they coursed about,
>   And shouted as they ran—
> Turning to mirth all things of earth,
>   As only boyhood can ;
> But the Usher sat remote from all,
>   A melancholy man."

Our ushers were not melancholy men, but sometimes one
of them would bring a book to read while we played, and this
was sufficient to carry out the resemblance to the poem, and
summon up to my imagination this charming spot whenever
I read it.

In one corner of this field there was a rather deep circular
hole, from which chalk was brought up as a top-dressing for
some of the poor gravel soil, and this was one of the instances
which led me to the belief that chalk was always somewhere
underground.   In this field I was once told that a wonderful
plant, the bee-orchis, was sometimes found, and my father
used to talk of it as a great rarity.   Once, during the time
we lived at Hertford, some one showed us the flower, and I
remember looking at it as something so strange as to be
almost uncanny, but as I never found one myself I did not
think more of it.

Just over the boundary wall of our school playground,
and continuing along the side of the churchyard, and then
across the fields for a long distance southward, was a dry,
irregular ditch or channel cut in the gravel by flood-water
after heavy rains.   In places this would be very deep—six or
eight feet or more, in others shallow, and in some places there
were vertical drops where regular little waterfalls occurred
after storms.   The whole appearance of this channel was
very strange and mysterious, as there was nothing like it
anywhere else.   We called it the Gulps or Gulphs, but it is
now marked on the ordnance maps as Hag's Dell, showing
that it was looked upon as a mysterious phenomenon by
those who gave it the name.   This also was a kind of playground, and we sometimes spent a whole afternoon wandering
about it.   In the neighbourhood of Morgan's Walk, however,
there were many interesting spots, among others, some old
hedgerows which had been so undermined in a chalky slope
as to form complete overhanging caves, one of which I and
two of my companions made our own, and stored it with a
few necessaries, such as bits of candle, a tinder-box with flint-
steel and matches, and a few provisions, such as potatoes,
which we could roast in our fire, and play at being brigands.

It was in a rather out-of-the-way spot, and quite concealed from ordinary passers-by, and during all the time that we frequented it, we were never disturbed by visitors.

Among the interesting places in the town itself were the castle and the Bluecoat School. The castle was a modern building in the castellated style, but it stood in spacious grounds of about four acres near the middle of the town, with the river flowing through a part of it, and with about two hundred yards of the old defensive wall still remaining in a very complete state. During a short period the family of some of our schoolfellows lived in the castle, and we occasionally went there to play with them, and enjoyed scrambling along the top of the old wall, which, having a parapet still left, was quite practicable and safe. The moat which formerly surrounded it, and was connected with the river, had been long filled up and formed into gardens, which sloped down from the outside of the wall. The original castle was built by Edward the Elder to protect the town against the Danes.

The Bluecoat School was a branch of the celebrated school of the same name, or more properly, Christ's Hospital, in London. It stood at the upper end of Fore Street, opposite where the London Road branched off. Enclosed by lofty iron railings and gates was an oblong playground, about four hundred feet long by a hundred feet wide, bounded on each side by low buildings, forming offices, schoolrooms, and dormitories, while at the end were the large dining-hall and schoolrooms, and in front, near the great gates, the master's residence. On the gate pillars stood two nearly life-size figures of boys in the costume of the school—long blue coat and yellow petticoat, with breeches and yellow stockings, a dress which was quite familiar to us. Occasionally we went to see the boys dine in the grand dining-hall, where the old-world style of everything was of great interest. At the ringing of an outside bell the boys, 250 in all, came in, and seated themselves at the long rows of tables. Then one of the older boys mounted a sort of pulpit and read a long grace, followed by a hymn, in which the boys joined. Then the serving began, a number of the boys taking this duty by turns. Hot meat and vegetables were served on flat wooden platters instead of plates, and I used to pity the boys for not having any place for gravy, which to me was (and still is) the chief luxury of hot meat. What was still more amusing to us was that in place of mugs there were little wooden flagons with wooden hoops and handles, in which they had, I think, beer. If I remember rightly, during the meal the boy in the pulpit read a chapter from the Bible, and at the end there was another grace and hymn. All was carried out with great regularity and very little noise, and the crowds of brightly clad boys, who had red leather belts over their blue coats, and whose yellow stockings were well visible, together with the fine, lofty hall, had a very pleasing effect.

Among the other features of interest in the town were All Saints' Church, adjoining the Grammar School. I used to wonder at what seemed to me a curious and rather dangerous plan of groups of four very slender pillars instead of one large one to support the arches on each side of the nave. I did not know then that these were characteristic of the Early English Gothic, but are not common in our churches. Another feature of this church was its peal of ten bells, which were not only uncommonly numerous, but were of very fine tones, so that when they were well rung, as they frequently were, they produced an exceedingly musical effect, which I have never heard equalled since. The church has since been burnt down and rebuilt, but whether the bells were saved I do not know.

Very conspicuous was the square, ugly brick Town Hall and Market Place at the bottom of Fore Street. This had, however, a large clock-face projecting outwards and supported by three or four pieces of wood which seemed to hold it quite detached from the building, and I used to wonder whether it was a huge watch with all the works inside it. What made this more curious (to me) was that it struck the hours and quarters on very loud and sweet-toned bells, which again I have never since heard surpassed. In this hall were the law-courts, where the Assizes were held, and to which I sometimes gained admittance, and heard a trial of some poor sheep-stealers, who in those days were liable to transportation for life, in order to protect the landed interest, which then ruled the country.

The elections for members of Parliament were at that time scenes of considerable show and excitement, and the members elected had to undergo the ceremony of being chaired, which consisted in being carried round the town on their supporters' shoulders seated in a chair highly decorated with rosettes and coloured ribbons. I well remember the election which took place after the Reform Bill of 1832 was passed, when Thomas Slingsby Duncombe was the Radical member, and was returned at the head of the poll. I saw him being chaired, and when he had been brought back to the door of his hotel, the chair was overturned, as was then the custom, and he had to jump out into his friends' arms to avoid an awkward fall. There was then a scramble for the ribbons and chair-coverings, which were carried away as trophies.

To celebrate the great national event—the passing of the Reform Bill—a banquet was given in the main street to all who chose to attend. It was summer time, and fine weather, and we went to see the feast, which was enjoyed by almost all the poorer people of the town on rows of tables which filled the street for a long distance.

In connection with the game of cricket, I may mention that in those days the players, whether professional or amateur, had none of the paraphernalia of padded leggings and gauntlets now worn ; while a suit of white duck, with an ordinary white or black top-hat, was the orthodox costume. This was the time when the practice of overhand bowling was just beginning, and there was much controversy as to whether or no it should be allowed. I once saw tried a curious bowling machine which it was thought might advantageously take the place of the human bowler. It was called a catapult, and was on the principle of the old instrument used for throwing stones into besieged cities. It consisted of a strong wooden frame about three feet high. On a cross-bar at top was a place for the ball, and this was struck by a knob on an upright arm, which was driven on to it by a powerful spring, something in the manner of a spring-trap. The upright arm was pulled back and held by a catch, which was released by pulling a cord. By slight alterations in the position of the ball and the force of the spring, the ball could be made to pitch on any spot desired, and could thus be slightly changed each time, as is the case with a good bowler. It seemed to answer very well, and it was thought that it might be used for practice where good bowlers were not available, but it never came into general use, and is now, perhaps, wholly forgotten.

CHAPTER IV

HERTFORD : MY SCHOOL LIFE

My recollections of life at our first house in St. Andrew's Street are very scanty. My father had about half a dozen small boys to teach, and we used to play together; but I think that when we had been there about a year or two, I went to the Grammar School with my brother John, and was at once set upon that most wearisome of tasks, the Latin grammar. It was soon after this that I had the first of the three serious illnesses which at different periods brought me within a few hours of death in the opinion of those around me. I know that it must have been after I went to the school by the way the illness began. We had school before breakfast, from half-past six to eight in summer, and as we had nearly half a mile to walk, it was necessary to be out of bed at six. One morning I got up and dressed as usual, went down the two flights of stairs, but when I got to the bottom, I suddenly felt so weak and faint and curiously ill all over that I could go no further, so I had to lie down on the bottom step, and was found there shortly afterwards by the servant coming down to light the fire. That was the beginning of a severe attack of scarlet fever, and I remember little more but heat and horrid dreams till one evening when all the family came to look at me, and I had something given me to drink all night. I was told afterwards that the doctor said this was the crisis, that I was to have port wine in tea-spoonfuls at short intervals, and that if I was not dead before morning I might recover.

For some weeks after this I lived a very enjoyable life in
bed, having tea and toast, puddings, grapes, and other luxuries
till I was well again. Then, before going back to Latin
grammar and other studies of the period, a little incident
or interlude occurred which I am unable to place at any
other period. How it came about I do not at all remember,
but a gentleman farmer from Norfolk must have come to see
us about some business, possibly connected with my sister
and her desired occupation as a governess, and seeing me,
and perhaps hearing of my recent illness, offered to take me
home with him for a visit to play with his boy of about my
age, and to go to Cromer, where his wife, with her sister and
son, were going for change of air. As it was thought that the
change would do me good, and I was delighted at the idea of
going to such a nice seaside place as Cromer, his offer was
kindly accepted. As it happened we did not go to Cromer,
but my visit was, so far as I remember, an enjoyable one.
We went by coach to Ely, where we stayed the night at a
large inn almost joining the cathedral. No doubt we had had
dinner on the way, and I had tea on our arrival, but my host,
whose name I cannot remember, dined with a large party of
gentlemen—probably a farmers' dinner—about six o'clock,
and he told me to walk about and see the shops or wait in
the hall, and I should come in for dessert. So for more than
an hour I wandered up and down the street near the hotel
and past the great entrance to the cathedral. At last a
servant came and called me in, and my friend bade me sit
beside him, and introduced me to the company as a real
Wallace—"Scots wae hae wi' Wallace bled," he added, I
suppose to show what he meant. Then I had fruit of many
kinds, including fine grapes, and a glass of wine, and after an
hour more went to bed.

In the morning, after breakfast, we started in a chaise
which had been sent from my friend's home overnight to
meet him, and we had a long drive to the farm, where we
arrived early in the afternoon, and found dinner ready for us.
There were, I think, two ladies, my friend's wife and her
sister, a boy about my own age, and I think the lady's

brother, who had come some miles on a pony to meet us, and rode back alongside of the carriage.

Of this visit I remember very little except one or two incidents. On the very day of our arrival, I think about tea-time, soon after I and my boy-friend had come in, Mrs. —— became very excited, and then went off into violent hysterics, and was obliged to be taken upstairs to bed. Whether this had anything to do with putting off the visit to Cromer, or some other domestic affairs, I never heard. However, next day all was right again, and I was treated very kindly, as if to show that I had nothing to do with it. I recall the house as a rather long white building with green outside shutters, with a lawn and flower-beds in front, and a kitchen garden and large orchard on one side. In the fields around were some fine trees, and I think there was a pond or a stream near the house and a small village not far off. I and my companion played and roamed about where we liked, but what most struck me was the fruit-gathering in the large orchard, which began the very day after our arrival. I had never seen so many apples before. They were piled in great heaps on the ground, while men and boys went up the trees with ladders and gathered those from the higher branches into baskets. Of course, my little friend knew the best trees, and we ate as many as we liked. Sometimes we went out for drives, or were taken to visit at houses near, or visitors came to tea; but how long I stayed there, or how I returned, I have no recollection, but the main features of the visit as here related have always remained clearly impressed upon my memory.

It may be well here to give a brief outline of my school life at Hertford and of the schoolmaster who taught me. The school itself was built in the year 1617, when the school was founded. It consisted of one large room, with a large square window at each end and two on each side. In the centre of one side was a roomy porch, and opposite to it a projecting portion, with a staircase leading to two rooms above the schoolroom and partly in the roof. The school-room was fairly lofty. Along the sides were what were termed

THE GRAMMAR SCHOOL, HERTFORD.

(*From an engraving in Turner's "History."* 1830.)

porches—desks and seats against the wall with very solid, roughly carved ends of black oak, much cut with the initials or names of many generations of schoolboys. In the central space were two rows of desks with forms on each side. There was a master's desk at each end, and two others on the sides, and two open fireplaces equidistant from the ends. Every boy had a desk the sloping lid of which opened, to keep his school-books and anything else he liked, and between each pair of desks at the top was a leaden ink-pot, sunk in a hole in the middle rail of the desks. As we went to school even in winter at seven in the morning, and three days a week remained till five in the afternoon, some artificial lighting was necessary, and this was effected by the primitive method of every boy bringing his own candles or candle-ends with any kind of candlestick he liked. An empty ink-bottle was often used, or the candle was even stuck on to the desk with a little of its own grease. So that it enabled us to learn our lessons or do our sums, no one seemed to trouble about how we provided the light.

The school was reached by a path along the bottom of All Saints' Churchyard, and entered by a door in the wall which entirely surrounded the school playground and master's garden. Over this door was a Latin motto—

"Inter umbras Academi studere delectat."

This was appropriate, as the grounds were surrounded by trees, and at the north end of the main playground there were two very fine old elms, shown in the old engraving of the school here reproduced.

The headmaster in my time was a rather irascible little man named Clement Henry Crutwell. He limped very much owing to one leg being shorter than the other, and the foot I think permanently drawn up at the instep, but he was very active, used no stick, and could walk along as quickly and apparently as easily as most people. He was usually called by the boys Old Cruttle or Old Clemmy, and when he overheard these names used, which was not often, he would give us a short lecture on the impropriety and impoliteness of miscalling those in authority over us. He was a good master, inasmuch as he kept order in the school, and carried on the work of teaching about eighty boys by four masters, all in one room, with great regularity and with no marked inconvenience. Whatever might be the noise and games going on when he was absent, the moment his step was heard in the porch silence and order at once reigned.

Flogging with a cane was not uncommon for more serious offences, while for slighter ones he would box the ears pretty severely. If a boy did not obey his orders instantly, or repeated his offence soon afterwards, however trifling it might be, such as speaking to another boy or pinching him surreptitiously, he often, without another word, came down from his desk and gave the offender a resounding box on the ear. On one occasion I well remember his coming down to a rather small boy, giving him a slap on one side of his head which knocked him down flat on the seat, and when he slowly rose up, giving him another, which knocked him down on the other side. Caning was performed in the usual old-fashioned way by laying the boy across the desk, his hands being held on one side and his feet on the other, while the master, pulling the boy's trousers tight with one hand, laid on the cane with great vigour with the other. Mr. Crutwell always caned the boys himself, but the other masters administered minor punishments, such as slight ear-boxes, slapping the palm with a flat ruler, or rapping the knuckles with a round one. These punishments were usually deserved, though not always. A stupid boy, or one who had a bad verbal memory, was often punished for what was called invincible idleness when it was really congenital incapacity to learn what he took no interest in, or what often had no meaning for him. When the usual extra tasks or impositions failed with such a boy he was flogged, but I cannot remember whether in such cases his conduct was improved or whether he was given up as "a thoroughly lazy, bad boy, who was a disgrace to the school," and thereafter left to go his own way. Such boys were often very good playfellows, and the magisterial denunciations had little effect upon us.

Mr. Crutwell was, I suppose, a fairly good classical scholar, as he took the higher classes in Latin and Greek. I left school too young even to begin Greek, but the last year or two I was in the Latin class which was going through Virgil's "Æneid" with him. The system was very bad. The eight or ten boys in the class had an hour to prepare the translation, and they all sat together in a group opposite each other and close to Mr. Crutwell's desk, but under pretence of work there were always two or three of the boys who were full of talk and gossip and school stories, which kept us all employed and amused till within about a quarter of an hour of the time for being called up, when some one would remark, " I say, let's do our translation ; I don't know a word of it." Then the cleverest boy, or one who had already been through the book, would begin to translate, two or three others would have their dictionaries ready when he did not know the meaning of a word, and so we blundered through our forty or fifty lines. When we were called up, it was all a matter of chance whether we got through well or otherwise. If the master was in a good humour and the part we had to translate was specially interesting, he would help us on wherever we hesitated or blundered, and when we had got through the lesson, he would make a few remarks on the subject, and say, " Now I will read you the whole incident." He would then take out a trans- lation of the " Æneid" in verse by a relative of his own—an uncle, I think—and, beginning perhaps a page or two back, read us several pages, so that we could better appreciate what we had been trying to translate. I, for one, always enjoyed these readings, as the verse was clear and melo- dious, and gave an excellent idea of the poetry of the Latin writer. Sometimes our laziness and ignorance were found out, and we either had to stay in an hour and go over it again, or copy it out a dozen times, or some other stupid imposition. But as this only occurred now and then, of course it did not in the least affect our general mode of procedure when supposed to be learning our lesson. Mr. Crutwell read well, with a good emphasis and intonation, and I obtained a better idea of what Virgil really was from his readings than from the fragmentary translations we scrambled through.

The three assistant masters, then called ushers, were very distinct characters. The English and writing master, who also taught French, was a handsome, fair young man named Fitzjohn. He was something of a dandy, wearing white duck trousers in the summer, and always having a bright-coloured stiff stock, which was the fashionable necktie of the day. Those being *ante*-steel-pen days he had to make and mend our quill pens, and always had a sharp penknife. He was consequently the authority among the boys on the different knife-makers and the best kind of hones for keeping them sharp ; and when he declared, as I once heard him, that some knives required oil and others water on the stone to bring them to a proper edge, we marvelled at his knowledge. What raised him still higher in our estimation was that he was a fairly good cricketer, and, even more exciting, he was one of the County Yeomanry, and upon the days appointed for drill or inspection, when from his bedroom over the schoolroom he came down in his uniform with sword and spurs, and marched across the room, our admiration reached its height. Though rather contemptuous to the younger boys, he was, I think, a pretty fair teacher. I learnt French from him for about two or three years, and though he taught us nothing colloquially, and could not, I think, speak the language himself, yet I learnt enough to read any easy French book, whereas my six years' grinding at Latin only resulted in a scanty knowledge of the vocabulary and grammar, leaving me quite unable to construe a page from a Latin author with any approach to accuracy. Of course this was partly due to the fact that one language is much more difficult than the other, but more to the method of instruction. Had half the time been devoted to teaching us simple colloquial Latin thoroughly, I feel sure it would have been far more useful to those who left school early, and who had no special talent for languages. The only use Latin has been to me has been the enabling me to understand the specific descriptions of birds and insects in that tongue, and also to appreciate the derivation from Latin of many of our common English words. If the remaining time had been spent in learning German, the result would have been far more useful, but I do not think this language was taught in the school.

The second master, or head usher, was named Hill. He had the end desk opposite to Mr. Crutwell's, and was a rather hard man, who knocked the boys' knuckles with his ruler very severely. On one occasion I remember seeing a boy whose hand was not only black and swollen from blows, but had the skin cut, and was covered with blood. In this case I think a complaint was made by the boy's parents, and Mr. H. was informed privately that he must be more moderate in the future. I do not think I ever had any lessons with this master.

The youngest of the ushers was named Godwin, and was a nephew of Mr. Crutwell. He was rather a large-limbed, dark young man of eighteen or twenty. He was very good natured, and was much liked by the boys, in whose games he often took part. He was, I believe, studying the higher classics with his uncle with the idea of going to the University, but I never heard what became of him afterwards. He taught generally in the school, but the only recollection I have of him as a teacher was in one special case. Shortly before I left the school, I and a few others were put to translate one of the works of Cicero, and we were to be heard the lesson by Godwin. We had none of us any experience of this author before, having translated only Ovid and Virgil. We sat down and worked away with our dictionaries till we knew the meanings, or some of the meanings, of most of the words, but, somehow, could not fit them together to make sense. However, at last we thought we had got something of the meaning. We were called up, and the boy at the head of the class began his translation. When he got stuck Godwin asked the others if they could help him, and when we could not, he would tell us the meaning of some difficult word, and then tell the translator to go on. He went on bit by bit till we got to the end of a long sentence. Then Godwin asked us if we thought we had got it right. We said we didn't know.

Then he said, "Let's see; I will read it just as you have translated it." This he did, and then we could see that we had not made the least approach to anything that was intelligible. So we had to confess that we could only make nonsense of it. Then he began, and translated the whole passage correctly for us, using very nearly the same words as we had used, but arranging them in a very different order, and showing us that the very ideas involved and the whole construction of the sentence was totally different from anything we had imagined. He did all this in a good-humoured way, as if pitying our being put upon a task so much beyond us, and, so far as I now recollect, that was our last as well as our first attempt at translating Cicero. I felt, however, that if we had had Godwin for our Latin teacher from the beginning we should have had a much better chance of really learning the language, and, perhaps, getting to understand Cicero, and appreciate the beauty and force of his style.

Next to Latin grammar the most painful subject I learnt was geography, which ought to have been the most interesting. It consisted almost entirely in learning by heart the names of the chief towns, rivers, and mountains of the various countries from, I think, Pinnock's "School Geography," which gave the minimum of useful or interesting information. It was something like learning the multiplication table both in the painfulness of the process and the permanence of the results. The incessant grinding in both, week after week and year after year, resulted in my knowing both the product of any two numbers up to twelve, and the chief towns of any English county so thoroughly, that the result was automatic, and the name of Staffordshire brought into my memory Stafford, Litchfield, Leek, as surely and rapidly as eight times seven brought fifty-six. The labour and mental effort to one who like myself had little verbal memory was very painful, and though the result has been a somewhat useful acquisition during life, I cannot but think that the same amount of mental exertion wisely directed might have produced far greater and more generally useful results. When I had to learn the chief towns of the provinces of Poland, Russia,

Asia Minor, and other parts of Western Asia, with their almost unpronounceable names, I dreaded the approaching hour, as I was sure to be kept in for inability to repeat them, and it was sometimes only by several repetitions that I could attain even an approximate knowledge of them. No interesting facts were ever given in connection with these names, no accounts of the country by travellers were ever read, no good maps ever given us, nothing but the horrid stream of unintelligible place-names, to be learnt in their due order as belonging to a certain country.

History was very little better, being largely a matter of learning by heart names and dates, and reading the very baldest account of the doings of kings and queens, of wars, rebellions, and conquests. Whatever little knowledge of history I have ever acquired has been derived more from Shakespeare's plays and from good historical novels than from anything I learnt at school.

At one period when the family was temporarily broken up, for some reason I do not remember, I was for about half a year a boarder in Mr. Crutwell's house, in company with twenty or thirty other boys; and I will here give the routine of a pretty good boarding-school at that period.

Our breakfast at eight consisted of a mug of milk-and-water and a large and very thick slice of bread-and-butter. For the average boy this was as much as they could eat, a few could not eat so much, a few wanted more, and the former often gave their surplus to the latter. Any boy could have an egg or a slice of bacon cooked if he bought it himself or had it sent from home, but comparatively very few had such luxuries. Three times a week half the boys had a hot buttered roll instead of the bread-and-butter. These penny rolls were much larger than any I have seen in recent years, although this was in the corn-law days, and one of them was as much as any boy wanted. They were cut in two longitudinally and well buttered, and were served quite hot from the kitchen oven. Any boy who preferred it could have bread-and-butter instead, as a few did, and any bread-and-butter boy who had not much appetite could have a thin slice instead of a thick one by asking for it.

For dinner at one o'clock we had hot joints of meat and vegetables for five days, hot meat-pies on Saturdays made of remnants, with some fresh mutton or beef to make gravy, well seasoned, but always with a peculiar flavour, which I think must have been caused by the meat having been slightly salted or pickled to keep it good. Of course the boys used to turn up their noses at this dinner, but the pie was really very good, with a good substantial crust and abundance of gravy. On Sundays we had a cold joint of meat, with hot fruit-pies in the summer and plum-pudding in the winter, with usually some extra delicacy as custard or a salad. Every boy had half a pint of fairly good beer to drink, and any one who wished could have a second helping of meat, and there were always some who did so, though the first helping was very liberal.

At half-past five, I think, we had milk-and-water and bread-and-butter as at breakfast, from seven to eight we prepared lessons for the next day, and at eight we had supper, consisting of bread-and-cheese and, I think, another mug of beer. The house where the masters lived and where we had our meals and slept was in Fore Street, and was about two hundred feet away from the school; and the large schoolroom was the only place we had to go to in wet weather, when not at meals, but as we were comparatively few in number, it answered our purpose very well.

Occasionally Mr. Crutwell gave us a special treat on some public occasion or holiday. Once I remember he gave us all syllabub in his private garden, two cows being brought up for the occasion, and milked into a pail containing two or three bottles of wine and some sugar. Having been all regaled with this delicacy and plum cake, and having taken a walk round the garden, we retired to our playground rejoicing.

Our regular games were cricket, baseball, leapfrog, high and long jumps, and, in the winter, turnpikes with hoops. This latter was a means of enabling those who had no hoops to get the use of them. They kept turnpikes, formed by two bricks or stones placed the width of the foot apart, and the hoop-driver had to pass through without touching. If the hoop touched he gave it up, and kept the turnpike in his place. When there were turnpikes every five or ten yards all round the playground and a dozen or more hoops following each other pretty closely, the game was not devoid of its little excitements. We never played football (so far as I remember), which at that time was by no means such a common game as it is now. Among the smaller amusements which were always much liked were marbles and pegtops. Marbles were either a game of skill or a form of gambling. In the latter game a small hole was made against a wall, and each player in turn asked for a hand of two or four or even a higher number from some other boy; then with an equal number of his own he tried to pitch them into the hole, and if all or any even number remained in he won the whole, while if the number was odd he lost them. When a boy had lost all his stock of marbles he bought a halfpenny worth and went on playing, and in the end some would lose all the marbles they began with and several pence besides, while others would retire with their trouser-pockets almost bursting with marbles, and in addition several pence resulting from sales in their pockets. I well remember the excitement and fascination even of this very humble form of gambling play; how we would keep on to the very last moment in hopes of retrieving our losses or adding to our gains, then rush home to dinner, and return as quickly as possible to play again before school began. It was really gambling, and though perhaps it could not have been wholly forbidden, it might have been discouraged and made the text for some important teaching on the immorality of gaining only by another's loss. But at that time such ideas had hardly arisen in the minds of teachers.

Pegtops, whipping-tops, and humming-tops were all more or less appreciated, but pegtops were decidedly the most popular, and at certain times a large number of the boys would have them. We used to pride ourselves on being able to make our tops keep up as long as possible, and often

painted them in rings of bright colours, which showed beautifully while they were spinning. Those made of box-wood
and of rather large size were preferred, as their weight, and
the longer string that could be used, caused them to spin
longer. The individuality of tops was rather curious, as some
could only be made to spin by holding them with the peg
upwards, others with it downwards, while others would spin
when held in either position, and thrown almost anyhow.
When tops were in fashion they might have been made the
vehicle for very interesting teaching of mechanics, but that
again was quite beyond the range of the ordinary schoolmaster of the early part of the nineteenth century.

During my last year's residence at Hertford an arrangement was made by which, I suppose, the fees paid for my
schooling were remitted on condition that I assisted in the
school. I was a good writer and reader, and while continuing
my regular classes in Latin and algebra, I took the younger
boys in reading and dictation, arithmetic and writing.
Although I had no objection whatever to the work itself, the
anomalous position it gave me in the school—there being a
score of boys older than myself who were scholars only—was
exceedingly distasteful. It led to many disagreeables, and
subjected me to painful insinuations and annoying remarks.
I was especially sensitive to what all boys dislike—the being
placed in any exceptional position, or having to do anything
different from other boys, and not of my own choice. Every
time I entered the schoolroom I felt ashamed, and whether I
was engaged at my own lessons or occupied as a teacher, I
was equally uncomfortable. I cannot now remember all the
details of what was to me a constant humiliation, but I am
sure it must have been a time of very real mental anguish
from one result that persisted almost into middle life. For
at least twenty years after I left school, and I think even
longer, I was subject to frequently recurring dreams of still
having to go to school in the hybrid position of pupil and
teacher, aggravated by feeling myself taller, and at last a
man, and yet suffering over again with increased intensity

the shyness and sense of disgrace of my boyhood. In my dreams I hated to go ; when I reached the schoolhouse I dreaded to open the door, especially if a few minutes late, for then all eyes would be upon me. The trouble of not always knowing what to do came upon me with exaggerated force, and I used to open my desk and fumble about among its contents so as to hide my face as long as possible.

After some years the dream became still more painful by the thought occurring to me sometimes that I need *not* go, that I had *really* left school ; and yet the next time the dream came I could not resist the impulse to go, however much I dreaded it. At last a phase came in which I seemed to have nothing to do at the school, and my whole time there was spent in pretending to do something, such as mending pens or reading a school-book, all the while feeling that the boys were looking at me and wondering what I was there for. Then would come a struggle *not* to go. I would say to myself that I was sure I had left school, that I had nothing to do there, that if I never went again nothing would happen ; yet for a long time I always *did* go again. Then for a time I would dream that it was close to the holidays, or that the next day was breaking-up, and that I had better not go at all. Then I would remember that my books and slate and other things were in my desk, and that I *must* take them away. And after this for some years I would still occasion-ally dream that I had to go on this last day to carry away my books and take formal leave of Mr. Crutwell. After having got to this point even, the dream reappeared, and I went over the last school-day again and again ; and then the final stage came, in which I seemed to have the old impulse to go to school, even started on the way, and then remembered that I had *really* left, that I need *never* go any more, and with an infinite sense of relief turned back, and found myself in some quite different life.

Now, the very long persistence of such a dream as this shows, I think, how deeply impressionable is the mind at this period of boyhood, and how very difficult it is to get rid of painful impressions which have been almost daily repeated.

Whether or not this particular form of experience in my boyhood produced any permanent effect on my character I cannot
say, but the mere continuance of a painful dream for so many
years is in itself an evil, and must almost certainly have had
an injurious effect upon the bodily health. Even in my home-
life I was subject to impressions of the same general nature,
though far less severe. Many slight faults of conduct which had
been long overlooked were often suddenly noticed, and I was
ordered at once to change them. One such that I remember
was that I had been accustomed to use my spoon at table
with my left hand, when I was one day told to use my right.
No doubt I could have done this without much trouble, but
I seemed to feel that to make such a change would be singular,
would draw the attention of my brothers and sisters to me,
and would be a kind of confession of ignorance or clumsiness
which I could *not* make. I felt too much ashamed to do it.
I put down my spoon and waited, and when I thought no one
was looking, took it up again in the way forbidden. This was
said to be obstinacy, but to me it seemed something else which
I could hardly describe. However, the result was that I was
sent away from table up to my bedroom, and was ordered to
have my meals there till I would "do as I was bid." I forget
exactly how it ended, but I think I remained under this
punishment several days, and that it was only under the kind
persuasions and advice of my mother and sisters that I was at
length allowed to come down ; and this was the most terrible
ordeal of all, and when I actually took the spoon in my right
hand, I felt more hurt and ashamed than when I was sent
away from table. This is only an example of numbers of
little things of a similar character, which were treated in the
same rough and dogmatic manner, which was then almost
universal, and was thought to be the only way of training
children. *How*, exactly, to treat each case must depend upon
circumstances, but I think that a little mild ridicule would
have a better effect than compulsion. I might have been told
that, although *we* did not much care about it, other people
would think it very strange, and that *we* should then be
ashamed because people would say that we did not know

good manners. Or I might have been asked to practise it by myself, and try the experiment, using sometimes one hand and sometimes the other, till at last, when the holidays or my next birthday came, or I first had new clothes on, I was to complete the victory over myself by discarding the left-hand spoon altogether.

One other case of this kind hurt me dreadfully at the time, because it exposed me to what I thought was the ridicule or contempt of the whole school. Like most other boys I was reckless about my clothes, leaning my elbows on the desk till a hole was worn in my jacket, and, worse still, when cleaning my slate using my cuff to rub it dry. Slate sponges attached by a string were unknown to our school in those days. As new clothes were too costly to be had very often, my mother determined to save a jacket just taken for school wear by making covers for the sleeves, which I was to wear in school. These were made of black calico, reaching from the cuff to the elbow, and though I protested that I could not wear them, that I should be looked upon as a guy, and other equally valid reasons, they were one day put in my pocket, and I was told to put them on just before I entered the school. Of course I could *not* do it; so I brought them back and told my mother. Then, after another day or two of trial, one morning the dreaded thunderbolt fell upon me. On entering school I was called up to the master's desk, he produced the dreaded calico sleeves, and told me that my mother wished me to wear them to save my jacket, and told me to put them on. Of course I had to do so. They fitted very well, and felt quite comfortable, and I dare say did not look so very strange. I have no doubt also that most of the boys had a fellow-feeling for me, and thought it a shame to thus make me an exception to all the school. But to me it seemed a cruel disgrace, and I was miserable so long as I wore them. How long that was I cannot remember, but I do not think it was very long, perhaps a month or two, or till the beginning of the next holidays. But while it lasted it was, perhaps, the severest punishment I ever endured.

In an article on the civilizations of China and Japan in

The Independent Review (April, 1904), it is pointed out that the universal practice of "saving the face" of any kind of opponent rests upon the fundamental idea of the right of every individual to be treated with personal respect. With them this principle is taught from childhood, and pervades every class of society, while with us it is only recognized by the higher classes, and by them is rarely extended to inferiors or to children. The feeling that demands this recognition is certainly strong in many children, and those who have suffered under the failure of their elders to respect it, can well appreciate the agony of shame endured by the more civilized Eastern peoples, whose feelings are so often outraged by the total absence of all respect shown them by their European masters or conquerors. In thus recognizing the sanctity of this deepest of human feelings these people manifest a truer phase of civilization than we have attained to. Even savages often surpass us in this respect. They will often refuse to enter an empty house during the absence of the owner, even though something belonging to themselves may have been left in it; and when asked to call one of their sleeping companions to start on a journey, they will be careful not to touch him, and will positively refuse to shake him rudely, as an Englishman would have no scruple in doing.

CHAPTER V

As the period from the age of six to fourteen which I spent at Hertford was that of my whole home-life till I had a home of my own twenty-eight years later, and because it was in many ways more educational than the time I spent at school, I think it well to devote a separate chapter to a short account of it.

During the year or two spent at the first house we occupied in St. Andrew's Street very little occurred to impress itself upon my memory, partly, I think, because I was too young and had several playfellows of my own age, and partly, perhaps, because the very small house and yard at the back offered few facilities for home amusements. There was also at that time too much inequality between myself and my brother John for us to become such constant companions as we were a little later.

When we moved to the house beyond the Old Cross, nearly opposite to the lane leading to Hartham, the conditions were altogether more favourable. The house itself was a more commodious one, and besides a yard at one side, it had a small garden at the back with a flower border at each side, where I first became acquainted with some of our common garden flowers. The gable end of the house in the yard, facing nearly south, had few windows, and was covered over with an old vine which not only produced abundance of grapes, but enabled my father to make some gallons of wine from the thinnings. But the most interesting feature of the premises to us two boys was a small stable with a loft over it, which, not being used except to store garden-tools and odd lumber, we had practically to ourselves. The loft especially was most delightful to us. It was reached by steps formed by nailing battens across the upright framing of the stable, with a square opening in the floor above. It thus required a little practice to climb up and down easily and to get a safe landing at top, and doing this became so easy to us that we ran up and down it as easily as sailors run up the shrouds of a vessel. Then the loft itself, under the sloping roof, gloomy and nearly dark in the remote corners, was almost like a robbers' cave, while a door opening to the outside by which hay could be pitched up out of a cart, afforded us plenty of light when we required it, together with the novel sensation and spice of danger afforded by an opening down to the floor, yet eight or nine feet above the ground.

This place was our greatest delight, and almost all the hours of daylight we could spare from school and meals were spent in it. Here we accumulated all kinds of odds and ends that might be useful for our various games or occupations, and here we were able to hide many forbidden treasures such as gunpowder, with which we used to make wild-fires as well as more elaborate fireworks. John was of a more mechanical turn than myself, and he used to excel in making all the little toys and playthings in which boys then used to delight. I, of course, looked on admiringly, and helped him in any way I could. I also tried to imitate him, but only succeeded in some of the simpler operations. Our most valuable guide was the "Boy's Own Book," which told us how to make numbers of things boys never think of making now, partly because everything is made for them, and also because children get so many presents of elaborate or highly ornamented toys when very young, that by the time they are old enough to make anything for themselves they are quite *blasé*, and can only be satisfied by still more elaborate and expensive playthings. I think it may be interesting to give a short enumeration of the things which at this time John and I used to make for ourselves.

I may mention first that, owing to the very straightened

circumstances of the family during the whole of our life at Hertford, we were allowed an exceedingly scanty amount of pocket-money. Till I was ten years old or more I had only a penny a week regularly, while John may perhaps have had twopence, and it was very rarely that we got tips to the amount of the smaller silver coins. We were, therefore, obliged to save up for any little purchase required for our various occupations, as, for example, to procure the saltpetre and sulphur required for making fireworks; the charcoal we could make ourselves, and obtain the iron filings from some friendly whitesmith. The simplest fireworks to make were squibs, and in these we were quite successful, following the recipe in the "Boy's Own Book." The cases we made beforehand with a little copy-book paper and paste. Crackers were much more difficult, and the home-made ones were apt to go off all at once instead of making the regular succession of bangs which the shop article seemed never to fail in doing. But by perseverance some fairly good ones were made, though they could never be thoroughly trusted. Roman candles we were also tolerably successful with, though only the smallest size were within our means; and we even tried to construct the beautiful revolving Catherine-wheels, but these again would often stop in the middle, and refuse either to revolve properly or to burn more than half way.

In connection with fireworks, we were fond of making miniature cannon out of keys. For this purpose we begged of our friends any discarded box or other keys with rather large barrels, and by filing a touch-hole, filing off the handle, and mounting them on block carriages, we were able to fire off salutes or startle our sister or the servant to our great satisfaction. When, later, by some exchange with a fellow school-boy or in any other way, we got possession of one of the small brass cannons made for toys, our joy was great; and I remember our immense admiration at one of these brass cannon, about six inches long, in the possession of a friend, which would go off with a bang as loud as that of a large pistol. We also derived great pleasure by loading one of our weapons to the very muzzle, pressing it down into the ground

so that we could lay a train of powder to it about two feet
long, and then escape to a safe distance, and see it jump up
into the air with the force of the explosion.

On the fifth of November we always had a holiday, and
in the evening there was always in the playground a large
bonfire and a considerable display of fireworks by a professional, some of the wealthier of the boys' parents contributing
the outlay. On these occasions almost all the day-scholars
came, their pockets more or less filled with crackers and
squibs, to occupy the time before the more elaborate fireworks. The masters were all present to help keep order
and prevent accidents, and no boy was allowed to light
squib or cracker till about seven o'clock, when Mr. Crutwell
himself lighted the first squib, threw it in the air, and was
immediately followed by the boys in every part of the playground, which soon presented a very animated scene. Many
of the parents, relatives, and friends of the boys were also
present, so that the playground was quite crowded, yet though
the boys recklessly threw squibs and crackers in all directions,
no accidents of any importance happened. Now and then a
boy would have the squibs or crackers in his pocket exploded,
but I do not remember any injury being done in that way.
But shortly after I left, I think, a serious accident occurred,
by which some one was permanently injured, and after that I
believe the miscellaneous fireworks of the boys were no longer
allowed.

Among our favourite playthings were pop-guns and miniature spring-guns and pistols. Pop-guns were made of stout
pieces of elder-wood, which, when the pith is pushed out
has a perfectly smooth, glossy inner surface which made a
better pop than those bought at the toy-shop. Many a
pleasant walk we had to get good straight pieces of elder,
which, when cut to the proper length and a suitable strong
stick made to force out the pellets of well-chewed brown
paper or tow, would shoot them out with a report almost
equal to that of a small pistol.

Far more elaborate and ingenious, however, were the
spring-pistols which my brother made so well and finished

so beautifully that he often sold them for a shilling or more, and thus obtained funds for the purchase of tools or materials. For the stocks he would beg odd bits of mahogany or walnut or oak from a cabinet-maker's shop, and carve them out carefully with a pocket-knife to the exact shape of pistol or gun. The barrel was formed of a goose-quill or swan's-quill, carefully fastened into the hollow of the stock with waxed thread, and about an inch of the hinder part of this had the upper half cut away to allow the spring to act. In the straight part near the bend of the stock a hole was cut for the trigger, which was held in its place by a stout pin passing through it on which it could turn. The only other article needed was a piece of strong watch or clock-spring, of which we could get several at a watchmaker's for a penny. The piece of watch-spring being broken off the right length and the ends filed to a smooth edge, was tied on to the stock between the barrel and the trigger, curving upwards, and one end fitting into a notch at the top of the trigger, while the other end was bent round so that the end fitted into a small notch in the open part of the quill at its hinder end. It was then cocked, and a pea or shot being placed in front of the spring, a slight pressure on the trigger would release it and cause it to drive out the shot or pea with considerable velocity. My brother used to take great delight in making these little pistols, shaping the stocks very accurately, rubbing them smooth with sandpaper, and then oiling or varnishing them; while every part was finished off with the greatest neatness. I do not think there was any boy in the school who made them better than he did, and very few equalled him.

One of the most generally used articles of a boy's stock of playthings are balls, and as these are often lost and soon worn out we used to make them ourselves. An old bung cut nearly round formed the centre; this we surrounded with narrow strips of list, while for the outside we used coarse worsted thread tightly wound on, which formed a firm and elastic ball. We had two ways of covering the balls. One was to first quarter it tightly with fine string, and using this as a base, cover the whole with closely knitted string by means of a very simple loop-stitch. A much superior plan was to obtain from the tan-yard some partly tanned sac-shaped pieces of calf-skin which were of just the size required for a small-sized cricket-ball. These were stretched over the ball, stitched up closely on the one side, the joint rubbed down smooth, and by its partial contraction when drying an excellent leather-covered ball was made, which at first was hairy outside, but this soon wore off. In this way, at a cost of about twopence or threepence, we had as good a ball as one which cost us a shilling to buy, and which served us well for our boyish games at cricket.

Other house occupations which employed much of our spare time in wet weather and in winter were the making of cherry-stone chains and bread-seals. For the former we collected some hundreds of cherry-stones in the season. These, with much labour and scraping of fingers, were ground down on each side till only a ring of suitable thickness was left. The rings were then soaked in water for some days, which both cleaned and softened them, so that with a sharp pen-knife they could be cut through, and by carefully expanding them the next ring could be slipped in, the joint closing up so as to be scarcely, if at all, visible. When nicely cleaned, and if made from stones of nearly uniform size, these chains made very pretty and useful watch-guards, or even necklaces for little girls of our acquaintance.

Bread-seals were easier to make, and were more interesting in their results. In those ante-penny-postage days envelopes were unknown, as one of the rules of the post-office was that each letter must consist of a single sheet, any separate piece of paper either enclosed or outside constituting it a double letter with double postage. Almost every letter, therefore, was sealed, and many of them had either coats-of-arms, crests, heads, or mottoes, so that besides the contents, which were, perhaps, only of importance to the recipient, the seal would often interest the whole family. In such a case we begged for the seal to be carefully cut round so that we might make a copy of it. To do this we required only a piece of the crumb of new bread, and with cleanly washed hands we worked this up with our fingers till it formed a compact stiff mass. Before doing this, we begged a little bright water-colour, carmine or Prussian blue, from our sisters, and also, I think, a very small portion of gum. When all was thoroughly incorporated so that the whole lump was quite uniform in colour and texture, we divided it into balls about the size of a large marble, and carefully pressed them on to the seals, at the same time squeezing the bread up between our fingers into a conical shape to form the upper part of the seal serving as a handle and suspender. Each seal was then carefully put away to dry for some days, when it got sufficiently hard to be safely removed. It was then carefully trimmed round with a sharp pen-knife, and accurately shaped to resemble the usual form of the gold or silver seals which most persons carried on their watch-chains to seal their letters. The seal itself would be perfectly reproduced with the glossy surface of the original, and when still more hardened by thorough drying, would make a beautiful impression in sealing-wax. In this way we used to get quite a collection of ornamental seals, which, if carefully preserved, would last for years.

Almost all the above amusements and occupations were carried on in the stable and loft already described, during the two or three years we lived there. After that my brother John went to London, and was apprenticed to a builder to learn carpentry and joinery. When left alone at home, my younger brother being still too young for a playmate, I gave up most of these occupations, and began to develop a taste for reading. I still had one or two favourite companions with whom I used to go for long walks in the country round, amusing ourselves in gravel or chalk pits, jumping over streams, and cutting fantastic walking-sticks out of the woods; but nothing afterwards seemed to make up for the quiet hours spent with my brother in the delightful privacy of the loft which we had all to ourselves. The nearest approach to it was about a year later when, for some family reason that I quite forget, I was left to board with Miss Davies at

All Saints' Vicarage, then used as a post-office, a large rambling old house with a large garden, in which there was among other fruit an apple tree which bore delicious ribston-pippins, of which I was allowed to eat as many as I liked of the windfalls. In this house there was a loft in the roof, which I was told was full of old furniture and other things, so I one day asked if I might go up into it. Miss Davies, who was very kind though melancholy, said I might. So I went up, and found all kinds of old broken or moth-eaten furniture, broken lamps, candlesticks, and all the refuse of a house where a family have lived for many years. But among these interesting things I hit upon two veritable treasures from my point of view. One was a very good, almost new, cricket-bat, of a size just suitable to me ; and the other was still more surprising and attractive to me, being a very large, almost gigantic, box-wood pegtop, bigger than any I had seen. It seemed to me then almost incredible that such treasures could have been ranked as lumber, and purposely left in that old attic. I thought some one must surely have put them there for safety, and would soon come and claim them. I therefore waited a few days till Miss Davies seemed rather more communicative than usual, when I said to her, " I found something very nice in the lumber-room." " Oh, indeed ; and what is it ? " said she, " I did not know there was anything nice there." " May I go and fetch them for you to see ? " said I ; and she said I might. So I rushed off, and brought down the top and the bat, and said, " I found these up there ; do you know whose they are ? " She looked at them, and said, " They must have belonged to ——," mentioning a name which I have forgotten. " They have been there a good many years." Then, as I looked at them longingly, she said, "You can have them if you like "—as if they were of not the least value. I felt as if I had had a fortune left me. The top was the admiration of the whole school. No one had so large a top or had even seen one so large, yet I was quite able to spin it properly, my hands being rather large for my age. This occurred in the winter, and when the cricket season came, I equally enjoyed my bat, which at once elevated me to the rank of the few bigger boys who had bats of their own.

But even these rapturous delights were not so enduring, and certainly not so educational, as those derived from making as well as possessing toys and playthings, and the year or two I spent with my brother in these pleasant occupations were certainly the most interesting and perhaps the most permanently useful of my whole early boyhood. They enabled me to appreciate the pleasure and utility of doing for one's self everything that one is able to do, and this has been a constant source of healthy and enjoyable occupation during my whole life. It led, I have no doubt, to my brother being apprenticed to a carpenter and builder, where he became a first-rate workman ; and from him later on I learnt to use the simpler tools. During my whole life I have kept a few such tools by me, and have always taken a pleasure in doing the various little repairs continually needed in a house and garden. I therefore look with compassion on the present generation of children and schoolboys who, from their earliest years, are overloaded with toys, so elaborately constructed and so highly finished that the very idea of making any toys for themselves seems absurd. And these purchased toys do not give anything like the enduring pleasure derived from the process of making and improving as well as afterwards using ; while it leads to the great majority of men growing up without any idea of doing the simplest mechanical work required in their own homes.

It was during our residence at this house near the Old Cross that, I think, my father enjoyed his life more than anywhere else at Hertford. Not only had he a small piece of garden and the fine grape-vine already mentioned, but there was a roomy brew-house with a large copper, which enabled him to brew a barrel of beer as well as make elder-wine and grape-wine, bottle gooseberries, and other such work as he took great pleasure in doing. When here also, I think, he hired a small garden about half a mile off, where he could grow vegetables and small fruit, and where he spent a few

hours of every fine day. And these various occupations were
an additional source of interest and instruction to us boys.
It was here, however, that our elder sister died of consumption in the year 1832, a little before she attained her twenty-
second year. This was a severe loss to my father and mother,
though I was not of an age to feel it much. I think it was
soon afterwards that my remaining sister went to live at
Hoddesdon, four miles away, as governess to two girls in a
gentleman's family there. These girls were somewhere near
my age, or a little older, and occasionally in the summer
my brother and I were invited to dine and spend the afternoon with them, which we greatly enjoyed, as there was a
large garden, and beyond it a large grass orchard full of
apple and other fruit trees. We also enjoyed the walk there,
and back in the evening, through the picturesque country I
have already described. My sister lived in this family for
two or three years, and was on terms of affection with the two
girls till they were married.

In the year 1834, I think, my sister went to a French
school in Lille in order to perfect herself in conversation, in
view of becoming a governess or keeping a school. But the
following year the misfortune occurred that still further
reduced the family income. Mr. Wilson, who had married
my mother's only sister, was one of the executors of her
father's will, and as he was a lawyer (the other executor
being a clergyman), and his own wife and her sister were the
only legatees, he naturally had the sole management of the
property. Owing to a series of events which we were only
very imperfectly acquainted with, he became bankrupt in this
year, and his own wife and large family were at once reduced
from a condition of comfort and even affluence to poverty,
almost as great as our own. But we children also suffered, for
legacies of £100 each to my father's family, to be paid to us
as we came of age, together with a considerable sum that had
reverted to my mother on the death of her stepmother in
1828, had remained in Mr. Wilson's hands as trustee, and
was all involved in the bankruptcy. He did all he possibly

could for us, and ultimately, I believe, repaid a considerable
part of the money, but while the legal proceedings were in
progress, and they lasted full three years, it was necessary for
us to reduce expenses as much as possible. We had to leave
our comfortable house and garden, and for a time had the
use of half the rambling old house near All Saints' Church
already mentioned.

Before this, I think, my brother John had gone to London
to be apprenticed, and the family at home consisted only of
myself and my younger brother Herbert till my sister returned
from France. It must have been about this time that I was
sent for a few months as a boarder at the Grammar School, as
already stated ; but this whole period of my life is very indistinct. I am sure, however, that we moved to the next house
in St. Andrew's Street early in 1836, because on May 15 of
that year an annular eclipse of the sun occurred, visible in
England, and I well remember the whole family coming out
with smoked glasses into the narrow yard at the side of the
house in order to see it. I was rather disappointed, as it only
produced a peculiar gloom such as often occurs before a
thunderstorm. While we were here a brewery was being
built at the bottom of the yard, and while inspecting it and
inquiring what the various tanks, boilers, etc., were for, I
learnt that the word " water " was tabooed in a brewery ; that
it must always be spoken of as " liquor," and any workman
or outsider mentioning " water " is immediately fined or called
upon to stand a gallon of beer, or more if he can afford it.

At midsummer, I think, we again moved to a part of a
house next to St. Andrew's Church, where we again had the
Silk family for neighbours in the larger half of the house.
They also had most of the garden, on the lawn of which was
a fine old mulberry tree, which in the late summer was so
laden with fruit that the ground was covered beneath it, and
I and my friend George used to climb up into the tree,
where we could gather the largest and ripest fruit and feast
luxuriously.

This was the last house we occupied in Hertford, the
family moving to Hoddesdon some time in 1837, to a pretty

but very small red-brick house called Rawdon Cottage, while
I went to London and stayed at Mr. Webster's with my
brother John, preparatory to going with my eldest brother
William to learn land-surveying.

During the time I lived at Hertford I was subject to
influences which did more for my real education than the
mere verbal training I received at school. My father belonged
to a book club, through which we had a constant stream of
interesting books, many of which he used to read aloud in
the evening. Among these I remember Mungo Park's travels
and those of Denham and Clapperton in West Africa. We
also had *Hood's Comic Annual* for successive years, and I
well remember my delight with " The Pugsley Papers " and
" A Tale of the Great Plague," while as we lived first at a
No. 1, I associated Hood's " Number One " with our house,
and learnt the verses by heart when I was about seven years
old. Ever since those early experiences I have been an
admirer of Hood in all his various moods, from his inimitable
mixture of pun and pathos in his " Sea Spell," to the exquisite
poetry of " The Haunted House," " The Elm Tree," and
" The Bridge of Sighs."
We also had some good old standard works in the house,
" Fairy Tales," " Gulliver's Travels," " Robinson Crusoe," and
the " Pilgrim's Progress," all of which I read over again and
again with constant pleasure. We also had " The Lady of
the Lake," " The Vicar of Wakefield," and some others ; and
among the books from the club I well remember my father
reading to us Defoe's wonderful " History of the Great
Plague." We also had a few highly educational toys, among
which were large dissected maps of England and of Europe,
which we only had out as a special treat now and then, and
which besides having the constant charm of a puzzle, gave us
a better knowledge of topographical geography than all our
school teaching, and also gave me that love of good maps
which has continued with me throughout life. Another
valuable toy was a model of a bridge in wood, the separate
stones constituting the arch of which could be built up on a

light centre, showing beautifully the principle of the arch, and how, when the keystone was inserted the centre supports could be removed and a considerable weight supported upon it. This also was a constant source of pleasure and instruction to us, and one that seems to be not now included among instructive toys.

I think it was soon after we went to the Old Cross house that my father became librarian to a fairly good proprietary town library, to which he went for three or four hours every afternoon to give out and receive books and keep everything in order. After my brother John left home and I lost my chief playmate and instructor, this library was a great resource for me, as it contained a large collection of all the standard novels of the day. Every wet Saturday afternoon I spent there; and on Tuesdays and Thursdays, which were our four-o'clock days, I usually spent an hour there instead of stopping to play or going straight home. Sometimes I helped my father a little in arranging or getting down books, but I had most of the time for reading, squatting down on the floor in a corner, where I was quite out of the way. It was here that I read all Fenimore Cooper's novels, a great many of James's, and Harrison Ainsworth's "Rookwood," that fine highwayman's story containing a vivid account of Dick Turpin's Ride to York. It was here, too, I read the earlier stories of Marryat and Bulwer, Godwin's "Caleb Williams," Warren's "Diary of a Physician," and such older works as "Don Quixote," Smollett's "Roderick Random," "Peregrine Pickle," and "Humphry Clinker," Fielding's "Tom Jones," and Miss Burney's "Evelina." I also read, partially or completely, Milton's "Paradise Lost," Pope's "Iliad," Spenser's "Faërie Queene," and Dante's "Inferno," a good deal of Byron and Scott, some of the *Spectator* and *Rambler*, Southey's "Curse of Kehama," and, in fact, almost any book that I heard spoken of as celebrated or interesting. At this time "Pickwick" was coming out in monthly parts, and I had the opportunity of reading bits of it, but I do not think I read it through till a considerably

later period. I heard it a good deal talked about, and it
occasioned quite an excitement among the masters in the
Grammar School. Walton's "Angler" was a favourite of
my father's, and I well remember a wood-cut illustration of
Dove Dale with greatly exaggerated rocks and pinnacles,
which made me long to see such a strange and picturesque
spot—a longing which I only gratified about a dozen years
ago, finding it more exquisitely beautiful than I had imagined
it to be, even if not quite so fantastic.

I may now say a few words about our home-life as regards
meals and other small matters, because I think its simplicity
was perhaps better for children than what is common now.
Till we reached the age of ten or twelve we never had tea or
coffee, our breakfast consisting of bread-and-milk and our
tea of milk-and-water with bread-and-butter. Toast, cake,
muffins, and such luxuries were only indulged in on festive
occasions. At our one-o'clock dinner we began with pudding
and finished with meat and vegetables. During this period
we made our own bread, and good wholesome bread it was,
made with brewer's yeast (which I often went for to the
brewery), and sent to the nearest baker to be baked, as were
most of our baked pies and puddings. Kitcheners were almost
unknown then, and meat was roasted before the open fire with
a clock-work jack, dripping-pan, and large tinned screen to
reflect the heat and to warm plates and dishes.

A few words about the cost of living will not be out of
place here, and will serve to correct some erroneous ideas on
the subject. Tea was about double the price it is now, but
coffee and cocoa were about the same as at present; and
these latter were commonly used for breakfast, while tea was
only taken at tea, and then only by the older members of the
family. Sugar was also more than twice as dear, but milk,
eggs, and butter were all cheaper. Although this was in the
corn-law days I doubt if our bread was any dearer than it is
now, and it was certainly much better. It was ground in the
mills of the town from wheat grown in the country round, and
the large size of the penny rolls, which I have already

mentioned, shows that there cannot have been much difference
of price to the retail buyer, who was then usually one or two
steps nearer to the actual corn-grower than he is now.   Meat
also was cheaper than now.   The price of the best beef was
sixpence to sevenpence a pound; while mutton was sevenpence to eightpence for the best joints, but for ordinary
parts much less.   In the country gleaning was a universal
practice, and numbers of cottagers thus got a portion of their
bread; while a much larger proportion than now lived in the
country and had large gardens or a few acres of land.   My
mother often took me with her when visiting such poor
cottagers as were known to her, and my impression is that
there was very little difference in the kind and degree of the
rural poverty of that day and this; and a few years later, as
I shall show, the same may be said of the skilled mechanic.
As a prime factor in this question, it must always be remembered that rent, both in villages and towns, was in most
cases less than half what it is at present, and this more than
compensated for the few cheaper articles of food and clothing
to-day.

My father and mother were old-fashioned religious people
belonging to the Church of England, and, as a rule, we all
went to church twice on Sundays, usually in the morning and
evening.   We also had to learn a collect every Sunday morning, and were periodically examined in our catechism.   On
very wet evenings my father read us a chapter from the Bible
and a sermon instead of the usual service.   Among our friends,
however, were some Dissenters, and a good many Quakers,
who were very numerous in Hertford; and on rare occasions
we were taken to one of their chapels instead of to church,
and the variety alone made this quite a treat.   We were
generally advised when some " friend " was expected to speak,
and it was on such occasions that we visited the Friends'
Meeting House, though I remember one occasion when, during
the whole time of the meeting, there was complete silence.
And when any brother or sister *was* " moved to speak," it
was usually very dull and wearisome; and after having attended
two or three times, and witnessed the novelty of the men and

women sitting on opposite sides of the room, and there being no pulpit and no clergyman and no singing, we did not care to go again. But the Dissenters' chapel was always a welcome change, and we went there not unfrequently to the evening service. The extempore prayers, the frequent singing, and the usually more vigorous and exciting style of preaching was to me far preferable to the monotony of the Church service; and it was there only that, at one period of my life, I felt something of religious fervour, derived chiefly from the more picturesque and impassioned of the hymns. As, however, there was no sufficient basis of intelligible fact or connected reasoning to satisfy my intellect, this feeling soon left me, and has never returned.

Among our Quaker friends were two or three to whose houses we were occasionally invited, and I remember being greatly impressed by the excessive cleanliness and neatness of everything about their houses and gardens, corresponding to the delicate colouring and simple style of their clothing. At that time every Quaker lady wore the plainest of dresses, but of the softest shades of brown or lilac, while the men all wore the plain cutaway coat with upright collar, also of some shade of brown, which, with the low broad-brimmed beaver hat of the best quality, gave them a very distinctive and old-world appearance. They also invariably used "thee" and "thou" instead of "you" in ordinary conversation, which added to the conviction that they were a people apart, who had many habits and qualities that might well be imitated by their neighbours of other religious denominations.

CHAPTER VI

LONDON WORKERS, SECULARISTS AND OWENITES

HAVING finally left school at Christmas, 1836, I think it was early in 1837 that I was sent to London to live at Mr. Webster's in Robert Street, Hampstead Road, where my brother John was apprenticed. My father and mother were then about to move to the small cottage at Hoddesdon, and it was convenient for me to be out of the way till my brother William could arrange to have me with him to learn land-surveying. As I shared my brother's bedroom and bed, I was no trouble, and I suppose I was boarded at a very low rate. As the few months I spent here at the most impressionable age had some influence in moulding my character, and also furnished me with information which I could have obtained in no other way, I devote the present chapter to giving a short account of it.

Mr. Webster was a small master builder, who had a work shop in a yard about five minutes' walk from the house, where he constantly employed eight or ten men preparing all the joinery work for the houses he built. At that time there were no great steam-factories for making doors and windows, working mouldings, etc., everything being done by hand, except in the case of the large builders and contractors, who had planing and sawing-mills of their own. Here in the yard was a sawpit in which two men, the top- and bottom-sawyers, were always at work cutting up imported balks of timber into the sizes required, while another oldish man was at work day after day planing up floor-boards. In the shop itself windows and doors, cupboards, staircases, and other joiner's work was always going on, and the men employed all lived in the small streets surrounding the shop.   The working hours were from six to half-past five, with one and a half hours out for meals, leaving a working day of ten hours.

Having nothing else to do, I used to spend the greater part of my time in the shop, seeing the men work, doing little jobs occasionally, and listening to their conversation.   These were no doubt an average sample of London mechanics, and were on the whole quite as respectable a set of men as any in a similar position to-day.   I soon became quite at home in the shop, and got to know the peculiarities of each of the men.   I heard their talk together, their jokes and chaff, their wishes and their ideas, and all those little touches of character which come out in the familiar intercourse of the workshop. My general impression is that there was very little swearing among them, much less than became common thirty years later, and perhaps about as much as among a similar class of men to-day.   Neither was there much coarseness or indecency in their talk, far less indeed than I met with among professional young men a few years afterwards.   One of the best of the workmen was a very loose character—a kind of Lothario or Don Juan by his own account—who would often talk about his adventures, and boast of them as the very essence of his life.   He was a very good and amusing talker, and helped to make the time pass in the monotony of the shop ; but occasionally, when he became too explicit or too boastful, the foreman, who was a rather serious though very agreeable man, would gently call him to order, and repudiate altogether his praises of the joys of immorality.   But I never once heard such foul language as was not uncommonly used among themselves by young men of a much higher class and much more education.

Of course, I heard incidentally a good deal about how they lived, and knew exactly what they earned, and I am thus enabled to correct some very erroneous statements which have been made of late years as to the condition of artisans in the early part of the nineteenth century, before the repeal of the corn-laws. Perhaps the most glaring and the most numerous of these errors are due to Sir Robert Giffen, who, being considered an official statistical authority, continues to be quoted to the present day as if his statements were to be absolutely relied on. More often quoted than any other of his writings is his " Progress of the Working Classes in the last Half Century," given as a Presidential Address to the Statistical Society in 1883, and issued as a pamphlet, price threepence, in 1884, at the request of several friends, including Mr. Gladstone, who styled it "a masterly paper." It would occupy a whole chapter to expose the errors and the fallacies that pervade this paper, and I must therefore confine myself to two points only, that of the rise of wages and of the food of skilled artisans.

Mr. Giffen gives the weekly wages of carpenters at Manchester as 24s. fifty years ago and 34s. in 1883, an increase of 42 per cent., but he omits to give prices for London. In the Report of the Industrial Remuneration Conference, Mr. J. G. Hutchinson gives the wages at Greenwich in 1832 as 32s. 6d., and in 1876 as 39s. 8d., a rise of only 22 per cent. Again, Mrs. Ellis, a Huddersfield pattern-weaver, told the conference that Mr. Giffen's statements in the same table, of the earnings of her fellow-workers, were grossly inaccurate. He gave them as 25s. a week against 16s. fifty years earlier, whereas they were only earning an average of 20s. in 1883. The wages where my brother worked were 30s. a week for all the men employed. We see, therefore, that Mr. Giffen's general statement that wages have risen "in most cases from 50 to 100 per cent." is open to the gravest doubt; while even if it were nearly accurate, it would not by any means prove what he claims—that these workers are *very much* better off than they were fifty years earlier. He certainly saves himself, verbally, by terming it an "apparent rise," but he never attempts to get at the real rise, and throughout his argument hardly refers to this point again. Yet it is a most important one, on account of the fact which he notices, that, at the date of his paper as now, in all the building trades wages are reckoned and paid by *the hour*, instead of by *the day* as at the

earlier period, when also men were rarely discharged except at
the week end. Then, again, Mr. Giffen speaks of the shorter
hours of work which from "one or two scattered notices" he
estimates at nearly 20 per cent., and then adds, " The workman gets from 50 to 100 per cent. more money for 20 per
cent. less work ; in round figures, he has gained from 70 to
120 per cent. in fifty years in money return." What a conclusion for a statistician, from a very limited comparison of
wages obtained almost wholly from the masters, and from
"one or two scattered notices" as regards hours of work !

But it is when he deals with the real value or purchasing
power of this greatly exaggerated increase of wages that we
find the grossest errors and the wildest declamation. After
just remarking that " sugar and such articles " have decreased
greatly in price, that clothing is also cheaper, and that though
house-rent has gone up, "it cannot have gone up so much as
to neutralize to any serious extent the great rise in the money
wages of the workman," he admits that the increase in the
price of meat is considerable. And then comes this amazing
statement : "The truth is, however, that meat fifty years ago
was not an article of the workman's diet as it has since
become. He had little more concern with its price than with
the price of diamonds."

I was so perfectly astounded at this statement that I at
once made a few inquiries. A very intelligent man, a printer
in the City, gave me facts from his own observation. About
the time referred to, his father kept a public-house in or near
Greenwich, much frequented by mechanics and other workmen, who came there in considerable numbers to have their
dinner. He assured me that almost without exception they
had fresh meat, which they either brought ready cooked, or
had purchased on their way to work and cooked in a
frying-pan or gridiron at the kitchen fire, many of them
bringing large chops or steaks of good quality. Remembering the cheapness of meat when I was a boy, and remembering also the well-to-do appearance of the carpenters in Mr.
Webster's shop, I wrote to ask my brother how they lived
during the twelve years he was in London, the last six working

as a journeyman in large shops and living on journeyman's wages. His statement is as follows:—

"Having been personally associated with the workers in the building trade about half a century ago (from 1835 to 1845), I feel qualified to describe the social condition of skilled mechanics at that period, more especially that of the carpenters and joiners. At that time every kind of work was done by hand, no machines except hand-tools were ever used, even boards of all thicknesses being sawn on the premises by hand labour out of thick planks from Northern Europe or Canada.

"The wages of good workmen were 5s. a day of ten hours; and 6d. an hour was added or deducted for any variation from that time. No wages were paid except for a fair amount of work, and if the work was temporarily suspended by rain or otherwise, no compensation was given or expected. All the joiner's work was done in shops, generally well lighted and with good sanitary conditions; nothing but the rough carpenter's work was done in buildings before the roof was on. Working hours were from 6 a.m. to 5.30 p.m., with an hour and a half out for breakfast and dinner. Men were paid weekly on Saturday evening, and were generally discharged at that time, and the last two hours and a half were allowed for grinding tools.

"The best workmen were seldom discharged unless in very dull times. At many shops men often worked for years without ever losing time except through sickness or accident; but, of course, these were the very best men. There were always some out of work, especially in winter or in times of depression.

"As regards their social condition, the skilled workman with his 30s. a week, if a single man of steady and frugal habits, could save half his wages and have proper food, lodging, and clothing suitable to his position. His furnished lodging of one room would cost 4s. a week, and his three meals a day, taken at the eating-houses and coffee-shops, would not cost more than 8s. a week; his working clothes were cheap, and he would have one superior suit for Sundays and holidays. Of course, if he were of a gay disposition, he would spend more and save less, but that would not be the indispensable outlay of a working man.

"In the case of a married man with a family, it would, of course, be more difficult to save money, but I have known men live well and respectably, bring up a family, and put by regularly for the expected 'rainy day,' and eventually build their own house, and start in business, in a small way at first, and become masters and gain a competence; but these are exceptional cases.

"The generality of carpenters and joiners with a family would live in lodgings of two or three rooms with their own furniture (much of which the man could make in his spare time in the evening), paying 5s. or 6s. a week, and with a careful and industrious wife could live well on their wages, clothe and educate their children, and still have something to put by. I have *never known* a carpenter in work, whether married or single, that did not have *a good dinner of meat and vegetables* every day, and on Sundays something extra ; they always had beer for dinner and often at their work about ten o'clock, and sometimes in the afternoon.

"As near as I can recollect the prices of provisions were for meat from 6d. to 8d. a pound, bread 7d. the four-pound loaf, butter 10d., cheese 8d., and sugar 6d. to 9d. The brick-layers had about the same wages as the carpenters, but owing to lost time during bad weather, they were generally not so well off, or generally so well housed and fed, but I never heard or knew of any destitution or want among them. Of the social condition of the plasterers, painters, and other house finishers I know less, but all appeared well satisfied with their condition, and, at all events, no general dissatis-faction was expressed."

It is, I think, quite clear from this statement of my brother's that the standard of comfort of the skilled artisan was as high fifty years ago as it is now, notwithstanding his somewhat lower wages and his working ten instead of nine hours a day.

There being no railways and many more small employers, he seldom spent anything in going to and from his work; while, as access to the country was then easier, his holidays cost him less, with more enjoyment, than going by rail to some place fifty miles away. It is also absolutely certain that the food of the workman was quite as good as it is now or even better, and that meat and beer formed regular articles of consumption by the average mechanic.

Now, these almost incredible errors as to matters of fact teach us that Government officials are quite unfitted to deal with such questions as these, mainly because they know nothing at first hand of the lives of the workers and thus omit to take account of some of the most essential factors in the problem at issue.

Thus Mr. Giffen slurs over and minimizes the universal increase of rent. In the report already quoted, Miss Edith Simcox gives the results of two inquiries into the poorer districts of Westminster. A communication to the Statistical Society in 1840 showed that at that time somewhat less than a quarter of the wages went to pay rents; while a somewhat similar inquiry in 1884 by the *Pall Mall Gazette* showed that in another part of Westminster rents were on the average, for the same accommodation, nearly three times as much as those recorded forty years before. Combining these two results, it is clear that, even if workmen have smaller or fewer rooms than at the earlier period, they must still pay nearly twice as much rent, and this enormous increase will absorb a large portion, and in some cases the whole of the increase in wages.

Another point which Mr. Giffen omits to notice and allow for is the fact, well known to all workmen who remember the earlier period, that the decreased cost of clothing is quite illusory; the badness of the materials, made for show rather than for wear, render them really dearer. At the early period referred to shoddy was not invented, and paper as part of the soles in workmen's boots was unknown. The corduroys and fustians then generally worn by mechanics would last twice or thrice as long as the cheaper articles now sold under the same name. Boots were then all good leather and hand-sewn, and though not so highly finished and a little dearer than the cheapest kinds now made, would outlast two or three pairs of the latter. At about the same period my strong surveying boots cost 14*s*. a pair, but were really better in quality than what I should pay 20*s*. for now. The general result was, that the workman's clothing cost him rather less then than they do at the present day.

Another point Mr. Giffen overlooks which is of consider-able importance. In the earlier period referred to almost all workshops and factories were much smaller than they are now, and employed each a much smaller number of men, who were therefore able to live within about half a mile or less of their work. If they were sent to work at a distance they went in their master's time, or if by omnibus at their master's expense. Now, however, the hundreds of men in each large builder's or contractor's shops frequently live a mile or several miles away, and can only reach the shop when work begins either by a long and hurried walk or by paying tram or rail-way fare to shorten the distance. Under average circum-stances, having often to lose time waiting for train or tram, and having a walk at both ends from home to station and from station to work, each often half a mile or more, the loss of time morning and evening fully makes up for any shorten-ing of actual working hours, while the daily fares are a not unimportant deduction from the increased wages. Taking all these things into consideration, we see clearly how it was that the mechanic of the thirties and forties of the last century was able to afford quite as much meat as his successor of to-day, and was, on the whole, quite as well off.

As my brother was, at the time I am now speaking of, nearly nineteen and a very good workman, he had complete liberty in the evenings after seven o'clock, the only limitation being that he was back about ten ; while on special occasions he was allowed to take the door-key. He often took me with him on fine evenings to some of the best business streets in

London to enjoy the shops, and especially to see anything of particular interest exhibited in them. Among these objects was one of the earliest of the large plate-glass windows now so universal, which, though of quite moderate size, perhaps five feet high by four or five wide, was at that time a wonder. I also remember some curious clocks so constructed as to look like perpetual motion, which greatly interested and often puzzled us. But our evenings were most frequently spent at what was then termed a "Hall of Science," situated in John Street, Tottenham Court Road (now altered to Whitfield Street). It was really a kind of club or mechanics' institute for advanced thinkers among workmen, and especially for the followers of Robert Owen, the founder of the Socialist move-ment in England. Here we sometimes heard lectures on Owen's doctrines, or on the principles of secularism or agnos-ticism, as it is now called; at other times we read papers or books, or played draughts, dominoes, or bagatelle, and coffee was also supplied to any who wished for it. It was here that I first made acquaintance with Owen's writings, and especially with the wonderful and beneficent work he had carried on for many years at New Lanark. I also received my first knowledge of the arguments of sceptics, and read among other books Paine's "Age of Reason."

It must have been in one of the books or papers I read here that I met with what I dare say is a very old dilemma as to the origin of evil. It runs thus: "Is God able to pre-vent evil but not willing? Then he is not benevolent. Is he willing but not able? Then he is not omnipotent. Is he both able and willing? Whence then is evil?" This struck me very much, and it seemed quite unanswerable, and when at home a year or two afterwards, I took the opportunity one day to repeat it to my father, rather expecting he would be very much shocked at my acquaintance with any such infidel literature. But he merely remarked that such problems were mysteries which the wisest cannot understand, and seemed disinclined to any discussion of the subject. This, of course, did not satisfy me, and if the argument did not really touch the question of the existence of God, it did seem to prove that the orthodox ideas as to His nature and powers cannot be accepted.

I was also greatly impressed by a tract on " Consistency," written by Robert Dale Owen, the eldest son of Robert Owen, and as a writer superior in style and ability to his father. The chief subject of it was to exhibit the horrible doctrine of eternal punishment as then commonly taught from thousands of pulpits by both the Church of England and Dissenters, and to argue that if those who taught and those who accepted such dogmas thoroughly believed them and realized their horror, all worldly pleasures and occupations would give way to the continual and strenuous effort to escape such a fate. I remember one illustration quoted from a sermon, to enable persons to realize to some extent what eternal punishment meant. After the most terrible description had been given of the unimaginable torments of hell-fire, we were told to suppose that the whole earth was a mass of fine sand, and that at the end of a thousand years one single grain of this sand flew away into space. Then—we were told—let us try to imagine the slow procession of the ages, while grain by grain the earth diminished, but still remained apparently as large as ever,— and still the torments went on. Then let us carry on the imagination through thousands of millions of millions of ages, till at last the globe could be seen to be a little smaller—and then on and on, and on for other and yet other myriads of ages, till after periods which to finite beings would seem almost infinite the last grain flew away, and the whole material of the globe was dissipated in space. And then, asked the preacher, is the sinner any nearer the end of his punishment? No! for his punishment is to be infinite, and after thousands of such globes had been in the same way dissipated, his torments are still to go on and on for ever! I myself had heard such horrible sermons as these in one of the churches in Hertford, and a lady we knew well had been so affected by them that she had tried to commit suicide. I therefore thoroughly agreed with Mr. Dale Owen's conclusion, that the orthodox religion of the day was degrading and hideous, and that the only true and wholly beneficial religion was that which inculcated the service of humanity, and whose only dogma was the brotherhood of man. Thus was laid the foundation of my religious scepticism.

Similarly, my introduction to advanced political views, founded on the philosophy of human nature, was due to the writings and teachings of Robert Owen and some of his disciples. His great fundamental principle, on which all his teaching and all his practice were founded was that the character of every individual is formed *for* and not *by* himself, first by heredity, which gives him his natural disposition with all its powers and tendencies, its good and bad qualities; and, secondly, by environment, including education and surroundings from earliest infancy, which always modifies the original character for better or for worse. Of course, this was a theory of pure determinism, and was wholly opposed to the ordinary views, both of religious teachers and of governments, that, whatever the natural character, whatever the environment during childhood and youth, whatever the direct teaching, all men *could* be good if they liked, all *could* act virtuously, all *could* obey the laws, and if they wilfully transgressed any of these laws or customs of their rulers and teachers, the only way to deal with them was to punish them, again and again, under the idea that they could thus be *deterred* from future transgression. The utter failure of this doctrine, which has been followed in practice during the whole period of human history, seems to have produced hardly any effect on our systems of criminal law or of general education; and though other writers have exposed the error, and are still exposing it, yet no one saw so clearly as Owen did how to put his views into practice; no one, perhaps, in private life has ever had such opportunities of carrying out his principles; no one has ever shown so much ingenuity, so much insight into character, so much organizing power; and no one has ever produced such striking results in the face of enormous difficulties as he produced during the twenty-six years of his management of New Lanark.

Of course, it was objected that Owen's principles were
erroneous and immoral because they wholly denied free-will,
because he advocated the abolition of rewards and punishments as both unjust and unnecessary, and because, it was
argued, to act on such a system would lead to a pandemonium
of vice and crime. The reply to this is that, acting on the
principle of absolute free-will, every government has alike
failed to abolish, or even to any considerable degree to
diminish, discontent, misery, disease, vice, and crime; and
that, on the other hand, Owen *did*, by acting on the principle
of the formation of character enunciated by him, transform
a discontented, unhealthy, vicious, and wholly antagonistic
population of 2500 persons to an enthusiastically favourable,
contented, happy, healthy, and comparatively moral community, without ever having recourse to any legal punishment whatever, and without, so far as appears, discharging
any individual for robbery, idleness, or neglect of duty; and
all this was effected while increasing the efficiency of the
whole manufacturing establishment, paying a liberal interest
on the capital invested, and even producing a large annual
surplus of profits which, in the four years 1809–13, averaged
£40,000 a year, and only in the succeeding period, when the
new shareholders agreed to limit their interest to 5 per cent.
per annum, was this surplus devoted to education and the
general well-being of the community.

In view of such an astounding success as this, what is the
use of quibbling about the exact amount of free-will human
beings possess? Owen contended, and proved by a grand
experiment, that environment greatly modifies character, that
no character is so bad that it may not be greatly improved
by a really good environment acting upon it from early
infancy, and that society has the power of creating such an
environment. Now, the will is undoubtedly a function of the
character of which it is the active and outward expression;
and if the character is enormously improved, the *will*, resulting in actions whether mental or physical, is necessarily improved with it. To urge that the will is, and remains through
life, absolutely uninfluenced by character, environment,

or education; or to claim, on the other hand, that it is wholly and absolutely determined by them—seem to me to be propositions which are alike essentially unthinkable and also entirely opposed to experience. To my mind both factors necessarily enter into the determination of conduct as well as into the development of character, and, for the purposes of social life and happiness, a partial determinism, as developed and practised by Owen, is the only safe guide to action, because over it alone have we almost complete control. Heredity, through which it is now known that ancestral characteristics are continually reappearing, gives that infinite diversity of character which is the very salt of social life ; by environment, including education, we can so modify and improve that character as to bring it into harmony with the possessor's actual surroundings, and thus fit him for performing some useful and enjoyable function in the great social organism.

Although most people have heard of New Lanark, few have any idea of Owen's work there or of the means by which he gradually overcame opposition and achieved the most remarkable results. It will, therefore, not be out of place to give a short account of his methods as explained in his autobiography ; and it will also be advisable to give a very brief sketch of the early life of one of the most remarkable, most original, and, in many respects, most truly admirable characters which has adorned the nineteenth century.

Robert Owen was born in 1771, and brought up in Newtown, a small town in Montgomeryshire, North Wales. His father was a saddler by trade ; his mother a farmer's daughter. He was sent to the town school when about five years old, where the teaching was limited to what are now termed the three R.'s, and he learnt so quickly that when about seven years old the schoolmaster took him as an usher to teach the younger children, and for the next two years he learnt nothing more at school except how to teach. This, however, he appears to have taught himself to some purpose, as his after-life shows. At nine he entered the shop of a draper and haberdasher, a friend of his father's, where he went daily for a year, but taking his meals at home. He was a great reader, and being well known to all the inhabitants, and evidently much liked and admired, he had free access to all the libraries in the place, including those of the clergyman, doctor, lawyer, etc., and he says that he generally read a volume every day. He also thought much about all that he read, and at one time, having read many religious books, he wrote three sermons, which he afterwards destroyed. He also learnt dancing, of which he was very fond, and this led him to observe the characters of boys and girls, and also had an important influence on his views and practice of education.

At the age of ten, at his own request, he went to London, where an elder brother was engaged in a saddler's shop. Through his father's introductions and the recommendation of the draper in Newtown, he soon obtained an engagement with a haberdasher at Stamford, who had a large business in the finest qualities of goods, which he supplied to all the nobility and gentry in the country round. The boy Owen was to have his board, lodging, and washing, no salary the first year, £8 the second, and £10 the third, and he tells us that from the time of entering this house he supported himself, and never applied for or received any pecuniary aid from his parents. Here he remained three years, and the hours of business being comparatively short, by getting up early he was able to read five hours a day. He also learnt here to distinguish the different qualities of all the finest fabrics, which was of great use to him in after-life.

He then returned to London, and after a visit to his family in Wales, entered a large ready-money shop on Old London Bridge, where he had £25 a year, but was at work for fifteen or sixteen hours a day ; so after a year he obtained another situation in a large shop in Manchester at a salary of £40 a year. Here he remained till he was eighteen, and a circumstance occurred which changed the whole course of his life.

A mechanic named Jones supplied the firm with wire frames for ladies' bonnets, of which large numbers were sold.

He brought a supply weekly, and it was Owen's duty to receive
them from him, and being an intelligent man, they had some
conversation together. Jones was full of the wonderful
improvements then being made in machinery for cotton-
spinning. He had seen some of these machines at work, and
was sure he could make them and work them if he had a
little capital. At last he persuaded Owen to lend him £100
(borrowed from his brother in London), for which he was to
have half the profits of the work. Owen accordingly left
his employer after due notice, and rented a suitable machine
shop, in which about forty men were soon employed making
the newly invented "mules" for spinning cotton. Jones
superintended the work, and Owen kept the accounts, paid
the men, and saw that regular hours were worked; he being
the first to enter and the last to leave the workshop. The
"mules" were sold as quickly as made, and thus the small
capital was made to serve ; but Owen soon saw that Jones had
no business capacity, whereas Owen was, as he afterwards
proved, one of the greatest organizers who ever lived. He,
therefore, watched the work closely, learnt all he could about
it, and when an offer was made by another person with some
capital to buy him out, he gladly accepted the offer which they
made him, of six of the mule machines, a reel, and a making-
up machine with which to pack the skeins of yarn into bundles
for sale. He, however, only received three mules with the
two other machines, and immediately hired an empty building, set them up in one of the rooms, bought the cotton rovings
ready for spinning, and hired three men to work the machines.
The finished yarn was spun in hanks of one hundred and forty
yards each, the hanks made up into bundles of five pounds
weight, and wrapped neatly in paper, all which work was done
by himself, and he then sold it to the agent of some Glasgow
manufacturers of British muslins, then quite a new business.
In this way he found he could make a clear profit of £6 a
week.

A few months later he accidentally heard that a wealthy
manufacturer, Mr. Drinkwater, had advertised for a manager
for some new spinning-mills which he had just built and filled

with the best machinery under the management of Mr. Lee, a civil engineer, who had unexpectedly left him, he himself knowing nothing of the business. Owen applied for the post, being then barely twenty years old, and looking younger. He asked £300 a year salary ; and after a few inquiries as to character, seeing his little factory of three mules, and examining his books, Mr. Drinkwater engaged him, and about a week afterwards he was called upon to take charge of a large factory employing about five hundred workpeople. The former manager had left the day before, Mr. Drinkwater did not come to introduce him, and he was simply sent there as the new manager. His business was to purchase the raw material, to make the machines, for the mill was not nearly completed ; to manufacture the yarn, and to sell it ; to keep the accounts, pay the wages, and take the whole responsibility of the first *fine* cotton-spinning establishment by machinery that had ever been erected. Hitherto his life had been spent in retail shops, where he had learnt the qualities of various fabrics, and how to buy and sell, but till his short experience with Jones and with his three spinning-mules, he had never even seen any textile machinery or learnt anything about its construction.

He describes how he suddenly found himself in the midst of five hundred men, women, and children, who were busily occupied with machinery, much of which he had scarcely seen, and never in their regular connection so as to manufacture from the raw cotton to the finished thread. We can well understand his feelings, and how he said to himself, " How came I here ? And how is it possible I can manage these people and this business ? " His description of how he *did* manage it, without ever showing his complete ignorance ; how he not only superintended the completion of the mill and carried on the whole thing successfully, but in a very short time noticed imperfections in the thread, found out the defect in the machinery or in the mode of working that led to these imperfections, and then had these defects remedied ; how the quality and selling value of the output steadily advanced ; how the organization of the whole mill was perfected, and yet the

workpeople were satisfied with the various new rules and
regulations he adopted; and how, during the four years he
remained there, he continually improved the output; how his
salary was raised by agreement to £500 a year, to be followed
the next year by his becoming a partner with one-fourth share
in the whole concern—is one of the most interesting and
remarkable incidents in modern biographical literature.

Owing to family arrangements Mr. Drinkwater wished
Owen to withdraw from the partnership, but begged him to
remain as manager, and name his own salary. This he
declined, soon found another offer, built new mills, and
carried them on successfully for several years, till, in the year
1800, he became partner and sole manager of the New
Lanark mills, and married the daughter of Mr. Dale, the
former proprietor.

Gradually, for many years, he had been elaborating his
theory of human nature, and longing for an opportunity
of putting his ideas in practice. And now he had got his
opportunity. He had an extensive factory and workshops, with
a village of about two thousand inhabitants all employed in
the works, which, with about two hundred acres of surrounding
land, belonged to the company. The character of the workers
at New Lanark is thus described by Mr. W. L. Sargant in
his work "Robert Owen and his Social Philosophy," when
describing the establishment of the mills about fifteen years
before Owen acquired them: "To obtain a supply of adult
labourers a village was built round the works, and the houses
were let at a low rent; but the business was so unpopular
that few, except the bad, the unemployed, and the destitute,
would settle there. Even of such ragged labourers the numbers
were insufficient; and these, when they had learned their
trade and become valuable, were self-willed and insubordinate."
Besides these, there were about five hundred children, chiefly
obtained from the workhouses of Edinburgh and other large
towns, who were apprenticed for seven years from the age of
six to eight, and these were lodged and boarded in a large
building erected for the purpose by the former owner, Mr.
Dale, and was well managed. But these poor children had

to work from six in the morning to seven in the evening (with
an hour and three-quarters for meals) ; and it was only after
this task was over that instruction began. The poor children
hated their slavery ; many absconded ; some were stunted,
and even dwarfed in stature ; and when their apprenticeship
expired at the ages of thirteen to fifteen, they commonly went
off to Glasgow or Edinburgh, with no natural guardians, and
trained for swelling the mass of vice and misery in the towns.
" The condition of the families who had immigrated to the
village was also very lamentable. The people lived almost
without control in habits of vice, idleness, poverty, debt, and
destitution. Some were drunk for weeks together. Thieving
was general, and went on to a ruinous extent. . . . There was
also a considerable drawback to the comfort of the people in
the high price and bad quality of the commodities supplied in
the village."

When Owen told his intimate friends who knew all these
facts that he hoped to reform these people by a system of
justice and kindness, and gradually to discontinue all punishment, they naturally laughed at him for a wild enthusiast;
yet he ultimately succeeded to such an extent that hardly any
one credited the accounts of it without personal inspection,
and its fame spread over the whole civilized world. He had,
besides the conditions already stated, two other great difficulties to overcome. The whole of the workers and overseers
were strongly antagonistic to him as being an Englishman,
whose speech they could hardly understand, and who, they
believed, was sent to get more money for the owners and
more work out of themselves. They, therefore, opposed all
he did by every means that ingenuity could devise, and though
he soon introduced more order and regularity in the work
and improved the quality of the yarn produced, they saw in
all this nothing but the acts of a tool of the mill-owners somewhat cleverer, and therefore more to be dreaded, than those
who had preceded him. An equally fierce opposition was
made to any improvement in the condition of the houses
and streets as to dirt, ventilation, drainage, etc. He vainly
tried to assure the more intelligent of the overseers and

workmen that his object was to improve their condition, to make them more healthy and happier and better off than they were. This was incredible to them, and for two years he made very little progress.

His second great difficulty was that his partners were business men, who expected him to carry on the works on ordinary business principles, so as to obtain for them at least as large returns as any other factories in the country. Generally, he was absolute and sole manager, but he knew that he could not make any large or extensive alterations till he had obtained a surplus revenue beyond what was expected. For the first two years he limited his improvements to the factory itself and its management, and to endeavours, mostly in vain, to obtain the confidence of the workers.

One thing, however, he did for the benefit of the workers which had some effect in disarming their enmity and suspicions. Instead of the retail shops where inferior articles were sold at credit for very high prices, he established stores and shops where every article of daily consumption was supplied at wholesale prices, adding only the cost of management. The result was that by paying ready money the people got far better quality at full 25 per cent. less than before; and the result soon became visible in their superior dress, improved health, and in the general comfort of their houses.

But what at length satisfied them that their manager was really their friend was his conduct when a great temporary scarcity of cotton and its rapid rise in price caused most of the mills to be shut, and reduced the workers to the greatest distress. But though Owen shut up the mills he continued to pay every worker full wages for the whole of the four months during which the scarcity lasted, employing them in thoroughly cleaning the mills and machinery, repairing the houses, etc. This cost £7000, which he paid on his own responsibility; but it so completely gained the confidence of the people that he was afterwards able to carry out improvements without serious obstruction. Being wholly

opposed to infant labour he allowed all arrangements with
the guardians to expire, built a number of better houses,
and thus obtained families of workers to take the place of
the children ; but difficulties with the partners arose, the
property was sold to a fresh set of partners, Owen being still
the largest shareholder and manager, and a few years later
again sold to Owen and a few of his personal friends, who
agreed to allow him to manage the property, and to expend
all profits above 5 per cent. for the benefit of the workers.
Among his co-shareholders were Jeremy Bentham, with Joseph
Foster and William Allen, well-known Quakers. It may be
here stated that the property was purchased of Mr. Dale for
£60,000, and was sold to Owen and his friends in 1814 for
£114,100. This great increase of value was due in part to
the large profits made by cotton mills generally at this period,
and partly to Owen's skilful management and judicious
expenditure.

He was now at last able to carry out his plans for the
education of the children, none of whom he would allow to
enter the mills as workers till they were ten years old. He
built handsome and roomy schools, playrooms and lecture-
rooms for infants from two to six, and for the older children
from six to ten years old ; and he obtained the best masters
for the latter. The infant schools were superintended by
himself, and managed by teachers he himself selected for
their manifest love of children. His instructions to them
were " that they were on no account ever to beat any one of
the children, or to threaten them in any manner in word or
action, or to use abusive terms, but were always to speak to
them with a pleasant countenance, and in a kind manner and
tone of voice ; that they should tell the infants and children
that they must on all occasions do all they could to make
their playfellows happy ; and that the older ones, from five to
six years of age, should take especial care of the younger
ones, and should assist to teach them to make each other
happy." And these instructions, he assures us, were strictly
followed by the man and woman he chose as infant-school
master and mistress.

No books were to be used ; but the children "were to be taught the uses and nature or qualities of the common things around them, by familiar conversation when the children's curiosity was excited so as to induce them to ask questions respecting them." The schoolrooms were furnished with paintings of natural objects, and the children were also taught dancing, singing, and military evolutions, which they greatly enjoyed. The children were never kept at any one occupation or amusement till they were fatigued, and were taken much into the open air and into the surrounding country, where they were taught something about every natural object. Here we see all the essential features of the educational systems of Pestalozzi and Frœbel, worked out by his own observations of child-nature from his own childhood onward, and put into practice on the first opportunity with a completeness and success that was most remarkable.

He tells us that his numerous visitors, latterly numbering two thousand every year, were more amazed and delighted with the schools than with any other part of the establishment ; and that during the visit of "a lady of the highest rank of our own nobility—after inspecting the dancing, the music, and all the other lessons and exercises out-of-doors, of the infants and children in their playground, while attentively witnessing their kindness of manner to each other, their unaffected, unrestrained, joyous happiness, and remembering their efficiency in their indoor exercises—this lady said to me with tears in her eyes, 'Mr. Owen, I would give any money if my children could be made like these.' And truly those who were trained from infancy through these schools were by far the most attractive, and the best and happiest human beings, I have ever seen. Their manner was unaffectedly graceful, and, when spoken to by strangers, naturally polite, with great innocent simplicity. The total absence of all fear, and full confidence in and affection for their teachers, with the never-ceasing expression of perfect happiness, gave these children of working cotton-spinners a character for their age superior to any I have yet seen." It was also noted how this training improved the physical appearance of the children, and many visitors declared that they had never seen so many beautiful girls and boys as in the schools at New Lanark.

The effect of his system on the adult workers was hardly less remarkable. To stop the continued pilfering of bobbins and other small articles used in the mills, he invented a system (unfortunately not explained) by which the many thousands of these articles which passed from hand to hand daily were so recorded automatically that the loss of one by any particular worker could be always detected. In this way robbery, large or small, was always discovered, *but no one was ever punished for it.* The certainty of discovery, however, prevented its being attempted, and it very soon ceased altogether.

Equally novel and ingenious was his method of avoiding the necessity for punishment, or even for a word of censure, for the many petty offences or infractions of rules that are inevitable in every large establishment. Owen calls it "the silent monitor," but the workers called it the "telegraph." Each superintendent of a department had a character-book, in which the daily conduct of every worker was set down by marks for each of the ordinary offences, neglect of work, swearing, etc., which when summed up gave a result in four degrees—bad, indifferent, good, excellent. For every individual there was a small wooden, four-sided tally, the sides being coloured black, blue, yellow, and white, corresponding to the above degrees of conduct. This tally was fixed at each one's work-place, with the indicative colour outward, so that as Owen or his representative passed down the shops at any time during the day, he could note at a glance the conduct of each one during the preceding day, and thus get both a general and a detailed view of the behaviour of the workers. If any one thought they were unfairly treated they could complain to him, but in hardly any cases did this happen. He tells us, " As I passed through all the rooms, and the workers observed me always to look at these telegraphs— and when black I merely looked at the person, and then at the colour—but never said a word to one of them by way of blame. At first," he says, "a large proportion daily were black and blue, few yellow, and scarcely any white. Gradually the blacks were changed for blue, the blues for yellow, and the yellows for white. Soon after the adoption of this telegraph I could at once see by the expression of countenance what was the colour which was shown. As there were four colours there were four different expressions of countenance, most evident to me as I passed along the rooms. . . . Never perhaps in the history of the human race has so simple a device created in so short a period so much order, virtue, goodness, and happiness, out of so much ignorance, error, and misery. And for many years the permanent daily conduct of a very large majority of those who were employed deserved, and had, No. 1 placed as their character on the books of the company."

To show that Owen did not exaggerate the improved condition of New Lanark, it will be well to give the estimates of experienced and independent visitors. In 1819 the town of Leeds sent a deputation, consisting of Mr. Edward Baines, Mr. Robert Oastler, and Mr. John Cawood, to report on the character and condition of the workers at New Lanark. They spent four days in a careful inspection and examination of the whole establishment, and the following are a few extracts from their general report. Speaking first of the children in the schools, from two to ten years of age, they say, " They appear like one well-regulated family, united together by the ties of the closest affection. We heard no quarrels from the youngest to the eldest ; and so strongly impressed are they with the conviction that to be happy themselves it is necessary to make those happy by whom they are surrounded, that they had no strife but in offices of kindness."

"The next class of the population in the Lanark establishment consists of boys and girls between ten and seventeen years of age. These are all employed in the mill, and in the evening from seven to half-past eight o'clock they pursue their education. The deportment of these young people is very exemplary. In business they are regular and diligent, and in their manners they are mild and engaging."

"In the adult inhabitants of New Lanark we saw much
to commend. In general they appeared clean, healthy, and
sober. Intoxication, the parent of so many vices and so
much misery, is indeed almost unknown here. The consequence is that they are well clad, well fed, and their dwellings
are inviting. . . . In this well-regulated colony, where almost
everything is made that is wanted by either the manufactory
or its inhabitants, no cursing or swearing is anywhere to be
heard. There are no quarrelsome men or brawling women."

Every visitor to New Lanark who published any account
of his observations seems to have agreed as to the exceptional
health, good conduct, and well-being of the entire population ;
while residents in the vicinity, as well as the ruling authorities
of the district, bore witness that vice and crime were almost
wholly unknown. And it must be remembered that this was
all effected upon the chance population found there, which
was certainly no better if no worse than the usual lowest
class of manufacturing operatives at that period. There
appears to have been not a single case of an individual or a
family being expelled for bad conduct ; so that we are compelled to trace the marvellous improvement that occurred
entirely to the *partial* application of Owen's principles of
human nature, most patiently and skilfully applied by himself. They were necessarily only a partial application,
because a large number of the adults had not received the
education and training from infancy which was essential for
producing their full beneficial results. Again, the whole
establishment was a manufactory, the property of private
capitalists, and the adult population suffered all the disadvantages of having to work for long hours at a monotonous
employment and at low rates of wages, circumstances wholly
antagonistic to any full and healthy and elevated existence.
Owen used always to declare that the beneficial results at
which all visitors were so much astonished were only one-
tenth part of what *could* and *would* be produced if his
principles were fully applied. If the labour of such a community, or of groups of such communities, had been directed
with equal skill to produce primarily the necessaries and

comforts of life for its own inhabitants, with a surplus of such goods as they could produce most advantageously for themselves, in order by their sale in the surrounding district to be able to supply themselves with such native or foreign products as they required, then each worker would have been able to enjoy the benefits of change of occupation, always having some alternation of outdoor as well as indoor work ; the hours of labour might be greatly reduced, and all the refinements of life might have been procured and enjoyed by them.

On considering the whole course of Owen's life, the one great error he committed was to give up the New Lanark property and management, and spend his large fortune in the endeavour to found communities in various countries of chance assemblages of adults, which his own principles should have shown him were doomed to failure. He always maintained that a true system of education from infancy to manhood was *essential* to the best formation of character. His infant schools had only been about ten years in existence, when, owing to some difficulties with his Quaker partners, who had always objected to the dancing and drill, he gave up the management into their hands.

This was a weakness due to his amiable temper, which could not bear to be the cause of difference with his friends. Under the circumstances he might well have refused to give up an establishment which was wholly his own creation, and whose splendid success was unequalled in the world. He possessed nearly half the shares, and the profits were so large that he could soon have paid off the remainder, and become the sole owner. If they had absolutely refused to sell, he might have sold his interest and started another community on improved lines, to which it is almost certain the whole of the inhabitants of New Lanark would have voluntarily removed in order to be under his beneficent rule. He would thus have had all the advantages of not losing the young people he had so thoroughly trained, and might have gone on during his life extending the establishment till it became almost wholly self-supporting, and ultimately, when the majority of the inhabitants had been trained from childhood under his supervision, self-governing also. Had he done this, his beautiful system of education, and the admirable social organization founded on his far-seeing and fundamentally true philosophy of human nature, might still have existed, as a beacon-light guiding us towards a better state of industrial organization. In that case we should not have now found ourselves, after another century of continuous increase of wealth and command over nature, with a much greater mass of want and misery in our midst than when he first so clearly showed the means of abolishing them.

Notwithstanding this one fatal error, an error due to the sensitive nobility of his character and to his optimistic belief in the power of truth to make its way against all adverse forces, Robert Owen will ever be remembered as one of the wisest, noblest, and most practical of philanthropists, as well as one of the best and most lovable of men.

I have a recollection of having once heard him give a short address at this "Hall of Science," and that I was struck by his tall spare figure, very lofty head, and highly benevolent countenance and mode of speaking. Although later in life my very scanty knowledge of his work was not sufficient to prevent my adopting the individualist views of Herbert Spencer and of the political economists, I have always looked upon Owen as my first teacher in the philosophy of human nature and my first guide through the labyrinth of social science. He influenced my character more than I then knew, and now that I have read his life and most of his works, I am fully convinced that he was the greatest of social reformers and the real founder of modern Socialism. For these reasons I trust that my readers will not consider the space I have here devoted to an outline of his great work at New Lanark is more than the subject deserves.

The preceding sketch of his life and work is founded upon his "Life" written by himself, and accompanied by such a mass of confirmatory reports and correspondence as to show that it can be thoroughly relied on. It has, however,

long been out of print, and very few people have read it or
even heard of it, and it is for this reason that I have given
this brief outline of its contents. The fine obituary notice
of Owen by his contemporary and friend, Mr. G. J. Holyoake,
together with the book on his life and times by his fellow-
worker, Lloyd Jones, show that I have in no way exaggerated
either his character or his achievements.

CHAPTER VII

BEDFORDSHIRE : SURVEYING

IT was, I think, early in the summer of 1837 that I went with my brother William into Bedfordshire to begin my education as a land-surveyor. The first work we had was to survey the parish of Higham Gobion for the commutation of the tithes. It was a small parish of about a thousand acres, with the church, vicarage, and a good farmhouse on the highest ground, and a few labourers' cottages scattered about, but nothing that could be called a village. The whole parish was one large farm ; the land was almost all arable and the fields very large, so that it was a simple piece of work. We took up our quarters at the Coach and Horses public-house in the village of Barton-in-the-Clay, six miles north of Luton, on the coach-road to Bedford. We were nearly a mile from the nearest part of the parish, but it was the most convenient place we could get.

An intelligent young labourer was hired to draw the chain in measuring, while I carried a flag or measuring-rod and stuck in pegs or cut triangular holes in the grass, where required, to form marks for future reference. We carried bill-hooks for cutting rods and pegs, as well as for clearing away branches that obstructed the view, and for cutting gaps in the hedges on the main lines of the survey, in order to lay them out perfectly straight. We started work after an early breakfast, and usually took with us a good supply of bread-and-cheese and half a gallon of beer, and about one o'clock sat down under the shelter of a hedge to enjoy our lunch.

My brother was a great smoker, and always had his pipe after lunch (and often before breakfast), and, of course, the chain-bearer smoked too. It therefore occurred to me that I might as well learn the art, and for a few days tried a few whiffs. Then, going a little too far, I had such a violent attack of headache and vomiting that I was cured once and for ever from any desire to smoke, and although I afterwards lived for some years among Portuguese and Dutch, almost all of whom are smokers, I never felt any inclination to try again.

Three miles north of Barton was the small village of Silsoe adjoining Wrest Park, the seat of Earl Cowper, whose agent, Mr. Brown, was known to my brother, and had, I think, obtained for him the parish survey we were engaged upon. A young gentleman three or four years older than myself who was, I think, a pupil of Mr. Brown's, was sent by him to learn a little land-surveying with us, and was a pleasant companion for me, especially as we were often left alone, when my brother was called away on other business, sometimes for a week at a time. Although the country north of Barton was rather flat and uninteresting, to the south it was very picturesque, as it was only about half a mile from the range of the North Downs, which, though only rising about three hundred feet above Barton, yet were very irregular, jutting out into fine promontories or rounded knolls with very steep sides and with valleys running up between them. The most charming of these valleys was the nearest to us, opening behind the church. It was narrow, with abundance of grass and bushes on the sides of a rapid-flowing streamlet, which, about a quarter of a mile further, had its source in a copious spring gushing out from the foot of the chalk-hill. On the west side of this valley the steep slope was thickly covered with hazel and other bushes, as well as a good many trees, forming a hanging wood full of wild flowers, and offering a delightful shade in the heat of the afternoon. About a mile to the east there was an extensive old British earthwork called Ravensburgh Castle, beyond which was another wooded valley; between these was a tolerably level piece of upland where the villagers played cricket in the summer.

My friend, whose name I forget (we will call him Mr. A.), was a small-sized but active young fellow, very good-looking, and quite the dandy in his dress. He was proud of his attractions, and made friends with any of the good-looking village girls who would talk to him. One day we met a pretty rosy-cheeked girl about his own age—a small farmer's daughter—and after a few words, seeing she was not disinclined for a chat, he walked back with her, and I went home. When he returned, he boasted openly of having got her to promise to meet him again, but the landlord advised him to be careful not to let her father see him. A day or two after, as we were passing near the place, he saw the girl again, and I walked slowly on. I soon heard loud voices, and, looking back, saw the girl's father, a big, formidable-looking man, threatening the young Lothario with his stick, and shouting out that if he caught him there again with *his* girl, he would break every bone in his body. When the young gentleman came back he was not the least abashed, but told us the whole story very much as it had happened, and rather glorying in his boldness in not running away from so big and enraged a man, and intimating that he had assuaged his anger by civil words, and had come away with flying colours.

One day he and I went for a walk over the hills towards Hitchin, where on the ordnance map a small stream was named Roaring Meg, and we wanted to see why it was so called. We found a very steep and narrow valley something like that called the Devil's Dyke near Brighton ; but this was thickly wooded on both sides, and the little stream at the bottom, rushing over a pebbly bed, produced a roaring sound which could be heard at a considerable distance. This northern range of downs has the advantage over the south downs of having numerous springs and streams on both sides of it, and these are especially abundant around the ancient village of Toddington, five miles west of Barton, where the ordnance map shows about twenty springs, the sources of small streams, within a radius of two miles.

It was while living at Barton that I obtained my first

information that there was such a science as geology, and
that chalk was not *everywhere* found under the surface, as I
had hitherto supposed. My brother, like most land-surveyors,
was something of a geologist, and he showed me the fossil
oysters of the genus Gryphæa and the Belemnites, which we
had hitherto called "thunderbolts," and several other fossils
which were abundant in the chalk and gravel around Barton.
While here I acquired the rudiments of surveying and mapping, as well as calculating areas on the map by the rules of
trigonometry. This I found very interesting work, and it
was rendered more so by a large volume belonging to my
brother giving an account of the great Trigonometrical Survey
of England, with all the angles and the calculated lengths of
the sides of the triangles formed by the different stations on
hilltops, and by the various church spires and other conspicuous objects. The church spires of Barton and Higham
Gobion had been thus used, and the distance between them
accurately given ; and as the line from one to the other ran
diagonally across the middle of the parish we were surveying, this was made our chief base-line, and the distance as
measured found to agree very closely with that given in the
survey. This volume was eagerly read by me, as it gave an
account of all the instruments used, including the great theo-
dolite three feet in diameter for measuring the angles of the
larger triangles formed by distant mountain tops often twenty
or thirty miles apart, and in a few cases more than a hundred
miles ; the accurate measurement of the base-lines by steel
chains laid in wooden troughs, and carefully tightened by
exactly the same weight passing over a pulley, while the ends
were adjusted by means of microscopes ; the exact temperature being also taken by several thermometers in order to allow
for contraction or expansion of the chains ; and by all these
refinements several base-lines of seven or eight miles in length
were measured with extreme accuracy in distant parts of the
country. These base-lines were tested by repeated measurements in opposite directions, which were found to differ only
by about an inch, so that the mean of all the measurements
was probably correct to less than half that amount.

These bases were connected by the system of triangulation already referred to, the angles at all the stations being taken with the best available instruments and often repeated by different observers, while allowance had also to be made for height above the sea-level, to which all the distances had to be reduced. In this way, starting from any one base, the lengths of the sides of all the triangles were calculated, and ultimately the length of the other bases; and if there had been absolutely no error in any of the measurements of base-lines or of angles, the length of a base obtained by calculation would be the same as that by direct measurement. The results obtained showed a quite marvellous accuracy. Starting from the base measured on Salisbury Plain, the length of another base on the shore of Lough Foyle in the north of Ireland was calculated through the whole series of triangles connecting them, and this calculated length was found to differ from the measured length by only five inches and a fraction. The distance between these two base-lines is about three hundred and sixty miles.

These wonderfully accurate measurements and calculations impressed me greatly, and with my practical work at surveying and learning the use of that beautiful little instrument the pocket-sextant, opened my mind to the uses and practical applications of mathematics, of which at school I had been taught nothing whatever, although I had learnt some Euclid and algebra. This glimmer of light made me want to know more, and I obtained some of the cheap elementary books published by the Society for the Diffusion of Useful Knowledge. The first I got were on Mechanics and on Optics, and for some years I puzzled over these by myself, trying such simple experiments as I could, and gradually arriving at clear conceptions of the chief laws of elementary mechanics and of optical instruments. I thus laid the foundation for that interest in physical science and acquaintance with its general principles which have remained with me throughout my life.

It was here, too, that during my solitary rambles I first

began to feel the influence of nature and to wish to know more of the various flowers, shrubs, and trees I daily met with, but of which for the most part I did not even know the English names. At that time I hardly realized that there was such a science as systematic botany, that every flower and every meanest and most insignificant weed had been accurately described and classified, and that there was any kind of system or order in the endless variety of plants and animals which I knew existed. This wish to know the names of wild plants, to be able even to speak of them, and to learn anything that was known about them, had arisen from a chance remark I had overheard about a year before. A lady, who was governess in a Quaker family we knew at Hertford, was talking to some friends in the street when I and my father met them, and stayed a few moments to greet them. I then heard the lady say, "We found quite a rarity the other day—the Monotropa; it had not been found here before." This I pondered over, and wondered what the Monotropa was. All my father could tell me was that it was a rare plant; and I thought how nice it must be to know the names of rare plants when you found them. However, as I did not even know there were books that described every British plant, and as my brother appeared to take no interest in native plants or animals, except as fossils, nothing came of this desire for knowledge till a few years later.

Barton was a rather large straggling village of the old-fashioned, self-contained type, with a variety of small tradesmen and mechanics, many of whom lived in their own freehold or leasehold houses with fair-sized gardens. Our landlord was a young man fairly educated and intelligent. One of his brothers was a tailor, and made such good clothes that my brother remarked upon the excellent cut and finish of a suit worn by our host. Their eldest brother lived in a very good old roomy cottage in the village, and was, I think, a wheelwright, and I was sometimes asked to tea there, and found them very nice people, and there was a rather elderly unmarried sister who was very talkative and satirical. Most

of the villagers, and some of the farmers around, used to come
to the house we lived in, and among them was a painter and
glazier, who was married while I was there, and who was subjected to good-humoured banter when he came to the house
soon afterwards. These, with the necessary blacksmith and
carpenter, with a general shop or two and a fair number of
labourers, made up a little community, most of whom seemed
fairly well off.

Our landlord was a Radical, and took a newspaper called
The Constitutional, which was published at Birmingham, and
contained a great deal of very interesting matter. This was
about the time when the dean and chapter refused to allow a
monument to be erected to Byron in Westminster Abbey,
which excited much indignation among his admirers. One of
these wrote some lines on the subject which struck me as being
so worthy of the occasion that I learnt them by heart, and by
constant repetition (on sleepless nights) have never forgotten
them. They were printed in the newspaper without a signa-
ture, and I have never been able to learn who was the author
of them. I give them here to show the kind of poetry I
admired then and still enjoy—

> " Away with epitaph and sculptured bust !
> Leave these to decorate the mouldering dust
> Of him who needs such substitutes for fame—
> The chisel's pomp to deck a worthless name.
> Away with these ! A Byron needs them not ;
> Nature herself selects a deathless spot,
> A nation's heart : the Poet cannot die,
> His epitaph is Immortality.
> What are earth's mansions to a tomb like this ?
> When time hath swept into forgetfulness
> Wealth-blazoned halls and gorgeous cemeteries,
> The mouldering Abbey with its sculptured lies,
> His name, emblazoned in the wild, the free,
> The deep, the beautiful of earth, shall be
> A household word with millions. Dark and wild
> His song at times, his spirit was the child
> Of burning passion. Yet when he awoke
> From his dark hours of bondage, when he broke
> His cage and seized his harp, did he not make
> A peal of matchless melody and shake

The very earth with joy. Still thrills the heart
Of man at those sweet notes ; scared despots start
To curse them from their thrones ; they pierce the cell
And cheer the captive in his chains ; they tell
Lessons of life to struggling liberty.
Death mars the man but spares his memory,
Nor tears one laurel from his wreath of fame.
How many glorious thoughts of his we claim
Our heritage for ever ; beacon lights
To guide the barque of freedom through the nights
Of tyranny and woe, when not a star
Of hope looks down to glad the mariner :
Thoughts which must ever haunt us, like some dream
Of childhood which we ne'er forget, a gleam
Of sunshine flashing o'er life's troubled stream ! "

The last eight lines of this poem form a passage characterized by deep feeling and poetic beauty of a high order.
My brother was an admirer of Byron, and he used to say
that his description of Satan, in the "Vision of Judgment,"
was finer than anything in Milton. This poem, which is
essentially a satirical parody of Southey's poem with the
same title, yet contains some grand passages on behalf of
political and religious liberty. The lines my brother thought
so fine (and I agree with him) are the following :—

" But bringing up the rear of this bright host,
    A Spirit of a different aspect waved
His wings, like thunder-clouds above some coast
    Whose barren beach with frequent wrecks is paved ;
His brow was like the deep when tempest-tost ;
    Fierce and unfathomable thoughts engraved
Eternal wrath on his immortal face,
    And *where* he gazed a gloom pervaded space."

Those who only know Byron by his more romantic or
pathetic poems, and who may think the panegyric of the
anonymous writer in *The Constitutional* * to be overdrawn,
should read "The Age of Bronze," which is pervaded

* This newspaper—*The Constitutional*—appears to have existed only two
years. The *Daily News*, referring to a sale of Thackeray rarities last year, states
that he contributed several articles to that paper as Paris correspondent, and that,
in consequence, a set of the paper sold in 1899 for two hundred guineas. A
friend informs me that it does not exist in the Bodleian Library.

throughout with the detestation of war, with admiration of those who fought only for freedom, and with scorn and contempt for the majority of English landlords, who subordinated all ideas of justice or humanity to the keeping up of their rents. Even if it stood alone, this one poem would justify the poet as an upholder of the rights of man and as a truly ethical teacher.

Returning from this digression to the villagers who came within my range at the little tavern where we lodged, I had an opportunity of seeing a good deal of drunkenness, inevitably brought on by the fact that only in the public-house could any one with enforced leisure have the opportunity of meeting friends and acquaintances and of hearing whatever news was to be had. Sometimes a labourer out of work, and having perhaps a week's wages in his pocket, would have a pint of beer in the morning, and while waiting alone for some one to come in, would, of course, require another to pass away the time ; and sometimes, if a young unmarried man, he would remain quietly drinking beer the whole day long. On one such occasion the landlord told me that a man had consumed twenty-two pints of beer during the day. At that time there was no temperance party, no body of people who thought drinking intoxicants altogether wrong ; while deliberately aiding a man to get drunk was often a mere amusement. My brother was a great smoker but a small drinker, and he used to say that as he neither drank nor expectorated while smoking it did him no harm—a view which seems very doubtful. He was, however, accustomed to take a glass of spirits and water in the evening, and usually kept a gallon jar of gin in a cupboard by the fireplace, not only for his own use, but to have something besides beer to offer any friend who called. He had several acquaintances at Silsoe, the architect of the mansion then being built for Earl Cowper being an old friend of about his own age, a Mr. Clephan. One day, I remember, a young farmer whose acquaintance we had made while surveying gave us a call, and my brother hospitably invited him to take a glass of gin, which he accepted. He was rather a weak young man and had already drunk a good deal of beer, and soon became talkative, and as my brother asked him to take more gin, he did so, and at last he became quite incoherent and so troublesome, though perfectly good-natured, that we had to ask the landlord to take charge of him till he was able to go home. But his speech and actions were so ludicrous that all present were kept in a roar of laughter, and everybody seemed to think it an excellent and quite harmless bit of fun.

When I was alone at Barton I used frequently to sit in the tap-room with the tradesmen and labourers for a little conversation or to hear their songs or ballads, which I have never had such an opportunity of hearing elsewhere. Some of these were coarse, but not as a rule more so than among men of a much higher class, while purely sentimental songs or old ballads were very frequent, and were quite as much appreciated. I regret that I did not write down all that I heard here, but at that time I did not know that there would be any purpose in doing so, and I cannot remember the actual words of any of them. One that was occasionally sung was the old Masonic Hymn, beginning—

"Come all you freemasons that dwell around the globe,
    That wear the badge of innocence, I mean the royal robe,
    Which Noah he did wear when in the ark he stood,
    When the world was destroyed by a deluging flood "—

but I think it was never sung in its complete form. The well-known poacher's song with its musical refrain—

"Oh ! 'tis my delight of a shiny night, in the season of the year,"

was also rather a favourite ; but there was one ballad about Bonaparte which was often called for, but of which I can remember nothing but a line beginning—

"Then upspoke young Napoleon."

It was a really good ballad, describing some incidents in Napoleon's early life, and was remarkable as treating him from quite a heroic point of view, so different from the enormous mass of gross and stupid caricature and abuse which prevailed during the epoch of his military successes throughout Europe.

As there was no work of importance after the maps and reference books of the parish we had been surveying were completed and delivered, and winter was approaching, I went home for a short holiday. My father and mother and my younger brother were then living in Hoddesdon, and as there was no direct conveyance I made the journey on foot. It was, I think, the end of November, and as the distance was about thirty miles, and I was not very strong, I took two days, sleeping on the way at a roadside public-house. I went through Hitchin and Stevenage, and near the former place passed a quarry of a reddish chalk almost as hard as marble, which was used for building. This surprised me, as I had hitherto only seen the soft varieties of chalk, and had been accustomed to look upon it as more earth than stone. The only other thing that greatly interested me was a little beyond Stevenage, where, on a grassy strip by the roadside, were six ancient barrows or tumuli, which I carefully inspected; and whenever I have since travelled by the Great Northern Railway, I have looked out for these six tumuli, near to which the line passes.

Where I slept the night I forget, but its results were long remembered, for I was given a bed which I presume had been occupied by some tramp, and I found that I had brought away with me two different kinds of body-lice, one of which took me a long time and the application of special ointments to get rid of. This was the only time in my life that I suffered from these noisome insects.

After a few weeks at home at Hoddesdon, I went back to Barton, where we had some work till after Christmas. On New Year's Day, 1838, the first section of the London and Birmingham Railway was opened to Tring, and I and my brother took advantage of it to go up to London, where he

had some business. We stayed at a quiet hotel in Lamb's
Conduit Street, and the next day I walked to Hoddesdon for
a short holiday. My brother while in London obtained the
survey for tithe commutation of a parish in Bedfordshire,
where I was to meet him on the 14th or 15th of January,
at the village of Turvey, eight miles beyond Bedford.

CHAPTER VIII

BEDFORDSHIRE : TURVEY

I HAD first to go back to Barton to pay a few bills and pack up the books, instruments, etc., we had left there to be sent by carrier's waggon. I therefore left home on the 12th, and I think walked back to Barton, and the next day did what was required, took leave of my friends there, and on the morning of the 14th, after an early breakfast, started to walk to Turvey through Bedford, a distance of about twenty miles.

The reason I am able, without any diary, note, or letter to refer to, to fix the date of this particular walk is rather a curious one. While I was at home, or shortly before, a new almanack had appeared, which professed to predict the weather on every day of the year, on scientific principles, and the first week was said to be wonderfully correct. I was so much interested in this, and talked so much about it, that my mother procured it for me just before I left home as a New Year's present. It was called "Murphy's Weather Almanack," and was published, I think, at a shilling. The first three days were marked " Fair, frost," and the next three "Change." This was, I believe, nearly correct, but how near I cannot remember. The next fortnight, however, impressed itself upon my memory, partly because I had the book and marked it day by day, and partly on account of the remarkable weather and its exact fulfilment. From the 7th to the 13th every day was set down as " Fair, frost," and so it was. Then came the 14th, marked "Change ; " then again "Fair, frost," every day to the 20th, which was marked " Lowest temperature;" after which the indications were change, followed by rain.

Now, as the 14th was the day of my walk to Bedford and Turvey, I was rather anxious, and when I got up in the morning and saw that the sky was clear, I thought the almanack was wrong, and was glad of it; but as soon as I began my journey I found the air milder and the roads decidedly softer than the day before, and this soon increased, till by midday there was a regular thaw, which made the roads quite soft, but as there had been no snow not disagreeably wet. I had, therefore, a very pleasant walk. I dined at Bedford, and reached Turvey before dark.

For the next six days we were at work laying out the main lines for the survey of the parish, cutting hedges, ranging flags, ascertaining boundaries, and beginning the actual measurements, and every day the frost continued exactly as predicted by Murphy, culminating in the greatest cold on the 20th, after which there was a break.

I may here state that the rest of the year was very inaccurate, though there were certain striking coincidences. The hottest day was nearly, or quite, correct. In August nine days consecutively were exactly as predicted, and in December the very mild weather and fine Christmas Day was correct.

But the perfect accuracy of the fourteen consecutive days with the break on one day of an otherwise continuous frost, and that day being fixed on my memory by the circumstance of my having then to walk twenty miles, forced me to the conclusion that there must have been "something in it"—that this could not have been attained by pure guess-work, even once in a year, and though the most striking, it was not by any means the only success. My copy of the almanack disappeared half a century ago, but wishing to refresh my memory of the circumstances, and to fix definitely the year and day of my journey, I applied to the Meteorological Society to lend me the almanack if they possessed it. They very courteously obliged me, sending me the five years, 1838 to 1842, all that ever appeared, bound together. I then found that my

memory of the weather for a week before and after my walk
had been quite correct and as I have stated here, and I also
had the advantage of examining the succeeding years, with
notes of the actual weather in a considerable proportion of
the days entered in a space left for the purpose by the owner
of this copy. The place of observation, however, is not given,
and it is obvious that, as the weather is usually very different
in widely separated parts of the country, only those features
of it can have any chance of being predicted which are
common to the greater part of our island, and are persistent for
a considerable period. Looking over these records from this
point of view, I find the following points worthy of notice :—

In 1839 the lowest winter temperature was predicted for
January 9, and this was correct.

In 1840 sixteen days of frost were predicted in February ;
eleven of these are noted, and all are on the right days. In
March only seven days' rain were predicted, and it is noted as
a very dry month throughout. April was predicted to be a
mild and fine month, and it was so, though the *days* of rain,
etc., did not agree. In May the prediction was two days rain,
thirteen days changeable, the rest fair. Rain was noted
on nine days, the rest being fine and mild. June was about
equally correct. In the winter frost was predicted for the last
two weeks of the year, which was correct.

In 1841 March was predicted to be a fine, dry, and mild
month, which was correct. There was nothing very marked
in the rest of the year.

In 1842 frost was predicted for several days at the end of
January and the first week in February, which was correct.
April was foretold to have only four days' rain, and the
remark of the observer is, " A very dry month." May was to
have five days rainy and three changeable, and it is noted as
having had " rain on nine days," and as being " a very fine
month." In August rain was announced for six days only,
and the remark is, " Splendid August weather." Then at the
end came a great failure, for the last half of December was
predicted to be fine and frosty, but turned out to be " very
mild and rainy."

Thus ended the "Weather Almanack," and I am not aware whether the writer ever disclosed the exact method by which he arrived at his predictions. In each of the issues he had a somewhat lengthy introduction, the first of which purported to explain the principles of his system. But it was so exceedingly general and vague that it seemed more intended to conceal than to explain. It appears to me almost certain that the author must have had access to some old weather records for a long succession of years, and finding that very similar weather occurred at each recurring lunar cycle of nineteen years, he simply predicted day by day what the weather had been nineteen years before. This method has been recently applied by means of a longer cycle, which leads to a more accurate correspondence of the positions of the sun and moon, and has been said to produce very striking results. If that was really his method, his successes, though very partial, were yet, I think, sufficient to prove that the larger and more lasting phases of the weather in our latitudes are to a considerable extent dependent on the relative positions of the moon and sun, and that the moon really is, as has been so long and so generally believed, one of the factors in determining our very excentric weather phenomena.

Another curious little personal incident connected with this winter's frost may here be noted. One day I was out on the frozen meadows across the river Ouse, assisting in marking out one of our main lines which had to cross the windings of the river, when I saw a pleasant-looking young man coming towards me carrying a double-barrelled gun. When he was a few yards off, two very large birds, looking like wild geese, came flying towards us, and as they passed overhead at a moderate height, he threw up his gun, fired both barrels, and brought them both to the ground. Of course I went up to look at them, and found they were a fine pair of wild swans, the male being about five feet long from beak to end of tail. "That was a good shot," I remarked; to which he replied, "Oh! you can't miss them, they are as big as a barn door." Afterwards I found that this was young Mr. Higgins, of

Turvey Abbey, his father being one of the principal land-owners in the parish ; and in making out the reference books which give the owners of all the separate farms, etc., we found that he himself owned some property, and that his name was H. H. Higgins. This interested me, because one of my schoolfellow's initials had been H. H. H., *his* name being Henry Holman Hogsflesh, and I thought it curious that I should so soon again come across another H. H. H., and this made me remember the name of Mr. Higgins, which I might otherwise have totally forgotten.

More than half a century later (in November, 1889), I was invited to Liverpool to give some lectures, and some time before the date fixed upon I received a very kind letter from the Rev. H. H. Higgins inviting me to dine with him on my arrival, and offering to assist me in every way he could. I declined the invitation, but told him what hotel I was going to, and said that I should be glad to see him. His letter recalled to me my acquaintance at Turvey, but I did not see how a Liverpool clergyman could have any close relationship to a wealthy Bedfordshire landowner. I found Mr. Higgins at the station with a carriage ready, and he told me that, as I did not wish to go out to dinner, he and some friends had taken the liberty of ordering a dinner at my hotel, and hoped I would dine with them. He was as pleasant as an old friend, and of course I accepted. He was a short, rubicund, exceedingly good-humoured and benevolent-looking man, apparently some years older than myself, and looking very like what young Mr. Higgins of Turvey might have grown into. He somehow reminded me of Chaucer's description of a priest—

> " A little round, fat, oily man of God
> Was one I chiefly marked among the fry,
> He had a rogueish twinkle in his eye "—

except that he could hardly be described as round, or fat, but simply "jolly" in person as in manner. So when his friends left about an hour after dinner, I asked him, if he had no engagement, to stay a little longer, as I wished to find out the mystery. He was an enthusiastic naturalist, and we talked of many things, and the conversation turning on the land question, he remarked that he was perhaps one of the poorest landowners in England, for that he was heir to a considerable landed estate from which he never received anything, and probably never should, owing to family circumstances, which he stated. I then asked him if he knew a place called Turvey, in Bedfordshire, to which he replied, "I ought to know it, for I was born there, and my father owned the estate there to which I am heir." I then felt pretty sure of my man, and asked him if he remembered, during a very hard frost about fifty years ago, shooting a pair of wild swans at Turvey. "Why, of course I do," said he. "But how do you know it?" "Because I was there at the time and saw you shoot them. Do not you remember a thin tall lad who came up to you and said, 'That was a good shot,' and you replied, 'Oh! you can't miss them, they are as big as a barn door'?" "No," he said, "I don't remember you at all, but that is just what I should have said." His delight was great, for his story of how he shot the two wild swans was not credited even by his own family, and he made me promise to go to his house after the lecture on the next night, and prove to them that he had not been romancing. And when I went, I was duly introduced to his grown-up sons and daughters as one who had been present at the shooting of the swans, which I had been the first to mention. That was a proud moment for the Rev. H. H. Higgins, and a very pleasant one to myself.

Let us now return to Turvey and my experiences there. We lived at the chief inn in the place—perhaps the only one except some small beer-shops—called The Tinker of Turvey. The painted sign was a man with a staff, a woman, and a dog, and we were told in the village that the tinker meant was John Bunyan. But recent inquiry by a friend both in Bedford and at Turvey shows that this is perhaps a mistake. In a little book, "Turvey and the Mordaunts," by G. F. W. Munby, Rector of Turvey, and Thomas Wright (of Olney), we are told that there is a very rare pamphlet in the British Museum, entitled, " The Tincker of Turvey, his merry pastime from Billingsgate to Gravesend. The Barge being freighted with mirth, and mann'd with Trotter the tincker, Yerker a cobbler, Thumper a smith, and other merry fellows, every one of them telling his tale" (dated, London, 1630, 4to). There is a verse on the signboard as follows :—

> " The Tinker of Turvey, his dog, and his staff,
> Old Nell with her Budget will make a man laugh."

This may, perhaps, be taken from the old pamphlet, which certainly proves that " The Tinker of Turvey" was a character known before Bunyan's time, and as the tales told by the tinker and his companions are said to be exceedingly coarse, they were probably well known in country places, and the name would seem appropriate for an inn in the village named. It is possible, however, that the sign may have been first painted at a later date, and as Bunyan would no doubt have been well known at Turvey, as at other villages round Bedford, where he was accustomed to preach, he may have been represented or caricatured as the Tinker of Turvey on the signboard.

In this inn we had the use of a large room on the ground-floor, also used as a dining-room for the rare visitors requiring that meal, and in the evening as a farmers' room, where two or three often dropped in for an hour or two, while once a week there was a regular farmers' club, at which from half a dozen to a dozen usually attended. While at Barton I had become well acquainted with the labourers, mechanics, and small village shopkeepers; I here had an equal opportunity of observing how well-to-do farmers occupied their leisure. These seemed to be rather a serious class, whose conversation was slow, and devoted mainly to their own business, especially as to the condition of their sheep, how their " tegs " were getting on, or of a fat sheep being cast—that is, turned over on its back, and vainly struggling to get up again, when, if not seen and helped, they sometimes died. Most of the time was spent in silent smoking or sipping their glasses of ale or of spirits and water. Sometimes the talk would be of hunting, or even of the county races when any one was present who had horses good enough to run. On one evening I heard an agricultural problem solved by an expert, and it is the only piece of definite information I ever heard given on these occasions. A young farmer was complaining of the poor crop of wheat he had got from one of his best fields, and said he could not make it out. One of the large farmers, who was looked up to as an authority, asked, "What did you do to the field?" "Well," said the young man, "I ploughed it" (a pause); "I ploughed it twice." "Ah!" said the expert, "that's where you lost your crop." The rest looked approval. Some said, "That's it;" others said, "Ah!" The young man said nothing, but looked gloomy. Evidently the oracle had spoken, and nothing more was to be said; but I have often wondered since if that really *was* the cause of the bad crop of wheat. There seem to be so many other things to be taken account of—the kind of seed used; the mode of sowing, whether broadcast or drilled; the quantity and kind of manure used; the condition of the soil as regards moisture, freedom from weeds, and many other matters;—all, one would think, equally important with the mere difference between one or two ploughings. I should have liked to have asked about this at the time, but I was too shy and afraid of exposing my ignorance.

The farmers here were very proud of their mutton, and one with whom we were especially friendly told us one day about a fine sheep he had killed the previous year—five years old, I think he said—and that he had kept one of the legs of mutton six months in his cellar, which was large and very cool. He assured us that it was perfectly sweet, and that he invited several of his friends to dinner, and they all agreed that they had never eaten such fine mutton in their lives. At the time I hardly believed this, holding the usual opinion that meat *necessarily* putrefied, but I have no doubt now that he was speaking the truth, and that much of our meat would be greatly improved in quality if we had suitable places in which to store it for a few weeks or months before cooking.

Soon after we came to Turvey a young gentleman from Bedford came to us to learn a little surveying. He was, I think, the son of an auctioneer or estate agent, and was about eighteen or twenty years old. As my brother was occasionally away for several days at a time when we sometimes had nothing to go on with, he would amuse himself fishing, of which he was very fond. Sometimes I went with him, but I usually preferred walking about the country, though I cannot remember that I had at this time any special interest in doing so. He often caught some large coarse fish, such as bream or pike, which were the commonest fish in the river, but were hardly worth eating. Towards the latter part of our survey in the spring months, my brother left us a portion of the work to do by ourselves when he was away for a week or two, and as we worked very hard, and seldom got home before six in the evening, we had an unusually good appetite for our evening meal, and sometimes astonished our hosts. One occasion of this kind I have never forgotten. They had provided for our dinner a sparerib of young pork—a very delicate dish but not very substantial—with potatoes. My friend first cut the joint in half, about three or four ribs in each, and said to me, " I know you like fat ; if I cut off this lean piece, will you have the rest? " I joyfully assented, as I was very fond of the picking on the bones. We soon finished our portions, and then he cut the lean off the rest of the joint, gave me the ribs, and we very soon left nothing but the clean-picked bones, half of which I put on his plate so that it might not be thought that I had eaten the whole joint myself. The servant looked astonished at the empty dish when she brought us in a rather small apple-pudding. This was cut in two, and was hardly as much as we should have liked ; and when the servant saw another empty dish she smiled, and told us that some people had been waiting for the rest of the pork and pudding, and now had nothing for dinner ; at which we smiled, and asked for bread-and-cheese to finish with.

When at home and spending the larger part of every day in the schoolroom, I had never liked fat, which often made

me ill. But exercise for about ten hours every day in the
open air had improved my digestion and my general health
so that I could eat most kinds of fat, and have been very fond
of it during my whole life.

During our stay here we made the acquaintance of some
pleasant people, and on Sundays we were often asked out
to tea, which I should have enjoyed more than I did had it
not been for my excessive shyness, which was at this time
aggravated by the fact that I was growing very rapidly, and
my clothes, besides being rather shabby, were somewhat too
small for me. Another drawback was that our residence at
any place was too short to become really at home with these
passing friends. I was therefore left mostly to the companion-
ship of our own temporary pupil, and he, like the majority
of the young men I met at this period of my life, was by no
means an edifying acquaintance. Sporting newspapers, which
were then far grosser than they are now, were, so far as I
remember, his chief reading, and he had a stock of songs and
recitations of the lowest and most vicious type, with which he
used occasionally to entertain me and any chance acquaintances. There was one paper which I used very frequently to
see about this time, and which I think must have been taken
at most of the country inns we frequented. It was called, if
I remember rightly, *The Satirist*, and was full of the very
grossest anecdotes of well-known public characters, trials for
the most disgraceful offences reported in all their details, and
full accounts of prize-fights, which were then very common.
It was a paper of a character totally unknown now, and as it
no doubt reflected the ideas and pandered to the tastes of
a very considerable portion of the public in all classes of
society, it is not very surprising that most of the young men
of the middle classes that came across my path should have
been rather disreputable in conversation, though, perhaps, not
always so in character.

But, notwithstanding that I was continually thrown into
such society from the time I left school, I do not think it
produced the least bad effect upon my character or habits
in after-life. This was partly owing to natural disposition,

which was reflective and imaginative, but more perhaps to the quiet and order of my home, where I never heard a rude word or an offensive expression. The effect of this was intensified by my extreme shyness, which made it impossible for me to use words or discuss subjects which were altogether foreign to my home-life, as a result of which I have never been able to use an oath, although I have frequently felt those impulses and passions which in many people can only find adequate expression in such language. This, I think, is a rather striking example of the effects of home influence during childhood, and of that *kind* of education on which Robert Owen depended for the general improvement of character and habits.

CHAPTER IX

BEDFORDSHIRE: SILSOE AND LEIGHTON BUZZARD

It was some time in May or June of 1838 that we left Turvey for Silsoe, where my brother had some temporary work. I walked there, starting very early—I think about four or five in the morning; and a few miles from the village a fine fox jumped over a bank into the road a few yards in front of me, trotted quietly over, and disappeared into a field or copse on the other side. Never before or since have I seen a wild fox so near or had such a good view of one. I breakfasted at Bedford, and then walked to Silsoe.

This very small village is an appanage of Wrest Park, the seat of Earl de Grey, and is about halfway between Luton and Bedford. It consisted of a large inn with a considerable posting business, a few small houses, cottages, and one or two shops, and, like most such villages, it is no larger to-day than it was then. We boarded at the inn kept by a Mr. Carter, whose wife and two daughters, nice well-educated people, took an active part in the management. At this time it was very full of visitors in consequence of the work of building a fine new mansion then in progress and nearing completion. The architect and his clerk of the works were usually there, as was Mr. Brown, a nephew of the agent, and the lively young gentleman, Mr. A., who had been with us at Barton. Besides these, there were others who came for short periods, among whom I particularly remember a grave middle-aged man in black, whose conversation with my brother showed literary tastes and good education, which caused me to be much surprised when I learned that he was there solely to make the working drawings for the handrails of the principal staircase, and to superintend their proper execution. I remember hearing this gentleman speaking in praise of James Silk Buckingham as one of the most remarkable men and prolific writers of the day. Some six years later, I think, I heard a lecture in London by J. S. Buckingham on some of his travels, and the impression made upon me then was, and still is, that he was the best lecturer I ever heard, the most fluent and interesting speaker.

Our work here was mainly copying maps or making surveys connected with the estate, and for this purpose we had the use of a small empty house nearly opposite the inn, where a large drawing-table and a few chairs and stools were all the furniture we required. Here we used sometimes to sit of a summer's evening with one or two friends for privacy and quiet conversation, Mr. Clephan, the architect, and his clerk being our most frequent companions. My brother supplied them with gin-and-water and pipes, and I sat by reading a book or listening to their discourse. Sometimes they would tell each other stories of odd incidents they had met with, or discuss problems in philosophy, science, or politics. When jovially inclined, the architect's clerk would sing songs, many of which were of such an outrageously gross character that my brother would beg him to be more cautious so as not to injure the morals of youth. At one time, when Mr. Clephan was away, there was a fire at a farm quite near us which burnt some stacks and outbuildings, and caused considerable excitement in the village. We only heard of it early in the morning when the local fire-engine had at length succeeded in putting it out. My brother wrote an account of this to Mr. Clephan, with humorous descriptions of the sayings and doings of the chief village characters, and, in reference to what we saw when it was nearly all over, he said, " It could best be described in a well-known line from the Latin grammar, 'Monstrum, horrendum, informe, ingens cui lumen ademptum,' which might be freely rendered, 'a horrid shapeless mass whose

glim the engines dowse.'" He used to show me any letters
he thought might interest me, and this "free translation"
took my schoolboy fancy so that it has stuck in my memory.

One day, having to drive over to Dunstable on some
business, my brother took me with him. When there, we
walked out to a deep cutting through the chalk about a mile
to the north-west, where the road was being improved by
further excavation to make the ascent easier. This was
the great mail-coach road to Birmingham and Holyhead, and
although the railway from London to Birmingham was then
making and partly finished, nobody seemed to imagine that
in twelve years more a railway would be opened the whole
distance, and, so far as the mails and all through traffic were
concerned, all such costly improvement of the high-roads would
be quite unnecessary.

My brother had some conversation with the engineer who
was inspecting the work, and took a lump of chalk home with
him to ascertain its specific gravity, as to which there was
some difference of opinion. While taking luncheon at the
hotel we met a gentleman of about my brother's age, who
turned out to be a surveyor, and who was also interested in
engineering and science generally; and after luncheon they
borrowed a small pair of scales and a large jug of water, and
by suspending the chalk by a thread below the scale-pan,
they weighed it in water, having first weighed it dry in
the ordinary way, and the weight in air, divided by the
difference between the weights in air and water, gives
the specific gravity sufficiently near for ordinary purposes.
This little experiment interested me greatly, and made me
wish to know something about mechanics and physics. Mr.
Matthews lived at Leighton Buzzard, where he carried on
the business of watch-and-clock maker as well as that of
engineer and surveyor. He had undertaken the survey of
the parish of Soulbury, but having too much other work to
attend to, he was looking out for some one to take it off his
hands. This matter was soon agreed upon, and a few weeks
afterwards we left Silsoe to begin the work.

The village of Soulbury is a very small one, though the

parish is rather large. It is only three miles from Leighton, and we obtained accommodation in the school-house, a rather large red-brick house, situated at the further end of the village, where three roads met. It was occupied only by the schoolmaster and his sister, who kept house for him, so we had the advantage of a little society in a rather lonely place. They were both young people and fairly educated, but, as I thought even then, rather commonplace. The chief business of the village girls hereabouts was straw-plaiting, which they did sitting at their cottage doors, or walking about in the garden or in the lanes near, which therefore did not interfere with their getting fresh air and healthy exercise, as do all forms of factory work. Now, owing to cheap imported plait, the only work is in hat and bonnet-sewing, which involves indoor work, and is therefore less healthy as a constant occupation.

The district was rather an interesting one. The parish was crossed about its centre by the small river Ouzel, a tributary of the Ouse, bordered by flat verdant meadows, beyond which the ground rose on both sides into low hills, which to the north-east reached five hundred feet above the sea, and being of a sand formation, were covered with heaths and woods of fir trees. Parallel with the river was the Grand Junction Canal, which at that time carried all the heavy goods from the manufacturing districts of the Midlands to London. Following the same general direction, but about half a mile west on higher ground, the London and Birmingham Railway was in course of construction, a good deal of the earthwork being completed, most of the bridges built or building, and the whole country enlivened by the work going on.

At the same time the canal had been improved at great cost to enable it to carry the increased trade that had been caused by the rapid growth of London and the prosperity of agriculture during the early portion of the nineteenth century. About thirty miles further on the watershed between the river-basins of the Ouse and Severn had to be crossed, a district of small rainfall and scanty streams, from

which the whole supply of the canal, both for its locks as well
as for evaporation and leakage, had to be drawn.   Whenever
there was a deficiency of water here to float the barges and
fill the locks, traffic was checked till the canal filled again ;
and this had become so serious that, for a considerable portion
of the canal, it had been found necessary to erect steam-
engines to pump up the water at every lock from the lower
to the higher level.   Sometimes there were two, three, or
more locks close together, and in these cases a more powerful
engine was erected to pump the water the greater height.
Up to this time I had never seen a steam-engine, and therefore took the greatest interest in examining these both at rest
and at work.   They had been all erected by the celebrated
firm of Boulton and Watt, and were all of the low-pressure
type then in use, with large cylinders, overhead beam, and
parallel motion, but each one having its special features, the
purport of which was explained to me by my brother, and
gave me my first insight into some of the more important
applications of the sciences of mechanics and physics.

Of course at that time nobody foresaw the rapid development of railways all over the country, or imagined that they
could ever compete with canals in carrying heavy goods.
Yet within two years after the completion of the line to Birmingham, the traffic of the canal had decreased to 1,000,000
tons, while it was 1,100,000 tons in 1837.   Afterwards it
began slowly to rise again, and had reached 1,627,000 tons
in 1900, an exceedingly small increase as compared with
that of the railway.   And this increase is wholly due to local
traffic between places adjacent to the canal.

In the northern part of the parish, which extended nearly
to the village of Great Brickhill, were some curious dry
valleys with flat bottoms, and sides clothed with fir woods,
a kind of country I had not yet seen, and which impressed
me as showing some connection between the geological formation of the country and its physical features, though it was
many years later when, by reading Lyell's " Principles of
Geology," I first understood *why* it should be so.   Another
interesting feature of the place, which no one then saw the

significance of, was a large mass of hard conglomerate rock, or pudding-stone, which lay in the centre of the spot where the three roads met in front of the house where we lodged. It was roughly about a yard in diameter and about the same height, and had probably at some remote period determined the position of the village and the meeting-point of the three roads. Being a kind of rock quite different from any found in that part of England, it was probably associated with some legend in early time, but it is in all probability a relic of the ice-age, and was brought by the glacier or ice-sheet that at one time extended over all midland England as far as the Thames valley. But at this time not a single British geologist knew anything about a glacial epoch, it being two years later, in 1840, when Louis Agassiz showed Dr. Buckland such striking indications of ice-action in Scotland as to convince him of the reality of such a development of glaciers in our own country at a very recent period.

When we had completed our field-work, we moved into Leighton Buzzard, and lodged in the house of a tin-and-copper-smith in the middle of the town, where we completed the mapping and other work of the survey. Our landlord was a little active man with black hair and eyes and dark complexion. He told us that whenever his trade was slack he could make small tin mugs at a penny each and earn a fair living, as there was an inexhaustible demand for them. He was a very intelligent man, and he made the same objection to the success of the railway that had been made by many mechanics and engineers before him. This was, that the hold of the engine on the rails would not be sufficient to draw heavy trucks or carriages—that, in fact, the wheels would whizz round instead of going on, as they do sometimes now when starting a heavy train on greasy rails. He and others did not allow sufficiently for the weight of modern engines, which gives such pressure on the wheels as to produce ample friction or adhesion between iron and iron, though apparently smooth and slippery. This question used to be discussed in the old *Mechanics' Magazine*, and it was again and again declared that, however powerful engines were made, they would be unable to draw very heavy loads on account of the want of adhesion ; and all kinds of suggestions were made to remedy this supposed difficulty, such as sprinkling sand in front of the wheels, making the tyres rough like files, etc., all of which were found to be quite unnecessary, owing to the apparently unforeseen fact that as engines became more powerful they became heavier.

On the heath about a mile and a half north of Leighton there was a tumulus, and I was very anxious to know if there was anybody or thing buried under it. The whitesmith was equally interested, and he agreed to go with me some morning very early when we should not be likely to be interfered with. So we started one morning about five, with a couple of spades, and began digging straight down in the middle of the tumulus. It was light sandy soil, easy to move, and we dug a good large hole till we got down about five feet deep, which was the height of the barrow, and then, having found nothing whatever for our trouble, we filled the hole up again, laid on the turf, and got back to breakfast, very tired, but glad to have done it, even though we had found nothing.

Having finished our plans of Soulbury, and made the three copies needed with their books of reference, with some other odd work, my brother took me up to London on Christmas Eve, travelling by coach to Berkhampstead, and thence on to London by the railway, which had been just opened. We went third class for economy, in open trucks identical with modern goods trucks, except that they had hinged doors, but with no seats whatever, so that any one tired of standing must sit upon the floor. Luckily it was mild weather, and the train did not go more than fifteen or twenty miles an hour, yet even at that pace the wind was very disagreeable. The next day we went home to Hoddesdon for a holiday. It had been settled that, as no more surveying work was in view, I should go back to Leighton to Mr. Matthews for a few months to see if I should like to learn the watch and clock-making business as well as surveying and general engineering; and as there seemed to be nothing else available I did so.

Mr. William Matthews was a man of about thirty. He had been married two years, and had a little girl under a year old. Both he and Mrs. Matthews were pleasant people, and I felt that I should be comfortable with them. He had been partly educated under Mr. Bevan, a civil engineer of some reputation, who had made experiments on the strength of materials, the holding power of glue and nails, etc., and had invented an improved slide-rule. My brother had one of these rules, which we found very useful in testing the areas of fields, which at that time we obtained by calculating the triangles into which each field was divided. To check these calculations we used the slide-rule, which at once showed if there were any error of importance in the result. This interested me, and I became expert in its use, and it also led me to the comprehension of the nature of logarithms, and of their use in various calculations. Mr. Matthews had also charge of the town gas-works, which involved some knowledge of practical chemistry, and a good deal of mechanical work. I spent about nine months in his house, and during that time learnt to take an ordinary watch to pieces, clean it properly, and put it together again, and the same with a clock; to do small repairs to jewellery; and to make some attempts at engraving initials on silver. I also saw the general routine of gas manufacture; but hardly any surveying, which was the work I liked best. I was, therefore, very glad when circumstances, not connected with myself, put an end to the arrangement. Mr. Matthews received the offer of a partnership on very favourable terms in an old-established wholesale watchmaking firm in the city of London. Although he would have much preferred the more varied interests of a country life, he could not give up the certainty of a good income with prospect of increase, and thus be able to provide for his wife and family. Fortunately, about the same time my brother had engaged to go to Kington, in Herefordshire, to assist the Messrs. Sayce, with whom he had

been articled, and who had a large business in the surrounding districts.

A younger brother of Mr. Matthews, who was an amateur
chemist, was to take over the management of the gas-works,
and this led to a thorough overhauling of the whole plant,
including the mains and street lamps, so that everything
should be handed over in good working order ; and though I
had generally to mind the shop while the master was away, I
heard every detail discussed in the evening, and sometimes
went out with them after closing hours, to examine some
street lamp or house connection that showed indication of a
leak or water stoppage.   Before quitting this episode in my
early life, I may just note that in after years we became almost
neighbours, first in North-West London, and afterwards at
Godalming, and kept up a neighbourly friendship for many
years.   A son, William Matthews, jun., was brought up to
watchmaking, with the prospect of succeeding his father as
head of the London firm ; but the business was distasteful to
him, and when he came of age he entered the office of a
building surveyor.   But the strain of London life, and an
insatiable love of work when work was to be had, undermined his health, and he died in middle age.   Mr. Matthews
himself was also an example of an intelligent man with considerable ability entirely lost in the narrow round of a small
old-fashioned city business, which absorbed all his energies,
and, combined with a habit of excessive snuff-taking, affected
both his mental faculties and his physical health.   I am,
therefore, thankful that circumstances allowed me to continue
in the more varied, more interesting, and more healthy
occupation of a land-surveyor.

This may be considered the first of several turning-points
of my life, at which, by circumstances beyond my own control, I have been insensibly directed into the course best
adapted to develop my special mental and physical activities.
It was the death at this particular period of the senior partner
in the city watchmaking firm, and his having offered to Mr.
Matthews the opportunity of being his successor on exceedingly advantageous terms, that prevented me from becoming

a mechanical tradesman in a country town, by which my life would almost certainly have been shortened and my mental development stunted by the monotony of my occupation. If I had completed the year with Mr. Matthews, I should have been formally apprenticed to him ; and if he had gone into the City business afterwards, I should either have been passed over to his successor at Leighton, or my training would have been completed in London. This latter, though perhaps better financially, would have been far worse for me mentally and physically, since this wholesale business was the most monotonous and mechanical possible, as I learned some years afterwards when I visited the London office. To my surprise I then found that the business, which brought in a clear profit of about £1200 a year, had no factory, no machinery, no sign of watchmaking except in a very small room behind the office, where a single workman examined and tested the various portions of the watches as they were brought in by the outside piece-workers, the whole business being thus carried on in two small rooms in Bunhill Row. The movements of the watches dealt in were purchased in Coventry, where the various kinds in general use were designed, the separate parts cast, machine-cut, and filed to their proper gauges, and put together. The mainsprings and balance-springs, chains, hands, dials, and cases were usually purchased separately ; and for each class of watch a fitter was employed, whose business it was to put the parts together, find out any small defects, and correct them by hand, while any larger defect in any particular part was sent back to the workman or manufacturer responsible for it. The man at the office made a final examination of the completed watches, tested their performance, corrected any minute defect that was discoverable, and finally, in consultation with one of the firm, determined the grade or quality of the watch and the consequent price. What I should have learnt there would have been how to fit a watch together, how to test it for definite defects, how to judge of the design and workmanship, how to keep accounts, pay the workmen, and probably to act as a traveller for the firm. But even if my health

would have stood the office-work I should never have succeeded as a man of business, for which I am not fitted by
nature.   I rather think that this particular firm was the last
which carried on business in so old-fashioned a way, as the
good-will was, I believe, sold some thirty years later, when
Mr. Matthews retired.   My short experience as a shopboy
and watchmaker, and the association with a man of Mr.
Matthews's extensive knowledge in certain departments of
mechanics and engineering, no doubt helped in the all-round
development of my character, although I did not learn anything of much practical use in my after-life.

CHAPTER X

IN the autumn of 1839 my brother came to Leighton to take me away, and in a day or two we started for Herefordshire, going by the recently opened railroad to Birmingham, where we visited an old friend of my brother's, a schoolmaster, whose name I forget, and who I remember showed us with some pride how his school was warmed by hot-water pipes, then somewhat unusual. We then went on by coach through Worcester to Kington, a small town of about two thousand inhabitants, only two miles from the boundary of Radnorshire. It is pleasantly situated in a hilly country, and has a small stream flowing through it. Just beyond the county boundary, on the road to Old and New Radnor, there is an isolated craggy hill called the Stanner Rocks, which, being a very hard kind of basalt very good for road-metal, was being continually cut away for that purpose. It was covered with scrubby wood, and was the most picturesque object in the immediately surrounding country.

We obtained board and lodging at the house of a gun-maker, Mr. Samuel Wright, a jolly little man, who reminded me of the portrait of the immortal Mr. Pickwick, and who, on account of his rotundity, was commonly known in the town as Alderman Wright. Mrs. Wright was, on the contrary, very thin and angular. They were equally different in their characters; he was very slow of speech, but very fond of telling stories of his early life, usually very commonplace, and told in such a way as to be dreadfully wearisome. After every few words he would

stop, to let them sink in, then utter a few more with another
stop, and all mixed up with so many " says I's " and "says
he's," and " that's to say's," and little digressions about other
people, that it was usually impossible to make out what he
was driving at. Mrs. Wright, on the other hand, was a great
and rather voluble talker, and she would often interpose with,
" Now, Samiwell, you don't tell that right," and, of course,
that would only lengthen out the story. She was a very
active woman, a great scrubber and cleaner, and unusually
fond of fresh air ; but these good qualities were sometimes
inconvenient, as we all sat in a small room behind the shop,
which had three or four doors in it, which we usually found
open, and had to shut every time we came in. There was,
in fact, such a constant draught in this room that I jokingly
suggested a small windmill being put up, which might be
used to grind coffee, but she always said that it was the
warmest room in the house. Mr. Wright also seemed to
enjoy fresh air and water to an unusual degree in those days,
for early every morning, winter and summer, he would come
down undressed into his little back yard, and there pour cold
water all over his body, then scrub himself with a rough
towel, put on his underclothing, and return upstairs to finish
his toilet. But Mrs. Wright was an excellent cook, and
gave us very good meals, and the alderman was very good-
natured, let me look on while he cleaned and repaired guns,
and once, when I went with some friends to shoot young
rooks, he lent me an excellent double-barrelled gun for the
occasion ; and these good qualities made up for the little
eccentricities of both of them, who, though so different in
some respects, were evidently very attached to each other,
and never quarrelled. Mrs. Wright used to be fond of saying
how dreadful it would be if Samiwell should die first after
they had lived together so many years.

Our employers, two brothers, were also well-contrasted
characters. The elder, Mr. Morris Sayce, was a rather tall,
grey-haired man of serious aspect and rather silent and un-
communicative manner. He, I believe, devoted himself
chiefly to valuations and estate agency. The younger partner,

Mr. William Sayce, was a small, active, dark-haired man, rather talkative and fond of a joke, and as he attended to the surveying business, we saw most of him, and found him a pleasant superior. Both were married and had families of grown-up sons and daughters. They were very hospitable, and we were several times invited to dine or to evening parties at their houses, where we met some of the chief people in the town.

The offices were situated in a small house in a rather narrow street, the ground-floor being occupied by the partners' private office and a clerk's room, while a large room above was the chief map-drawing room, containing a large table ten or twelve feet long by five or six wide, used for mounting drawing-paper on canvas for large maps, with some smaller tables and desks, while other rooms were used chiefly for writing or store-rooms. There were a good many employées besides ourselves. The chief draughtsman and head of the office in the absence of the principals, was named Stephen Pugh, a thorough Welshman in appearance and speech, and a very pleasant and good-natured man, rather fond of poetry and general literature. The next marked character was a rather tall Irishman, a surveyor, who had the unconscious humour of his race, and was besides looked upon as somewhat of a philosopher. One evening, I remember, after work was over at the office, he undertook to give us an address on Human Nature or some such subject, which consisted of a rather prosy exposition of the ideas of Aristotle and the mediæval schoolmen on human physiology, without the least conception of the science of the subject at the time he was speaking. There were also a copying clerk, and two or three articled pupils, one or two about my own age, who helped to keep the office lively. In a solitary letter, accidentally preserved, written at this time to my earliest friend, George Silk, I find the following passage which well expresses the pleasure I felt in getting back to land-surveying :—

" I think you would like land-surveying, about half indoors and half outdoors work. It is delightful on a fine summer's day to be (literally) cutting all over the country,

following the chain and admiring the beauties of nature,
breathing the fresh and pure air on the hills, or in the noontide
heat enjoying our luncheon of bread-and-cheese in a pleasant
valley by the side of a rippling brook. Sometimes, indeed, it
is not quite so pleasant on a cold winter's day to find yourself
on the top of a bare hill, not a house within a mile, and
the wind and sleet chilling you to the bone. But it is all
made up for in the evening; and those who are in the house
all day can have no idea of the pleasure there is in sitting
down to a good dinner and feeling hungry enough to eat plates,
dishes, and all."

Although he was at least ten years older than myself,
Stephen Pugh was my most congenial friend in the office.
When I was away surveying, and for a year or two after we
had left Kington altogether, he and I used to correspond, and
often wrote rhymed letters, which were, of course, very poor
doggerel. I have, however, always kept in my memory a
portion of *one* of Pugh's letters, partly perhaps on account of
its extravagant flattery of my attempts at verse, though I
always knew that I had no poetic faculty whatever. The
letter began by describing what each one in the office was
doing just as work was over one evening, with characteristic
remarks on the idiosyncrasy of each; it then went on—

> " The board was covered o'er with canvas white,
> And looked Llyn Glwdy on a moonlight night,
> *   *   *   *   *   *
> When to *my* hand there came what could be better
> Than your poetic, wise, and humorous letter.
> Like that good angel mentioned by Saint John
> Who ope'd seven seals, I quickly opened one,
> And glancing o'er the page found to my joy
> Spontaneous poetry without alloy.
> The youth, cried I, who built this lofty rhyme
> Will be remembered to the end of time,
> And countless generations yet unborn
> Will read his verse upon a summer's morn,
> And think of him in that peculiar way
> We think of Byron in the present day," etc.

Some time during the winter I went alone to correct an
old map of the parish of New Radnor. This required no

regular surveying, but only the insertion of any new roads, buildings, or divisions of fields, and taking out any that had been cleared away. As these changes are not numerous and the new fences were almost always straight lines, it was easy to mark on the map the two ends of such fences by measuring from the nearest fixed point with a ten or fifteen-link measuring-rod, and then drawing them in upon the plan. Sometimes the direction was checked by taking an angle with the pocket sextant at one or both ends, where one of these could not be seen from the other. As the whole plan was far too large to be taken into the field, tracings were made of portions about half a mile square, which were mounted on stiff paper or linen, and folded up in a loose cover for easy reference. In this way a whole parish of several thousand acres could be examined and corrected in a week or two, especially in a country like Wales, where, from a few elevated points, large tracts could be distinctly seen spread out below, and any difference from the old map be easily detected. I liked this kind of work very much, as I have always been partial to a certain amount of solitude, and am especially fond of rambling over a country new to me.

New Radnor, though formerly a town of some importance, was then, and I believe is still, a mere village, and a poor one, Presteign being the county town. It is situated on the southern border of Radnor Forest, a tract of bare mountains about twenty square miles in extent, the highest point being a little over two thousand feet above the sea. Over a good deal of this country I wandered for about a week, and enjoyed my work very much. One day, when I had a little time to spare, I went a mile or two out of my way to see a rather celebrated waterfall, called Water-break-its-neck. I descended into the valley and walked down it, as I knew the fall was on one side of it in a small lateral valley, but owing to the glare of the afternoon sun, I did not see the opening in the shadow, and came down to the end of the valley. But I determined to see it, so turned back as fast as I could, and soon found it just out of sight, owing to a curve of the lateral valley. It must be a fine fall when the stream is full, as it then probably shoots out clear of the rock. But when I saw it there was only a film of water covering the surface of the rock from top to bottom. This surface is formed by the regular weathering of slaty beds in fine layers; the upper part curves downward but the lower half is very nearly or quite vertical and of considerable width, and the whole fall, as seen from near the foot of it, is perhaps sixty feet high. In the valley above this fall is another somewhat more irregular, but I had not time to see this, as it was getting dark when I turned homewards.

The little inn at which I stayed was very quiet and comfortable. The landlord and his wife were both quiet and refined-looking people, not the least like the ordinary type of innkeepers. In the evening I sat with them in a parlour where friends and a superior class of visitors only were admitted; and while I was there the district exciseman lodged in the house while making his rounds among the surrounding villages. He was a brisk and intelligent man, and was in no way treated as an enemy, but rather as a confidential friend. One evening when he and the host with myself were alone together, something brought up the names of Heloise and Abelard, whereupon the exciseman told us the whole story of these unfortunate lovers in a way that showed he was well acquainted with their correspondence, from which he quoted some of the more interesting passages, apparently verbatim, and with sympathetic intonation. This is the only occasion on which I have heard the subject dealt with in conversation, or, in fact, any similar subject in a village inn and between landlord and exciseman.

Early the next year, I think about February, my brother and I went to do some surveying at Rhaidr-Gwy (now more commonly called Rhayader), a small town in Radnorshire on the Upper Wye, and only fifteen miles from its source in the Plynlymmon range. A young man from Carmarthenshire came to us here to learn surveying. He was one of the very loose young men with whom I was often associated, and I think as regards the filthiness of his language and of the stories with which he used frequently to regale us he surpassed all. However, he was in other respects a pleasant companion, being quite unconscious that his conversation was not appreciated, and to him I probably owe my life. One day, I think on a Sunday afternoon, we were walking together up a rocky and boggy valley which extended some miles to the west of the town. As we were strolling alone, picking our way among the rocks and bog, I inadvertently stepped upon one of those small bog eyeholes which abound in such places, and are very dangerous, being often deep enough to swallow up a man, or even a horse. One leg went in suddenly up to the hip, and I fell down, but fortunately with my other leg stretched out upon the surface. I was, however, in such a position that I could not rise, and had I been alone my efforts to extricate myself might easily have drawn my whole body into the bog, as I could feel no bottom to it. But my companion easily pulled me out, and we walked home, and thought little of it. It had, however, been a hard frost for some time, and the mud was ice-cold, and after a few days I developed a bad cough with loss of appetite and weakness. The local doctor, John Henry Heaton by name, was a friend of ours, and he gave me some medicine, but it did no good, and I got worse and worse, with no special pain, but with a disgust of food, and for more than a week I ate nothing but perhaps a small biscuit each day soaked in tea without milk, though always before and since I greatly disliked tea without milk. At length the doctor got frightened, and told my brother that he could do nothing for me, and that he could not be answerable for my life. He added that he knew but one man who could save me, a former teacher of his, Dr. Ramage, who was the only man who could cure serious lung disease, though he was considered a quack by his fellow practitioners.

As I got no better, a few days later we started for London, I think sleeping at Birmingham on the way. On going to Dr. Ramage, who tested my lungs, etc., he told my brother

that he was just in time, for that in a week more he could
probably not have saved me, as I had an extensive abscess
of the lungs. His treatment was very simple but most effective, and was the forerunner of that rational treatment by
which it is now known that most lung diseases are curable.
He ordered me to go home to Hoddesdon immediately, to
apply half a dozen leeches to my chest at a place he marked
with ink, and to take a bitter medicine he prescribed to give
me an appetite; but these were only preliminaries. The
essential thing was the use of a small bone breathing-tube,
which he told us where to buy, and which I was to use three
times a day for as many minutes as I could without fatigue;
that I was to eat and drink anything I fancied, be kept
warm, but when the weather was mild sit out-of-doors. I
was to come back to him in a week.

The effect of his treatment was immediate. I at once
began to eat, and though I could not breathe through the
tube for more than a minute at first, I was soon enabled to
increase it to three and then to five minutes. It was constructed with a valve so that the air entered freely, but passed
out slowly so that it was kept in the lungs for a few seconds
at each inspiration. When I paid my second visit to Dr.
Ramage, he told me that I was getting on well, and need not
come to him again, that I was to continue using the breathing-tube for five minutes three or four times a day. He also
strongly advised me, now I saw the effect of deep and regular
breathing, to practise breathing in the same way without the
tube, and especially to do so when at leisure, when lying
down, or leaning back in an easy-chair, and to be sure to fill
my lungs well and breathe out slowly. "The natural food
of the lungs," he said, "is fresh air. If people knew this, and
acted upon it, there would be no consumption, no lung disease." I have never forgotten this. I have practised it all
my life (at intervals), and do so still, and I am sure that I
owe my life to Dr. Ramage's treatment and advice.

In about two months I was well again, and went back to
Kington, and after a little office-work my brother and I went

to the little village of Llanbister, near the middle of Radnorshire, the nearest towns being Builth, in Breconshire, and
Newtown, in Montgomeryshire, both more than twelve miles
distant. This was a very large parish, being fifteen miles
long, but I think we could only have corrected the old map
or we should have been longer there than we really were.
Here, also, we had a young gentleman with us for a month
or two to practise surveying. He was, I think, a Welshman,
and a pleasant and tolerably respectable young man, but he
had one dreadful habit—excessive smoking. I have never
met a person so much a slave to the habit, and even if I had
had any inclination to try it again after my first failure, his
example would have cured me.

He prided himself on being a kind of champion smoker,
and assured us that he had once, for a wager, smoked a good-
sized china teapot full of tobacco through the spout. He
smoked several pipes of very strong tobacco during the day,
beginning directly after breakfast, and any idle moments
were occupied by smoking. The village being an excessively
small one, and the population of the parish very scattered,
there was only one public-house, where we were living, and
the landlady went every week to market to lay in a stock of
necessaries, including tobacco. One market day our friend
found himself without tobacco, and on asking for some, was
told there was none till the mistress came home in the evening. He was in despair; went to the only little village shop,
but they did not keep it; to the two or three houses in the
village, but none was to be found. He was the picture of
misery all day; he could eat no dinner; he wandered about,
saliva dropping from his mouth, and looking as if he were
insane. The tobacco did not come till about seven in the
evening. His relief was great and instantaneous, and after a
pipe he was able to eat some supper. Had the tobacco not
come he declared he would have died, and I believe he would
have had a serious illness. This terrible slavery to the
smoking habit gave the final blow to my disinclination to
tobacco, which has been rendered more easy to me by my
generally good appetite and my thorough enjoyment of

appetizing food and drinks. Of the latter, I took beer and wine in moderation during the first fifty years of my life, after which period I became practically a total abstainer for special hygienic reasons ; and my own experience and observation has led me to the conclusion that alcoholic drinks, taken constantly, are especially injurious in old age and shorten the lives of many persons.

It was during this early period of my life that, on two occasions only, I exceeded the limits of moderation, and both were due to my youthful shyness and dislike of appearing singular in society. One of these was at a dinner at Mr. Sayce's, where the wine-drinking was especially prolonged, and when at last we left the table, I felt my head dizzy and my steps a little uncertain. The other was at Rhayader at a time when my brother was away, and Dr. Heaton and another friend were dining at the inn together with myself. At dinner the doctor ordered a bottle of port wine and filled my glass with the others. After dinner, the bottle being emptied, the doctor said, "One bottle is a very small allowance for three. Let's have another." Of course, the friend agreed, and I said nothing, and was too shy to make an excuse and leave the table. Of this bottle I tried, weakly, to refuse any share, but the doctor insisted on giving me half a glass each round ; and when this bottle was empty, he ordered another, saying, "That's only one each," and I was compelled to have some of that too, but I drank as little as I could, and again felt very dizzy and uncomfortable. Before going the doctor said to the waiter, "We've had three bottles of port ; charge one to each of us." Of course, I dare not say a word ; and when our bill came in, and my brother saw the bottle of port wine charged which he had not ordered, he asked for an explanation, and when I told him the circumstances, he evidently thought I had done very wrong, but said nothing more about it, knowing, perhaps, the difficulties of a shy lad in the society of men. This little circumstance, perhaps more than anything else, led to my never again taking more wine than I felt inclined to take, and that was usually two or three glasses only.

Before we left Llanbister my cousin, Percy Wilson, who was preparing for ordination after taking his degree at Oxford, came to stay a short time with us, and partly to see again the estate of Abbey-Cwm-Hir, which his father had purchased in the days of his prosperity and which was only a few miles distant, being, in fact, an adjoining parish. I and he walked over to see it one day, and found it to be situated in a lonely wild valley bounded by lofty and rather picturesque mountains. It was a small country house built by my uncle, partly from the heaped-up ruins of the ancient Cistercian monastery, the lower portion of the church still remaining, the walls having the remains of clustered columns attached to them. It would have made a charming summer residence in a few years, when the shrubs and trees had grown, and the whole surroundings had been somewhat modified by judicious planting, especially as Mr. Wilson had purchased, I believe, the entire estate, comprising the greater part of the parish, and including the whole valley and its surrounding mountains.

Two pencil sketches by my brother, made in a surveyor's field-book while at this place, have been preserved and are here copied, as examples of his delicacy of touch and power of giving artistic effect to the simplest objects. The upper one is the village taken from the house we lodged in, showing the low church at the end of the street, and the queer little house just opposite us, occupied then by the village shoe-maker, but showing some architectural pretensions as compared with the usual cottages in a small Welsh village. The lower one is a small and lonely chapel in a remote part of the parish, to which the local builder has given character, while the dreary surroundings are well indicated in the sketch.

When we had finished at Llanbister, we went about ten miles south to a piece of work that was new to me—the making of a survey and plans for the enclosure of common lands. This was at Llandrindod Wells, where there was then a large extent of moor and mountain surrounded by scattered cottages with their gardens and small fields, which,

LLANBISTER, RADNORSHIRE.

(*Pencil sketch by W. G. Wallace.* 1840.)

"A LONELY CHAPEL."

(*Pencil sketch by W. G. Wallace.* 1840.)

[*To face p.* 150, VOL. I.

with their rights of common, enabled the occupants to keep
a horse, cow, or a few sheep, and thus make a living. All
this was now to be taken away from them, and the whole
of this open land divided among the landowners of the
parish or manor in proportion to the size or value of their
estates. To those that had much, much was to be given,
while from the poor their rights were taken away ; for
though nominally those that *owned* a little land had some
compensation, it was so small as to be of no use to them in
comparison with the grazing rights they before possessed.
In the case of all cottagers who were tenants or leaseholders,
it was simple robbery, as they had no compensation whatever,
and were left wholly dependent on farmers for employment.
And this was all done—as similar enclosures are almost
always done—under false pretences. The "General Inclosure
Act" states in its preamble, "Whereas it is expedient to
facilitate the inclosure and improvement of commons and
other lands now subject to the rights of property which
obstruct cultivation and the productive employment of labour,
be it enacted," etc. But in hundreds of cases, when the
commons, heaths, and mountains have been partitioned out
among the landowners, the land remains as little cultivated
as before. It is either thrown into adjacent farms as rough
pasture at a nominal rent, or is used for game-coverts, and
often continues in this waste and unproductive state for half
a century or more, till any portions of it are required for
railroads, or for building upon, when a price equal to that of
the best land in the district is often demanded and obtained.
I know of thousands of acres in many parts of the south of
England to which these remarks will apply, and if this is not
obtaining land under false pretences—a legalized robbery of
the poor for the aggrandisement of the rich, who were the
law-makers—words have no meaning.

In this particular case the same course has been pursued.
While writing these pages a friend was staying at Llandrindod
for his wife's health, and I took the opportunity of asking him
what was the present condition of the land more than sixty
years after its inclosure. He informs me that, by inquiries

among old inhabitants, he finds that at the time nothing whatever was done except to enclose the portions allotted to each landlord with turf banks or other rough fencing; and that to this day almost all the great boggy moor, with the mountain slopes and summits, have not been improved in any way, either by draining, cultivation, or planting, but is still wild, rough pasture. But about thirty years after the inclosure the railway from Shrewsbury through South Wales passed through the place, and immediately afterwards a few villas and boarding-houses were built, and some of the enclosed land was sold at building prices. This has gone on year by year, and though the resident population is still only about 2000, it is said that 10,000 visitors (more or less) come every summer, and the chief increase of houses has been for their accommodation. My friend tells me that, except close to the village and railway, the whole country which was enclosed— many hundreds of acres—is still bare and uncultivated, with hardly any animals to be seen upon it. Milk is scanty and poor, and the only butter is Cornish or Australian, so that the inclosure has not led to the supply of the simplest agricultural needs of the population. Even the piece of common that was reserved for the use of the inhabitants is now used for golf-links!

Here, then, as in so many other cases, the express purpose for which alone the legislature permitted the inclosure has not been fulfilled, and in equity the whole of the land, and the whole money proceeds of the sale of such portions as have been built upon, should revert to the public. The prices now realized by this almost worthless land, agriculturally, are enormous. In or near the village it sells for £1500 an acre, or even more, while quite outside these limits it is from £300 to £400. All this value is the creation of the community, and it has only been diverted to the pockets of private persons by false pretences. And to carry out this cruel robbery, how many of the poor have suffered? how many families have been reduced from comfort to penury, or have been forced to emigrate to the overcrowded towns and cities, while the old have been driven to the workhouse, have become law-created paupers?

In regard to this fundamental question of land ownership people are so blinded by custom and by the fact that it is sanctioned by the law, that it may be well for a moment to set these entirely on one side, and consider what would have been the proper, the equitable, and the most beneficial mode of dealing with our common and waste lands at the time of the last general Inclosure Act in the early years of the reign of Queen Victoria. Considering, then, that these unenclosed wastes were the last remnant of our country's land over which we, the public, had any opportunity of free passage to breathe pure air and enjoy the beauties of nature; considering that these wastes, although almost worthless agriculturally, were of especial value to the poor of the parishes or manors in which they were situated, not only giving them pasture for their few domestic animals, but in some cases peat for fuel and loppings of trees for fences or garden sticks ; considering that an acre or two of such land, when enclosed and cultivated, would give them, in return for the labour of themselves and their families during spare hours, a considerable portion of their subsistence, would enable them to create a home from which they could not be ejected by the will of any landlord or employer, and would thus raise them at once to a condition of comparative independence and security, abolishing the terrible spectre of the workhouse for their old age, which now haunts the peasant or labourer throughout life, and is the fundamental cause of that exodus to the towns about which so much nonsense is talked ; considering, further, that just in proportion as men rise in the social scale, these various uses of the waste lands become less and less vitally important, till, when we arrive at the country squire and great landowner, the only use of the enclosed common or moor is either to be used as a breeding ground for game, or to add to some of his farms a few acres of land at an almost nominal rent—considering all these circumstances, and further, that those who perform what is fundamentally the most important and the most beneficial of all work, the production of food, should be able to obtain at least the necessaries of life by that work, and secure a comfortable old age by their own fireside

—how would any lover of his country think that such lands *ought* to be dealt with in the best interests of the whole community?

Surely, that the very first thing to be done should be to provide that all workers upon the land, either directly or indirectly, should have plots of from one to five acres, in proportion to the amount of such waste and the needs of the inhabitants. The land thus allotted to be held by them in perpetuity, from the local authority, at a low rent such as any farmer would give for it as an addition to his farm. In cases where the amount of common land was very great in proportion to the population, some of the most suitable land might be reserved for a common pasture, for wood or fuel, or for recreation, and the remainder allotted to applicants from adjacent parishes where there was no common land.

If it is asked, how are the various landowners and owners of manorial rights to be compensated? there are two answers, either of which is sufficient. The first is, that they would be fully compensated by the increased well-being of the community around them. Whenever such secure holdings have been given by private owners—as in the cases of Lord Tollemache and Lord Carrington—pauperism has been abolished, and even poverty of any kind greatly diminished. And as landlords pay rates, and diminished rates mean increased value of farm land, and, therefore, increased rents, the landlords would be more than compensated even in money's worth. Again, where it has been fairly tried, the surrounding large farmers, though at first violently opposed to such small holdings on the ground that they would make the labourers too independent, ultimately acknowledge that it greatly benefits them, because it surrounds them with a permanent population of good and experienced labourers, who are always ready at hay and harvest time to work for good wages, and thus save crops and secure them in the best condition when they might otherwise be deteriorated by delay, or totally lost for want of labour at the critical moment during a wet summer. Such a constant supply of labour benefits every farmer, abolishes to a large extent agricultural depression, and thus secures payment of the landlord's rents—again increasing the money value of his property.

And if, notwithstanding these demonstrated benefits, landlords still claim their pound of flesh, the money value of public land, which only laws made by their own class have given them, we will make our counterclaim for the land-tax at 4s. in the pound, "on the full annual value," as solemnly agreed by Parliament when the various services due from landlords to the crown were abolished and the tax fixed at what was then considered a very low rate, in lieu of them. The last valuation made was in 1692, and, notwithstanding the continual increase in land values from that time, as well as the continual decrease in the purchasing power of money, the land-tax continued to be paid on that absurdly low valuation, which in the reign of George III. was made permanent. The arrears of land-tax now equitably due will amount to more than the value of all the agricultural land of our country at the present time, and as when public rights are in question there is no time limit, existing landlords would do well not to be too clamorous for their alleged rights of property, since it may turn out that those "rights" do not exist.

Another thing that should be attended to in all such inclosures of waste land is the preservation for the people at large of rights of way over it in various directions, both to afford ample means of enjoying the beauties of nature and also to given pedestrians short cuts to villages, hamlets, or railway stations. One of the greatest blessings that might be easily attained if the land were resumed by the people to be held for the common good, would be the establishment of ample footpaths along every railway in the kingdom, with sufficient bridges or subways for safe crossing ; and also (and more especially) along the banks of every river or brook, such paths to be diverted around any dwelling-house that may have gardens extending to the water's edge, all such paths to be made and kept in repair by the District Councils. Under the present system old paths are often closed, but we never hear of new ones being made, yet such are now more than ever necessary when most of our roads are rendered dangerous by motor-cars and cycles, and exceedingly disagreeeble and unhealthy to pedestrians by the clouds of gritty dust continually raised by these vehicles.

Returning now to the question of the rights of the people at large to a share in their native land, I would further point out that the inclosure of commons is only one of many acts of robbery that have been perpetrated by or for the landlords. If we go back no further than the reign of Henry VIII. we have the whole vast properties of the abbeys and monasteries confiscated by the king, and mostly given away to personal friends or powerful nobles, without any regard whatever to the rights of the poor. Most of these institutions took the place of our colleges, schools, and workhouses. The poor were relieved by them, and they served as a refuge for the wanderer and the fugitive. No provision was made for the fulfilment of these duties by the new owners, and the poor and needy were thus plundered and oppressed. Under the same king and his successors all the accumulated wealth of the parish churches, in gold and silver vessels, in costly vestments often adorned with jewels, in paintings by great masters, and in illuminated missals which were often priceless works of art, were systematically plundered, court favourites obtaining orders to sequestrate all such "popish ornaments," in a certain number of cases keeping the produce for themselves, while in others they were sold for the king's benefit. The property thus stolen the Rev. A. Jessopp estimates to have been many times greater than the value of all the abbeys and monasteries of the kingdom!

If we consider the nature of this long series of acts of plunder of the people's land and other property, we find in it every circumstance tending to aggravate the crime. It was robbery of the poor by the rich. It was robbery of the weak and helpless by the strong. And it had this worst feature that distinguishes robbery from mere confiscation—the plunder was divided among the robbers themselves. Yet again, it was a form of robbery specially forbidden by the religion of the robbers—a religion for which they professed the deepest reverence and of which they considered themselves the special defenders. They read in what they called *The Word of God*, "Woe unto them that join house to house, that lay field to field, till there be no place, that they may be placed alone in the midst of the earth!" Yet this is what they were, and are, constantly striving for, not by purchase only, but by open or secret robbery. Again, they read in their holy book, "The land shall not be sold for ever: for the land is Mine;" and at every fiftieth year all land was to return to the family that had sold it, so that no one could keep land beyond the year of jubilee, the reason being that no man or family should be permanently impoverished by the misdeeds of his ancestors. But this part of the law they *never* obey.

This all-embracing system of land-robbery, for which nothing is too great and nothing too small; which has absorbed meadow and forest, moor and mountain, which has appropriated most of our rivers and lakes and the fish that live in them; which often claims the very seashore and rocky coasts of our island home, fencing them off from the wayfarer who seeks the solace of their health-giving air and wild beauty, while making the peasant pay for his seaweed manure and the fisherman for his bait of shell-fish; which has desolated whole counties to replace men by sheep or cattle, and has destroyed fields and cottages to make a wilderness for deer and grouse; which has stolen the commons and filched the roadside wastes; which has driven the labouring poor into the cities, and has thus been the primary and chief cause of the lifelong misery, disease, and early death of thousands who might have lived lives of honest toil and comparative well-being had they been permitted free access to land in their native villages;—it is the advocates and beneficiaries of this inhuman system who, when a partial restitution of their unholy gains is proposed, are the loudest in their cries of "robbery"!

But all the robbery, all the spoliation, all the legal and illegal filching, has been on *their* side, and they still hold the stolen property. *They* made laws to legalize their actions, and, some day, we, the people, will make laws which will not only legalize but justify our process of restitution. It will justify it, because, unlike their laws, which always took from the poor to give to the rich—to the very class which made the laws—ours will only take from the superfluity of the rich, *not* to give to the poor or to any individuals, but to so administer as to enable every man to live by honest work, to restore to the whole people their birthright in their native soil, and to relieve all alike from a heavy burden of unnecessary and unjust taxation. *This* will be the true statesmanship of the future, and it will be justified alike by equity, by ethics, and by religion.

In the few preceding pages I have expressed the opinions which have been gradually formed as the result of the experience and study of my whole life. My first work on the subject was entitled "Land Nationalization: its Necessity and its Aims," and was published in the year 1882 ; and this, together with the various essays in the second volume of my "Studies Scientific and Social," published in 1900, may be taken as expressing the views I now hold, and as pointing out some of the fundamental conditions which I believe to be essential for the well-being of society.

But at the time of which I am now writing such ideas never entered my head. I certainly thought it a pity to enclose a wild, picturesque, boggy, and barren moor, but I took it for granted that there was *some* right and reason in it, instead of being, as it certainly was, both unjust, unwise, and cruel. But the surveying was interesting work, as every trickling stream, every tree, every mass of rock or boggy waterhole, had to be marked on the map in its true relative position, as well as the various footpaths or rough cart-roads that crossed the common in various directions.

At that time the medicinal springs, though they had been used from the time of the Romans, were only visited by a few Welsh or West of England people, and there was little accommodation for visitors, except in the small hotel where we lodged. One of our great luxuries here was the Welsh mutton fed on the neighbouring mountains, so small that a hind-quarter weighed only seven or eight pounds, but which, when hung a few days or a week, was most delicious eating. I agree with George Borrow in his praise of this dish.   In his " Wild Wales " he says, " As for the leg of mutton it was truly wonderful; nothing so good had I ever tasted in the shape of a leg of mutton.   The leg of mutton of Wales beats the leg of mutton of any other country, and I had never tasted a Welsh leg of mutton before.   Certainly I shall never forget that first Welsh leg of mutton which I tasted, rich but delicate, replete with juices derived from the aromatic herbs of the noble Berwyn mountain, cooked to a turn, and weighing just four pounds."   Well done, George Borrow !   You had a good taste in ale and mutton, and were not afraid to acknowledge it.

CHAPTER XI

It was in the summer or early autumn of 1841 that we left Kington for the survey of a parish a few miles beyond the town of Brecon. As there was no coach communication, and the distance was only about thirty miles, we determined to walk, and having sent our luggage by coach or waggon, we started about sunrise, and after two hours' walking stopped at a nice-looking roadside public-house for breakfast. Our meal consisted of a large basin of bread-and-milk with half a pint of good ale in it, and sugar to taste, which had been recommended to my brother as the best thing to walk on. I certainly enjoyed it very much. We then walked on through the little town of Hay, and soon after midday had dinner at a village inn and a good rest, as the day was very hot and the roads hilly. In the afternoon I became very tired, and while we were still some miles from Brecon, I felt quite exhausted with the heat and fatigue. At length I became so faint that I had to lie down in the road to prevent myself from losing consciousness and falling down. However, with the aid of repeated rests I struggled on, and we reached Brecon when it was nearly dark.

The next morning I felt all right again, and as we started for our destination, I was delighted with the grand view of the double-headed Beacons, the highest mountain in South Wales, which, though five miles away, seem to rise up abruptly into the clouds as viewed down the street by which we entered the town. On leaving the town we crossed a bridge over the little rocky stream, the Honddu, which here enters the Usk, and gives the Welsh name to the town of Brecon—Aber-honddu—*aber* meaning the confluence or meeting of waters. So, Aberystwith, which has retained its Welsh name, is situated where the little river Ystwith enters the sea.   While living in Radnorshire, where hardly any Welsh is spoken, I had begun to take an interest in the picturesque names which primitive people always give to localities.   The first of these to which my attention was called by my brother was Llanfihangel-nant-Melan, a village about ten miles west of Kington, the name meaning " the Church of St. Michael on Melan's brook."   So, Abbey-cum-hir is the Abbey in the long valley ; while the celebrated Vale of Llangollen is, according to George Borrow, named after Collen, an ancient British hero who became Abbot of Glastonbury, but afterwards retired into the valley named after him.

Our road lay along the north side of the valley of the Usk, but at some distance from the river, through a very picturesque country, crossing many small rivers, often looking down upon the river Usk, which I took special interest in as my native stream, here approaching its source, and with frequent views of the Beacons when nearer hills did not intervene to block the view.   After a pleasant walk of about six miles we reached the tiny village of Trallong, the parish we had to survey, and obtained lodgings in the house of a shoemaker, where we were very comfortable for some months. The house was pleasantly situated about two hundred and fifty feet above the river, with an uninterrupted view to the south-east over woody hills of moderate height to the fine range of the Great Forest, culminating in the double peaks of the Beacons, which were seen here fully separated with the narrow ridge connecting them.   At sunset they were often beautifully tinted, and my brother made a charming little water-colour sketch of them, which, with most of his best sketches, were placed in an album by my sister, and this was stolen or lost while she was moving in London.

The family here were rather interesting.   The father, a middle-aged man, could not speak a word of English.   His grown-up sons, who helped in the shoemaking, spoke but little. The wife, however, a delicate woman and a great invalid, though having to do all the work of the household, spoke English very well, and told us that she preferred it to Welsh, because it was less tiring, the Welsh having so many gutturals and sounds which require an effort to pronounce correctly. There were also two little girls who went to the village school, and who spoke English beautifully as compared with our village children, because they had learnt it from the schoolmaster and their mother. Of course, the whole conversation in the house was in Welsh, and I picked up a few common words and phrases, and could understand others, though, owing to my deficiency in linguistic faculty, I never learnt to speak the language.

The schoolmaster was an intelligent and well-educated man, and he often called in the evening to have a little conversation with my brother. But almost the only special fact I remember about him was his passion for cold water. Every morning of his life he walked to the river half a mile off to take a dip before breakfast, and in some frosty days in winter I often saw him returning when he had had to break the ice at the river's edge.

I looked daily at the Beacons with longing eyes, and on a fine autumn day one of the shoemaker's sons with a friend or two and myself started off to make the ascent. Though less than six miles from us in a straight line, we had to take a rather circuitous course over a range of hills, and then up to the head of a broad valley, which took us within a mile of the summit, making the distance about ten miles. But the day was gloriously fine, the country beautiful, and the view from the top very grand ; while the summit itself was so curious as greatly to surprise me, though I did not fully appreciate its very instructive teaching till some years later, after I had ascended many other mountains, had studied Lyell's " Principles of Geology," and had fully grasped the modern views on sub-aerial denudation. As Brecknockshire is comparatively little known, and few English tourists make the

ascent of the Beacons, a short account of them will be both
interesting and instructive.

The northern face of the mountain is very rocky and precipitous, while on the southern and western sides easy slopes
reach almost to the summit. The last few yards is, however,
rather steep, and at the very top there is a thick layer of peat,
which overhangs the rock a little. On surmounting this on
the west side the visitor finds himself in a nearly flat triangular
space, perhaps three or four acres in extent, bounded on the
north by a very steep rocky slope, and on the other sides
by steep but not difficult grass slopes. To the north-east he
sees the chief summit about a quarter of a mile distant and
nearly fifty feet higher, while connecting the two is a narrow
ridge or saddle-back, which descends about a hundred feet in
a regular curve, and then rises again, giving an easy access to
the higher peak. The top of this ridge is only a foot or two
wide and very steep on the northern slope, but the southern
slope is less precipitous, and about a hundred yards down it
there is a small spring where the visitor can get deliciously
cold and pure water. The north-eastern summit is also
triangular, a little larger than the other, and bounded by a
very dangerous precipice on the side towards Brecon, where
there is a nearly vertical slope of craggy rock for three or four
hundred feet and a very steep rocky slope for a thousand,
so that a fall is almost certainly fatal, and several such
accidents have occurred, especially when parties of young men
from Brecon make a holiday picnic to the summit.

What strikes the observant eye as especially interesting
is the circumstance that these two triangular patches, forming
the culminating points of South Wales, both slope to the
south-west, and by stooping down on either of them, and
looking towards the other, we find that their surfaces correspond so closely in direction and amount of slope, that they
impress one at once as being really portions of one continuous mountain summit. This becomes more certain when
we look at the whole mountain mass, of which they form a part,
known as the "Fforest Fawr," or great forest of Brecknock.
This extends about twenty miles from east to west and ten

or twelve miles from north to south; and in every part of it
the chief summits are from 2000 to 2500 feet high, while
near its western end, about twelve miles from the Beacons, is
the second highest summit, Van Voel, reaching 2632 feet.
Most of these mountains have rounded summits which are
smooth and covered with grassy or sedgy vegetation, but
many of them have some craggy slopes or precipices on their
northern faces.

Almost the whole of this region is of the Old Red Sandstone formation, which here consists of nearly horizontal
strata with a moderate dip to the south; and the whole of
the very numerous valleys with generally smooth and gradually
sloping sides which everywhere intersect it, must be all due
to sub-aerial denudation—that is, to rain, frost, and snow—the
débris due to which is carried away by the brooks and rivers.
The geologist looks upon the rounded summits of these mountains as indications of an extensive gently undulating plateau,
which had been slowly raised above the surface of the lakes
or inland seas in which they had been deposited, and subjected
to so little disturbance that the strata remain in a nearly
horizontal position. When from the summit of any of these
higher mountains we look over the wide parallel or radiating
valleys with the rounded grassy ridges, and consider that the
whole of the material that once filled all these valleys to the
level of the mountain-top has been washed away day by day
and year by year, by the very same agencies that after heavy
rain now render turbid every brooklet, stream, and river,
usually so clear and limpid, we obtain an excellent illustration
of how nature works in moulding the earth's surface by a
process so slow as to be to us almost imperceptible.

This process of denudation is rendered especially clear to
us by the singular formation of the twin summits of the Brecon
Beacons. Here we are able, as it were, to catch nature at
work. Owing to the rare occurrence of a nearly equal rate
of denudation in four or five directions around this highest
part of the original plateau, we have remaining for our inspection two little triangular patches of the original peat-
covered surface joined together by the narrow saddle, as

THE BEACONS.
(Looking south.)

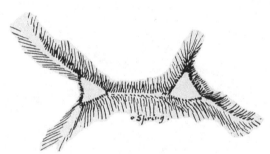

PLAN OF SUMMIT OF BEACONS.
(Looking north.)

SECTION THROUGH SUMMITS OF BEACONS.

[*To face p.* 165, VOL. I.

shown in the sketches opposite, showing a plan of the summits
and a section through them to explain how accurately the two
coincide in their slope with that of the original plateau.
Every year the frost loosens the rock on the northern precipices, every heavy rain washes down earth from the ridge,
while the gentler showers and mists penetrate the soil to the
rock surface, which they slowly decompose. Thus, year by
year, the flat portion of the summits becomes smaller, and a
few thousand years will probably suffice to eat them away
altogether, and leave rocky peaks more like that of Snowdon.
The formation, as we now find it, is, in my experience, unique
—that is, a mountain-top presenting two small patches of
almost level ground, evidently being the last remnant of the
great rolling plateau, out of which the whole range has been
excavated. Double-headed mountains are by no means uncommon, but they are usually peaked or irregular, and carved
out of inclined or twisted strata. The peculiarity of the
Beacons consists in the strata being nearly horizontal and
undisturbed, while the rock formation is not such as usually
to break away into vertical precipices. The original surface
must have had a very easy slope, while there were no meteorological conditions leading to great inequalities of weathering.
The thick covering of peat has also aided in the result by
preserving the original surface from being scored into gullies,
and thus more rapidly denuded.

After we had completed most of our work at Trallong we
had to go further up the valley to Devynock. This is an
enormous parish of more than twenty thousand acres, divided
into four townships or chapelries, the two eastern of which,
Maescar and Senni, we had to survey. In these mountain
districts, however, we only surveyed those small portions
where the new roads or new enclosures had been made,
the older maps being accepted as sufficiently accurate for
the large unenclosed areas of mountain land. We first went
to Senni Bridge, where both districts terminate in the Usk
valley; but after a short time I went to stay in a little public-
house at Senni in the midst of my work, while my brother

stayed at Devynock or at Trallong, which latter was quite as near for half the work.

On the other side of the river Usk there was a fine wooded rocky slope in which paths had been made near and above the river by some former resident owner, and this was a favourite walk on holidays. In the farmhouse adjacent a relative of the owner, a middle-aged man, who was apparently on the verge between eccentricity and madness, lived in retirement, and we heard a good deal of his strange ways, though they said he was quite harmless. He used to walk about a good deal with a pipe in his mouth and dressed in a game-keeper style, and he always stopped to make some remark, and then walked on without waiting for an answer. My brother made a rough pen-and-ink sketch of him, which has fortunately been preserved, and which is here reproduced, as it well represents his appearance and manner when meeting any one. Some of his sayings were not only wild but exceedingly coarse, others merely abrupt and strange. One day he would say, "Where's your pipe? Don't smoke? Then go home and begin if you want to be happy." Another time something like this, "Who are you? Come to look after me? They say I'm mad, but I ain't. I'm here to enjoy myself. Do as I like." One time when he met my brother, after some such rigmarole as the above, he ended with, "Shave your head and keep your toe-nails cut, and *you*'ll be all right."

When I went up to Senni Street (Heol Senni, as it is called in Welsh) I greatly enjoyed wandering over the pretty valley which extended a long way into the mountains, flowing over nearly level meadows and with an unusually twisted course. This I found was so erroneously mapped, the numerous bends having been inserted at random as if of no importance, that I had to survey its course afresh. Above the village there were several lateral tributaries descending in deep woody dingles, often very picturesque, and these had usually one or more waterfalls in their course, or deep rocky chasms ; and as these came upon me unexpectedly, and

OUR ECCENTRIC NEIGHBOUR AT DEVYNOCK.

(*From a sketch by W. G. Wallace.*)

[*To face p.* 166, VOL. I.

"MAEN LLIA," UPPER VALE OF NEATH.

[*To face p.* 167, VOL. I.

I had seen very few like them in Radnorshire, they were more especially attractive to me.

One Sunday afternoon I walked up the valley and over a mountain-ridge to the head waters of the Llia river, one of the tributaries of the river Neath, to see an ancient stone, named Maen Llia on the ordnance map. I was much pleased to find a huge erect slab of old red sandstone nearly twelve feet high, a photograph of which I am able to give through the kindness of Miss Florence Neale of Penarth. These strange relics of antiquity have always greatly interested me, and this being the first I had ever seen, produced an impression which is still clear and vivid.

The people here were all thoroughly Welsh, but the landlord of the inn, and a young man who lived with him, spoke English fairly well. Like most of the Welsh the landlord was very musical, and in the evenings he used to teach his little girl, about five years old, to sing, first exercising her in the notes, and then singing a Welsh hymn, which she followed with a tremendously powerful voice for so small a child. Her father was very proud of her, and said she would make a fine singer when she grew up.

While here, and also at Trallong, I went sometimes to church or chapel in order to hear the Welsh sermons, and also the Welsh Bible well read, and I was greatly struck with the grand sound of the language and the eloquence and earnestness of the preachers. The characteristic letters of the language are the guttural *ch*, the *dd* pronounced soft as "udh," the *ll* pronounced "llth." If the reader will endeavour to sound these letters he will have some idea of the effect of such passages as the following, when clearly and emphatically pronounced :—"Brenhin Brenhinoedd, ac Arglwydd Arglwyddi" ("King of Kings and Lord of Lords"). Again, "Ac a ymddiddanodd â mi, gan ddywedyd, Tyred, mi a ddangosaf i ti briodasferch" ("And talked with me, saying, Come hither, I will shew thee the bride"). These are passages from Revelation, but the following verse from the Psalms is still grander and more impressive :—

"Cyn gwneuthur y mynyddoedd, a llunio o honot y ddaear

a'r byd ; ti hefyd wyt Dduw, o dragywyddoldeb hyd dragywyddoldeb " (" Before the mountains were brought forth, or
ever Thou hadst formed the earth and the world, even from
everlasting to everlasting, Thou art God ").

The Welsh clergy are usually good readers and energetic
preachers, and seem to enjoy doing full justice to their rich
and expressive language, and even without being able to
follow their meaning it is a pleasure to listen to them.

Among the numerous Englishmen who visit Wales for business or pleasure, few are aware to what an extent this ancient
British form of speech is still in use among the people, how
many are still unable to speak English, and what an amount
of poetry and legend their language contains. Some account
of this literature is to be found in that very interesting book,
George Borrow's " Wild Wales," and he claims for Dafydd
ap Gwilym, a contemporary of our Chaucer, the position of
" the greatest poetical genius that has appeared in Europe
since the revival of literature." At the present day there are
no less than twenty weekly newspapers and about the same
number of monthly magazines published in the Welsh
language, besides one quarterly and two bi-monthly reviews.
Abstracts of the principal Acts of Parliament and Parliamentary papers are translated into Welsh, and one firm of
booksellers, Messrs. Hughes and Son, of Wrexham, issue a list
of more than three hundred Welsh books mostly published by
themselves. Another indication of the wide use of the Welsh
language and of the general education of the people, is the
fact that the British and Foreign Bible Society now sell
annually about 18,000 Bibles, 22,000 Testaments, and 10,000
special portions (as the Psalms, the Gospels, etc.) ; while the
total sale of the Welsh scriptures during the last century has
been 3½ millions. Considering that the total population of
Wales is only about 1½ millions, that two counties, Pembroke-
shire and Radnorshire, do not speak Welsh, and that the
great seaports and the mining districts contain large numbers
of English and foreign workmen, we have ample proof that
the Welsh are still a distinct nation with a peculiar language,
literature, and history, and that the claim which they are

now making for home rule, along with the other great sub-divisions of the British Islands, is thoroughly justified.

Our two other indigenous Celtic languages, Gaelic and Irish, or Erse, appear to have a far less vigorous literary existence. I am informed by the Secretary of the National Bible Society of Scotland that about three thousand Bibles and a little more than two thousand Testaments are sold yearly. The number of people who habitually speak Gaelic is, however, less than a quarter of a million, and the language seems to be kept up in a literary sense more by a few educated students and enthusiasts than to supply the needs of the people.

The Irish language is a form of Gaelic closely allied to that of Scotland, and there are still nearly a million people able to speak it, though only about one-tenth of that number use it exclusively. Owing to the prevalence of the Roman Catholic religion among the peasantry, very few copies of the Irish version of the Bible and Testament are now sold, and although the ancient literature was exceedingly rich and varied, any modern representative of it can hardly be said to exist. The strong vitality of the Welsh language as above sketched is therefore a very interesting feature of our country, and as it is undoubtedly suited to the genius of the people among whom it has survived, there seems to be no valid objection to its perpetuation. The familiar use of two languages does not appear to be in itself any disadvantage, while being able to appreciate and enjoy the literature of both must be a distinct addition to the pure intellectual pleasures of those who use them.

CHAPTER XII

SHROPSHIRE AND JACK MYTTON

AFTER having finished our work in Brecknockshire we returned to Kington for a few months, doing office-work and odd jobs of surveying in the surrounding country. Among these what most interested me was the country around Ludlow, in Shropshire, where there are beautiful valleys enclosed by steep low hills, often luxuriantly wooded, and watered by rapid streams of pure and sparkling water. I had by this time acquired some little knowledge of geology, and was interested in again being in an Old Red Sandstone country, which formation I had become well acquainted with in Brecknockshire, and which is so different from the Upper Silurian shales so prevalent in Radnorshire. In this country we were near the boundary of the two formations, and there were also occasional patches of limestone, and at every bit of rock that appeared during our work I used to stop a few moments to examine closely, and see which of the formations it belonged to. This was easily decided by the physical character of the rocks, which, though both varied considerably, had yet certain marked characteristics that distinguished them.

One day we were at work in a park near a country house named " Whittern," and my brother took a pencil sketch of it in his field-book. Just as he was finishing it the owner came out and talked with him, and seeing he was something of an artist, went to the house and brought out a portfolio of drawings in sepia, by his daughter, of views in the park and

"WHITTERN."

(An outdoor sketch by W. G. Wallace.   1842.)

[*To face p.* 170, VOL. I.

in the surrounding country. These seemed to me exceedingly well done and effective, and, of course, my brother
praised them, but, as I thought, only moderately, and as
" very good work for an amateur." I reproduce his sketch
on a reduced scale as showing his delicacy of touch even in
hasty out-of-door work, though, owing to the old yellowish
paper, the pencil marks come out very faint in the process
print.

While travelling by coach or staying at country inns in
Shropshire, we used to hear a good deal of talk about Jack
Mytton, of Halston, who had died a few years before, and
whose wild exploits were notorious all over the West of
England. He was a country gentleman of very old family,
and had inherited a landed estate bringing in about £10,000
a year, while having been a minor for eighteen years, there
was an accumulation of £60,000 when he came of age. In a
few years he spent all these savings, and continued to live
at such a rate that he had frequently to raise money.
All the grand oaks for which his estates were celebrated
were cut down, and it is said produced £70,000. About
half his property was entailed, but the other half was
sold at various times, and must have realized a very large
amount; while in the last years of his life, which he spent
either in prison for debt or in France, all the fine collection of
pictures, many by the old masters, and the whole contents of
his family mansion were sold, but did not suffice to pay his
debts or prevent his dying in prison. From the account given
by his intimate friend and biographer the total amount thus
wasted in about fifteen years could not have been much less
than half a million, but from the scanty details in his " Life "
it seems clear that he could not really have expended anything like this amount, but that his extreme good nature and
utter recklessness as to money led to his being robbed and
plundered in various ways by the numerous unscrupulous
persons who always congregate about such a character.

For those who have not read the account of his wasted
life one or two examples illustrative of his character may
be here given. Once, before he was of age, when dining out

in the country, he had driven over in a gig with a pair of horses tandem—his favourite style. On some of the party expressing the opinion that this was a very dangerous mode of driving, Mytton at once offered to bet the whole party £25 each that he would then and there drive his tandem across country to the turnpike road half a mile off, having to cross on the way a sunk fence three yards wide, a broad deep drain, and two stiff quickset hedges with ditches on the further side. All accepted the bet. It was a moonlight night, but twelve men with lanthorns accompanied the party in case of accidents. He got into and out of the sunk fence (I suppose what we call a Ha-ha) in safety, went at the drain at such a pace that both horses and gig cleared it, the jerk throwing Mytton on to the wheeler's back, from which he climbed up to his seat, drove on, and through the next two fences with apparent ease into the turnpike road without serious injury, thus winning this extraordinary wager.

He was as reckless of other person's lives and limbs as he was of his own, upsetting one friend purposely because he had just said that he had never been upset in his life, and jumping the leader over a turnpike gate to see whether he would take "timber," the gig being, of course, smashed, and Mytton with his friend being thrown out, but, strange to say, both uninjured.

He was a man of tremendous physical strength, and with a constitution that appeared able to withstand anything till he ruined it by excessive drinking. He was so devoted to sport of some kind or other that nothing came amiss to him, riding his horse upstairs, riding a bear into his drawing-room, crawling after wild ducks on the snow and ice stripped to his shirt, or shooting rats with a rifle. Several of these stories we heard told by the people we met, but there were many others of a nature which could not be printed, and which referred to the latter part of his life, when his wife had left him, and he had entered on that downhill course of reckless dissipation that culminated in his ruin and death.

Never was there a more glaring example of a man of

exceptional physical and mental qualities being ruined by the inheritance of great wealth and by a life of pleasure and excitement. Brought up from childhood on a great estate which he soon learnt would be his own ; surrounded by servants and flatterers, by horses and dogs, and seeing that hunting, racing, and shooting were the chief interests and occupations of those around him ; with an intense vitality and superb physique,—who can wonder at his after career ? At school he was allowed £400 a year, and it is said spent £800—alone enough to demoralize any youth of his disposition ; and as a natural sequence he was expelled, first from Westminster and then from Harrow. He was then placed with a private tutor for a year. He entered at both Universities but matriculated at neither ; and when nineteen became a cornet in the 7th Hussars, which he joined in France with the army of occupation after Waterloo. He quitted the army when of age, and settled at Halston.

Such having been his early life it would seem almost impossible that he could have profited much by his very fragmentary education ; yet his biographer assures us that he had a fair amount of classical knowledge, and throughout life would quote Greek and Latin authors with surprising readiness, and, moreover, would quote them correctly, and always knew when he made a mistake, repeating the passage again and again till he had it correct. Several examples are given when, in his later years, he quoted passages from Sophocles and Homer to illustrate his own domestic and personal misfortunes. But besides these literary tastes he was a man remarkable for many lovable characteristics and especially for a real sympathy for the feelings of others. After being arrested at Calais on bills he had accepted in favour of a person with whom he had had some dealings, as soon as he was released from prison by his solicitor paying the debt, he called upon his former creditor, not to upbraid him, but to walk with him arm-in-arm through the town, in order that the affair might not injure the creditor's character, he being a professional man. As his biographer says, few finer instances of generosity and good feeling are on record. It was this aspect of his character that led to his being so universally loved, that three thousand persons attended his funeral, with every mark of respect.

Here was a man whose qualities both of mind and body might have rendered him a good citizen, a happy man, and a cause of happiness to all around him, but whose nature was perverted by bad education and a wholly vicious environment. And such examples come before us continuously, exciting little attention and no serious thought. A few years back we had the champion plunger, who got rid of near a million in a very short time ; and within the last few years we have had in the bankruptcy court a young nobleman of historic lineage and great estates ; also a youth just come into a fortune of £12,000, who, while an undergraduate at Oxford, gave £5000 for four race-horses, which he had never seen, on the word of the seller about whom he knew nothing, spent over a thousand in training them, and in another year or two had got rid of the last of his thousands besides incurring a considerable amount of debt. But nobody seems to think that the great number of such cases always occurring, and which are probably increasing with the increasing numbers of great fortunes, really indicates a thoroughly rotten social system.

How often we hear the remark upon such cases, " He is nobody's enemy but his own." But this is totally untrue, and every such spendthrift is really a worse enemy of society than the professional burglar, because he lives in the midst of an ever-widening circle of parasites and dependents, whose idleness, vice, and profligacy are the direct creation of his misspent wealth. He is not only vicious himself, but he is a cause of vice in others. Perhaps worse even than the vice is the fact that among his host of dependents are many quite honest people, who live by the salaries they receive from him or the dealings they have with him, and the self-interest of these leads them to look leniently upon the whole system which gives them a livelihood. Innumerable vested interests thus grow up around all such great estates, and the more wastefully the owner spends his income the better it seems to be for all the tradesmen and mechanics in the district. But the fundamental evil is the kind of sanctity we attach to property, however accumulated and however spent. Hence no real reform is ever suggested ; and those who go to the root of the matter and see that the evil is in the very fact of inheritance itself, are scouted as socialists or something worse. The inability of ordinary political and social writers to follow out a principle is well shown in this matter. It is only a few years since Mr. Benjamin Kidd attracted much attention to the principle of " equality of opportunity " as the true basis of social reform, and many of the more advanced political writers at once accepted it as a sound principle and one that should be a guide for our future progress. Herbert Spencer, too, in his volume on " Justice," lays down the same principle, stating, as " the law of social justice " that " each individual ought to receive the benefits and evils of his own nature and consequent conduct ; neither being prevented from having whatever good his actions normally bring him, nor allowed to shoulder off on to other persons whatever ill is brought to him by his actions." This, too, has, so far as I am aware, never been criticized or objected to as unsound, and, in fact, the arguments by which it is supported are unanswerable. Yet no one among our politicians or ethical writers has openly adopted these principles as a guide for conduct in legislation, or has even seen to what they inevitably lead. Stranger still, neither Mr. Kidd nor Herbert Spencer followed out their own principle to its logical conclusion, which is, the absolute condemnation of unequal inheritance. Herbert Spencer even declares himself in favour of inheritance as a necessary corollary of the right of property rightfully acquired ; and he devotes a chapter to " The Rights of Gift and Bequest." But he apparently did not see, and did not discuss the effect of this in neutralizing his " law of social justice," which it does absolutely. I have myself fully shown this in a chapter on " True Individualism : the Essential Preliminary of a Real Social Advance " in my " Studies Scientific and Social."

It is in consequence of *not* going to the root of the matter, and *not* following an admitted principle to its logical conclusion, that the idea prevails that it is only the *misuse* of wealth that produces evil results. But a little consideration will show us that it is the inheritance of wealth that is wrong in itself, and that it necessarily produces evil. For if it is right, it implies that *inequality* of opportunity is right, and that "the law of social justice" as laid down by Herbert Spencer is *not a* just law. It implies that it is *right* for one set of individuals, thousands or millions in number, to be able to pass their whole lives without contributing anything to the well-being of the community of which they form a part, but on the contrary keeping hundreds, or perhaps thousands, of their fellow men and women wholly engaged in ministering to *their* wants, *their* luxuries, and *their* amusements. Taken as a whole, the people who thus live are no better in their nature—physical, moral, or intellectual—than other thousands who, having received no such inheritance of accumulated wealth, spend *their* whole lives in labour, often under exhausting, unhealthy, and life-shortening conditions, to produce the luxuries and enjoyments of others, but of which they themselves rarely or more often *never* partake. Even leaving out of consideration the absolute vices due to wealth on the one hand and to poverty on the other, and supposing both classes to pass fairly moral lives, who can doubt that *both* are injured morally, and that *both* are actually, though often unconsciously, the causes of ever-widening spheres of demoralization around them ? If there is one set of people who are tempted by their necessities to prey upon the rich, there is a perhaps more extensive class who are in the same way driven to prey upon the poor. And it is the very *system* that produces and encourages these terrible inequalities that has also led to the almost incredible result, that the ever-increasing power of man over the forces of nature, especially during the last hundred years, while rendering easily possible the production of all the necessaries, comforts, enjoyments, and wholesome luxuries of life for every individual, have yet, as John Stuart Mill declared, "not diminished the toil of any worker," but even, as there is ample evidence to prove, has greatly increased the total mass of human misery and want in every civilized country in the world.

And yet our rulers and our teachers—the legislature, the press, and the pulpit alike—shut their eyes to all this terrible demoralization in our midst, while devoting all their energies to increasing our already superfluous and injurious wealth-accumulations, and in compelling other peoples, against their will, to submit to our ignorant and often disastrous rule. As the great Russian teacher has well said, " They will do any-thing rather than get off the people's backs." And we, who adopt the principles of those great thinkers whom all delight to honour—Ruskin and Spencer—and urge the adoption of " equality of opportunity "—of equal education, equal nurture, an equal start in life—for all (implying the abolition of all inequality of inheritance) as the one Great Reform which will alone render all other reforms—all *general* social advance —possible, are either quietly ignored as idle dreamers, or openly declared to be " enemies of society."

These few remarks and ideas have been suggested to me by the life and death of Jack Mytton, and I trust that some of my readers may follow them up for the good of humanity.

CHAPTER XIII

GLAMORGANSHIRE : NEATH

IT was late in the autumn of 1841 that we finally bade adieu to Kington and the wild but not very picturesque Radnorshire mountains for the more varied and interesting county of Glamorgan. I have no distinct recollection of our journey, but I believe it was by coach through Hay and Brecon to Merthyr Tydvil, and thence by chaise to Neath. One solitary example of the rhyming letters I used to write has been preserved, giving my younger brother Herbert an account of our journey, of the country, and of our work, of which, though very poor doggerel, a sample may be given. After a few references to family matters, I proceed to description.

> " From Kington to this place we came
> By many a spot of ancient fame,
>   But now of small renown,
> O'er many a mountain dark and drear,
> And vales whose groves the parting year
>   Had tinged with mellow brown ;
> And as the morning sun arose
> New beauties round us to disclose,
>   We reached fair Brecon town ;
> Then crossed the Usk, my native stream,
>   A river clear and bright,
> Which showed a fair and much-lov'd scene
>   Unto my lingering sight."

We had to go to Glamorganshire to partially survey and make a corrected map of the parish of Cadoxton-juxta-Neath, which occupies the whole northern side of the Neath valley from opposite the town of Neath to the boundary of the county at Pont-Nedd-Fychan, a distance of nearly fifteen miles, with a width varying from two to three miles, the boundary running for the most part along the crest of the mountains that bound the valley on the north-west. We lodged and boarded at a farmhouse called Bryn-coch (Red Hill), situated on a rising ground about two miles north of the town. The farmer, David Rees, a rather rough, stout Welshman, was also bailiff of the Duffryn estate. His wife could not speak a word of English, but his two daughters spoke it very well, with the pretty rather formal style of those who have first learnt it at school. Here we stayed more than a year, living plainly but very well, and enjoying the luxuries of home-made bread, fresh butter and eggs, unlimited milk and cream, with cheese made from a mixture of cow's and sheep's milk, having a special flavour, which I soon got very fond of. In this part of Wales it is the custom to milk the ewes chiefly for the purpose of making this cheese, which is very much esteemed. Another delicacy we first became acquainted with here was the true Welsh flummery, called here "sucan blawd" (steeped meal), in other places "Llumruwd" (sour sediment), whence our English word "flummery." It is formed of the husks of the oatmeal roughly sifted out, soaked in water till it becomes sour, then strained and boiled, when it forms a pale brown sub-gelatinous mass, usually eaten with abundance of new milk. It is a very delicious and very nourishing food, and frequently forms the supper in farmhouses. Most people get very fond of it, and there is no dish known to English cookery that is at all like it; but I believe the Scotch "sowens" is a similar or identical preparation. This dish, with thin oatmeal cakes, home-made cheese, bacon, and sometimes hung beef, with potatoes and greens, and abundance of good milk, form the usual diet of the Welsh peasantry, and is certainly a very wholesome and nourishing combination. We, however, had also two other kinds of bread, both excellent, especially when made from new wheat. One was the ordinary huge loaves of farmhouse bread, the other what was called backstone bread—large flat cakes about a foot in diameter and an inch thick, baked over the fire on a large circular iron plate (formerly on a stone or slate, hence the name "bakestone" or "backstone"). This is excellent, either split open and buttered when hot, or the next day cut edgeways into slices of bread-and-butter, a delicacy fit for any lady's afternoon tea.

A little rocky stream bordered by trees and bushes ran through the farm, and was one of my favourite haunts. There was one little sequestered pool about twenty feet long into which the water fell over a ledge about a foot high. This pool was seven or eight feet deep, but shallowed at the further end, and thus formed a delightful bathing-place. Ever since my early escape from drowning at Hertford, I had been rather shy of the water, and had not learned to swim ; but here the distance was so short that I determined to try, and soon got to enjoy it so much that every fine warm day I used to go and plunge head first off my ledge and swim in five or six strokes to the shallow water. In this very limited sphere of action I gained some amount of confidence in the water, and afterwards should probably have been able to swim a dozen or twenty yards, so as to reach the bank of a moderate-sized river, or sustain myself till some neighbouring boat came to my assistance. But I have never needed even this moderate amount of effort to save my life, and have never had either the opportunity or inclination to become a practised swimmer. This was partly due to a physical deficiency which I was unable to overcome. My legs are unusually long for my height, and the bones are unusually large. The result is that they persistently sink in the water, bringing me into a nearly vertical position, and their weight renders it almost impossible to keep my mouth above water. This is the case even in salt water, and being also rather deficient in strength of muscle, I became disinclined to practise what I felt to be beyond my powers.

The parish being so extensive we had to stay at many different points for convenience of the survey, and one of these was about five miles up the Dulais valley, where we stayed at a small beershop in the hamlet of Crynant. I was often here alone for weeks together, and saw a good deal of the labourers and farmers, few of whom could speak any English. The landlady here brewed her own beer in very primitive fashion in a large iron pot or cauldron in the wash-house, and had it ready for sale in a few days—a rather thick and sweetish liquor, but very palatable. The malt and hops were bought in small quantities as wanted, and brewing took place weekly, or even oftener, when there was a brisk demand.

In my bedroom there was a very large old oak chest, which I had not taken the trouble to look in, and one morning very early I heard my door open very slowly and quietly. I wondered what was coming. A man came in, cautiously looking to see if I was asleep. I wondered if he was a robber or a murderer, but lay quite still. He moved very slowly to the big chest, lifted the lid, put in his arm, groped about a little, and then drew out a large piece of hung beef! The chest contained a large quantity bedded in oatmeal. My mind was relieved, and I slept on till breakfast time.

A young Englishman who was a servant in a gentleman's house near used to come to the beershop occasionally, and would sometimes give me local information or interpret for me with the landlady when no one else was at home. He seemed to speak Welsh quite fluently, yet to my great astonishment he told me he had only been in Wales three or four months, and could not read or write. He said he picked up the language by constantly talking to the people, and I have noticed elsewhere that persons who are thus illiterate learn languages by ear with great rapidity. It no doubt arises from the fact that, having no other mental occupations and no means of acquiring information but through conversation, their whole mental capacities are concentrated on the one object of learning to speak to the people. Some natural faculty of verbal memory must no doubt exist, but when this is present in even a moderate degree the results are often very striking. Somewhat analogous cases are those of teaching the deaf and dumb the gesture language, lip-reading, and even articulate speech which they cannot themselves hear, and the still more marvellous cases of Laura Bridgman and Helen Keller, in which was added blindness, so that the sense of touch was alone available for receiving ideas. The effect in developing the mind and enabling the sufferers to live full, contented, and even happy lives has been most marvellous, and give us a wonderful example of the capacity of the mind for receiving the most abstract ideas through one sense alone. Such persons, without proper training, would be in danger of becoming idiotic or insane from the absence of all materials on which to exercise the larger portion of their higher mental faculties. It is observed that, when first being taught the connection of arbitrary signs with objects, they are docile but apathetic, not in the least understanding the purport of the training. But after a time, when they perceive that they are acquiring a means of communicating their own wishes and even ideas to others, and receiving ideas and knowledge of the outer world from them, their whole nature seems transformed, and the acquisition and extension of this knowledge becomes the great object and the great pleasure of their lives. It seems to occupy all their thoughts and employ all their faculties, and they make an amount of progress which astonishes their teachers and seems quite incredible to persons ordinarily constituted. It gives them, in fact, what every one needs, some useful or enjoyable occupation for body and mind, and is almost equivalent to furnishing them with the faculties they have lost. A similar explanation may be given of the comparatively rapid acquisition by the deaf and dumb of those difficult arts—lip-reading by watching the motion of the lips and face of the speaker, and intelligible speech by imitating the motions during speech of the lips, tongue, and larynx by using a combination of vision and touch. These give them new means of communication with their fellows, and their whole mental powers are therefore devoted to their acquisition. It is a new employment for their minds, equivalent to a new and very interesting game for children, and under such conditions learning becomes one of their greatest pleasures. The same principle applies to the rapid acquisition of a new language by the illiterate. Being debarred from reading and

writing, all their intellectual pleasures depend upon converse
with their fellows, and thus their thoughts and wishes are
intensely and continuously directed to the acquisition of the
means of doing so.

A mile further up the valley was a small gentleman's house
with about a hundred and fifty acres of land attached, owned
and occupied by a Mr. Worthington, his wife and wife's sister.
They had, I believe, come there not long before from Devonshire, and being refined and educated people, we were glad
to make their acquaintance, and soon became very friendly.
Mr. Worthington was a tall and rather handsome man
between fifty and sixty ; while his wife was perhaps fifteen
or twenty years younger, rather under middle size and very
quiet and agreeable ; while her sister was younger, smaller,
and more lively. They lent us books and magazines, and we
often went there to spend the evening. I do not think our
friend knew much about farming, but he had a kind of working
bailiff and two or three labourers to cultivate the land, which,
however, was mostly pasture. The place is called Gelli-duch-
lithe, the meaning of which is obscure. "The grove and the
wet moor" is not inappropriate, and seems more likely than
any connection with "llaeth" (milk), which implies good
land or rich pastures, which were decidedly absent.

Mr. Worthington was an eccentric but interesting man. He
played the violin beautifully, and when in the humour would
walk about the long sitting-room playing and talking at intervals. He discussed all kinds of subjects, mostly personal,
and he was, I think, the most openly egotistical man I ever met,
and I have met many. After playing a piece that was one of
his favourites, he would say to my brother, "Was not that
fine, Mr. Wallace ? There are not many amateurs could play
in that style, are there ?—or professionals either," he would
sometimes add. And after telling some anecdote in which
he was the principal personage, he would often finish up with,
"Don't I deserve praise for that, Mr. Wallace ?" On one
occasion, I remember, after telling us of how he befriended a
poor girl and resisted temptation, he concluded with, "Was

not that a noble act, Mr. Wallace ?" to which we, as visitors,
were, of course, bound to assent with as much appearance of
conviction as we could manage to express. These things were
a little trying, but he carried them off so well, so evidently
believed them himself, and spoke in so earnest and dignified
a manner, that had we been more intimate, and could have
permitted ourselves to laugh openly at his more extravagant
outbursts, we should have had a more thorough enjoyment
of his society.

Of course, such an appreciation of his own merits led to
his taking the blackest view of all who opposed him, and thus
led to what was in the nature of a tragedy for his wife as well
as for himself, and one in which we had to bear our part.
His property was bounded on one side by the little river
Dulais, which wound about in a narrow belt of level pasture,
and in places appeared to have changed its course, leaving
dry channels, which were occasionally filled during floods. It
was to one of these further channels that our friend claimed
that his property extended, founding his belief on the evidence
of some old people who remembered the river flowing in this
channel, some of whom also declared that the cattle and sheep
belonging to Gelli used to graze there. He would talk for
hours about it, maintaining that the old water-line was always
the boundary, and that the adjoining landlord, Lord ——,
was trying to rob him by the power of his wealth and influence.
The whole of the little pieces of land in dispute did not
amount to more than half an acre and were not worth more
than a few pounds, and his own lawyer tried to persuade him
that the issue was very doubtful, and that even if he won, the
bits of land were not worth either the cost or the worry. But
nothing would stop him, and by his orders an act of trespass
was committed on the land to which he thus formally laid
claim, and after much correspondence an action was commenced against him by Lord ——'s lawyers. Then we were
employed to make a plan of the pieces claimed, and the case
came on for trial at the Cardiff Assizes.

The partner of the London solicitor came down for the
case and engaged one of the most popular barristers, the best

having been secured by the other side. Our friend was persuaded not to be present, and I was engaged to attend and take full notes of the proceedings, which I copied out in the evening and sent off to him. I stayed at a hotel with the lawyer, and the town being very crowded, we shared the same bedroom and had our meals together. He was by no means sanguine of success, and the first day's proceedings made him less so, as the other side stated that they had documents that proved their case, and intimated that the defendant knew it. The first day was Friday or Saturday, and we returned to Gelli till the Monday, and in the interval there occurred a scene. The lawyer felt confident that his client had not produced all the deeds he possessed relating to the estate, and insisted on being shown every single document or he would give up the case. Very reluctantly they were produced, and after a close examination one was found which had a map of the farm showing the boundary as claimed by the other side. The lawyer was a little man and lame, while Mr. Worthington was tall, erect, and defiant ; but the former stood up, and, holding the document in his hand, blazed out against his client. "Mr. Worthington," he said, "you have behaved scandalously, foolishly, almost like a madman. You have deceived your own lawyer, and put him in the wrong. You have denied the possession of documents which you knew were dead against your claim. Had we known of the existence of this deed we would never have defended your case, and if I were acting for myself alone I would throw it up instantly. But Mr. ——, my partner, is an old friend of yourself and your family, and to save you from open disgrace the case must go on to the end. But I tell you now, you will lose it, and you deserve to lose it, for you have not acted honourably or even honestly."

All this was said with the greatest fire and energy, and Mr. Worthington was, for the first time in my experience, completely cowed. He vainly tried to interpose a word, to disclaim knowledge of the importance of this deed, etc., but the lawyer shook his fist at him, and thoroughly silenced him. Finally, he told him that he should now act without consulting him, and if Mr. Worthington interfered in any way he would throw up the case.

It turned out as the lawyer expected. The other side had deeds showing the same boundary as that which Mr. Worthington had concealed. Our evidence as to possession was weak. Our counsel appealed to the jury for a poor man struggling for his rights against the power of wealth. But the judge summed up against us on the evidence, and the other side won. Mr. Worthington had insisted upon hearing his counsel's speech, which evidently gave him hopes, and when the verdict was given he was overwhelmed, looked altogether dazed, and I thought he would have a fit. But we got him at once out of court, went back to the inn, and as soon as possible drove home together. As soon as he recovered himself somewhat, he exclaimed, " My counsel was a noble fellow, *he* upheld the right ; but we had an unjust judge, Mr. Wallace." I forgot to mention that Mr. Worthington wore a brown curly wig, which I had at first taken for his natural hair, and when he was much excited he would suddenly snatch it off his head, when he looked rather ludicrous. The costs which he had to pay were very heavy, and he had to sell Gelli to pay them, and soon afterwards left the district to return to Devonshire. I fancy he had before lost a good deal of property, and this last misfortune was almost ruin. After they left I do not think we ever heard of them again, though my brother may have done so.

After living about a year at Bryn-coch we moved a little nearer the town to the other side of the Clydach river, and lodged with an old colliery surveyor, Samuel Osgood, in the employment of Mr. Price, of the Neath Abbey Iron Works. The house was an old but roomy cottage, and we had a large bedroom and a room downstairs for an office and living room, while Mr. Osgood had another, and there was also a roomy kitchen. A tramway from some collieries to the works ran in front of the house at a little distance, and we had a good view of the town and up the vale of Neath. Behind us rose the Drymau Mountain, nearly seven hundred feet above us, the level top of which was frequented by peewits, and whose steep slopes were covered with trees and bushes. Here we lived till I left Neath a year later, and were on the whole very comfortable, though our first experience was a rather trying one. The bedroom we occupied had been unused for years, and though it had been cleaned for our use we found that every part of it, bedstead, floor, and walls, in every crack and cranny, harboured the *Cimex lectularius*, or bedbug, which attacked us by hundreds, and altogether banished sleep. This required prompt and thorough measures, and my brother at once took them. I was sent to the town for some ounces of corrosive sublimate; the old wooden bedstead was taken to pieces, and, with the chairs, tables, drawers, etc., taken outside. The poison was dissolved in a large pailful of water, and with this solution by means of a whitewasher's brush the whole of the floor was thoroughly soaked, so that the poison might penetrate every crevice, while the walls and ceiling were also washed over. The bedstead and furniture were all treated in the same way, and everything put back in its place by the evening. We did all the work ourselves, with the assistance of Mrs. Osgood and a servant girl, and so effectual was the treatment that for nearly a year that we lived there we were wholly unmolested by insect enemies.

Mr. and Mrs. Osgood were both natives of the ancient town of Bideford, Devon, which they continually referred to as the standard of both manners and morality, to the great disadvantage of the Welsh. They were both old, perhaps between sixty and seventy, and thought old fashions were the best. Mr. Osgood was an old-fashioned surveyor, and was also a pretty good mechanic. He prided himself upon his work, upon his plans of the colliery workings, and especially upon his drawings, which were all copies from prints, usually very common ones, but which he looked upon as works of high art. Among these, he was especially proud of a horse, in copying which in pen and ink he had so exaggerated the muscular development that it looked as if the skin had been taken off to exhibit the separate muscles for anatomical teaching. It was a powerful-looking horse in the attitude of a high-stepper, but so exaggerated and badly drawn as to be almost ludicrous. It was framed and hung in his room, and he always called visitors' attention to it, and told them that Mr. Price, the owner of the collieries, had said that he could never get a horse like that one, as if this were the highest commendation possible of his work.

About that time the method of measuring the acreage of fields on maps by means of tracing-paper divided into squares of one chain each, with a beam-compass to sum up each line of squares, had recently come into use by surveyors ; and Mr. Osgood amused himself by making a number of these compasses of various kinds of wood nicely finished and well polished, rather as examples of his skill than for any use he had for them, though he occasionally sold them to some of the local surveyors. He had these all suspended vertically on the wall instead of horizontally, as they are usually placed, and as they look best. While we were one day admiring the workmanship of an addition to the series, he remarked, " I dare say you don't know why I hang them up that way; very few people do." Of course, we acknowledged we did not know. " Well," said he, " it is very important. The air presses with a weight of fifteen pounds on every square inch, and if I hung them up level the pressure in the middle would very soon bend them, and they would be spoilt." My brother knew it was no good to try and show him his error, so merely said, " Yes, that's a very good idea of yours," and left the old man in the happy belief that he was quite scientific in his methods. My brother took a sketch of him enjoying his pipe and glass of toddy of an evening, which was a very good likeness, and which is here reproduced.

After we had completed the survey and maps of Cadoxton, which occupied us about six months, we had not much to do except small pieces of work of various kinds. One of these was to make a survey and take soundings of the river between the bridge and the sea, a distance of three or four miles, for a proposed scheme of improving the navigation, making docks, etc., which was partly carried out some years later. We also

SAMUEL OSGOOD.

(*From a sketch by W. G. Wallace.* 1843.)

[*To face p.* 188, VOL. I.

had a little architectural and engineering work, in designing
and superintending the erection of warehouses with powerful
cranes, which gave me some insight into practical building.
To assist in making working drawings and specifications, my
brother had purchased a well-known work, Bartholomew's
"Specifications for Practical Architecture." This book, though
mainly on a very dry and technical subject, contained an
introduction on the principles of Gothic architecture which
gave me ideas upon the subject of the greatest interest and
value, and which have enabled me often to form an independent judgment on modern imitations of Gothic or of any
other styles. Bartholomew was an enthusiast for Gothic,
which he maintained was the only true and scientific system
of architectural construction in existence. He showed how
all the most striking and ornamental features of Gothic architecture are essential to the stability of a large stone-built
structure—the lofty nave with its clerestory windows and
arched roof; the lateral aisles at a lower level, also with
arched roofs; the outer thrust of these arches supported by
deep buttresses on the ground, with arched or flying buttresses above; and these again rendered more secure by
being weighted down with rows of pinnacles, which add so
much to the beauty of Gothic buildings. He rendered his
argument more clear by giving a generalized cross-section of
a cathedral, and drawing within the buttresses the figure of a
man, with outstretched arms pushing against the upper arches
to resist their outward thrust, and being kept more steady by
a heavy load upon his head and shoulders representing the
pinnacle. This section and figure illuminated the whole construction of the masterpieces of the old architects so clearly
and forcibly, that though I have not seen the book since, I
have never forgotten it. It has furnished me with a standard
by which to judge all architecture, and has guided my taste
in such a small matter as the use of stone slabs over window
openings in brick buildings, thus concealing the structural
brick arch, and using stone as a beam, a purpose for which
iron or wood are better suited. It also made me a very
severe critic of modern imitations of Gothic in which we often

see buttresses and pinnacles for ornament alone, when the roof is wholly of wood and there is no outward thrust to be guarded against; while in some cases we see useless gargoyles, which in the old buildings stretched out to carry the water clear of the walls, but which are still sometimes imitated when the water is carried into drains by iron gutters and water pipes. I also learnt to appreciate the beautiful tracery of the large circular or pointed windows, whose harmonies and well-balanced curves and infinitely varied designs are a delight to the eye; while in most modern structures the attempts at imitating them are deplorable failures, being usually clumsy, unbalanced, and monotonous. One of the very few modern Gothic buildings in which the architect has caught the spirit of the old work is Barry's Houses of Parliament, which, whether in general effect or in its beautifully designed details, is a delight to the true lover of Gothic architecture. My brother had seen the exhibition of the competing designs, and he used always to speak of the unmistakable superiority of Barry over all the others.

Among our few intellectual friends here was the late Mr. Charles Hayward, a member of the Society of Friends (commonly called Quakers), as were Mr. Price of Neath Abbey, and our temporary landlord, Mr. Osgood. Mr. Hayward had a bookseller's shop in the town combined with that of a chemist and druggist, but he himself lived in a pretty cottage about half a mile out of the town, where he had two or three acres of land, kept a cow, and experimented in agriculture on a small scale; while his partner, Mr. Hunt, lived at the shop. A year or two later these gentlemen gave up the business and took a farm from Mr. Talbot of Margam Abbey, which they farmed successfully for some years, their chemical knowledge enabling them to purchase refuse materials from some of the manufacturers in the district which served as valuable manures. Later, Mr. Hayward took a larger farm near Dartmouth, where I had the pleasure of visiting him after my return from the East. A good many years later, when I lived at Godalming, he was again my neighbour, as after the death of his wife he came to live with his nephew,

C. F. Hayward, Esq., a well-known London architect, who
had a country house close by my cottage. Mr. Hayward
began life with nothing but a good education, industry, and
a love of knowledge. He is an example of the possibility
of success in farming without early training and with very
scanty capital. Of course, the period was a good one for
farmers, but it was not every one who could have made even
a bare living under such unfavourable conditions. After he
came to live at Godalming, when over seventy years of age, he
began to exercise his hitherto dormant faculty of water-colour
drawing. For this he made most of his own colours from
natural pigments, earthy or vegetable, and executed a number
of bold and effective landscapes, showing that if he had had
early training he might have excelled in this beautiful art. Mr.
Hayward was among my oldest and most esteemed friends.

During the larger portion of my residence at Neath we
had very little to do, and my brother was often away, either
seeking employment or engaged upon small matters of business in various parts of the country. I was thus left a good
deal to my own devices, and having no friends of my own
age I occupied myself with various pursuits in which I had
begun to take an interest. Having learnt the use of the
sextant in surveying, and my brother having a book on
Nautical Astronomy, I practised a few of the simpler observations. Among these were determining the meridian by equal
altitudes of the sun, and also by the pole-star at its upper or
lower culmination ; finding the latitude by the meridian
altitude of the sun, or of some of the principal stars ; and
making a rude sundial by erecting a gnomon towards the
pole. For these simple calculations I had Hannay and
Dietrichsen's Almanac, a copious publication which gave
all the important data in the Nautical Almanac, besides
much other interesting matter, useful for the astronomical
amateur or the ordinary navigator. I also tried to make a
telescope by purchasing a lens of about two feet focus at an
optician's in Swansea, fixing it in a paper tube and using the
eye-piece of a small opera glass. With it I was able to

observe the moon and Jupiter's satellites, and some of the
larger star-clusters; but, of course, very imperfectly. Yet it
served to increase my interest in astronomy, and to induce
me to study with some care the various methods of construction of the more important astronomical instruments; and
it also led me throughout my life to be deeply interested
in the grand onward march of astronomical discovery.

But what occupied me chiefly and became more and
more the solace and delight of my lonely rambles among the
moors and mountains, was my first introduction to the
variety, the beauty, and the mystery of nature as manifested
in the vegetable kingdom.

I have already mentioned the chance remark which gave
me the wish to know something about wild flowers, but
nothing came of it till 1841, when I heard of and obtained a
shilling paper-covered book published by the Society for the
Diffusion of Useful Knowledge, the title of which I forget,
but which contained an outline of the structure of plants and
a short description of their various parts and organs; and
also a good description of about a dozen of the most common
of the natural orders of British plants. Among these were
the Cruciferæ, Caryophylleæ, Leguminosæ, Rosaceæ, Umbelliferæ, Compositæ, Scrophularineæ, Labiatæ, Orchideæ, and
Glumaceæ. This little book was a revelation to me, and for
a year was my constant companion. On Sundays I would
stroll in the fields and woods, learning the various parts and
organs of any flowers I could gather, and then trying how
many of them belonged to any of the orders described in my
book. Great was my delight when I found that I could
identify a Crucifer, an Umbellifer, and a Labiate; and as one
after another the different orders were recognized, I began to
realize for the first time the order that underlay all the
variety of nature. When my brother was away and there
was no work to do, I would spend the greater part of the day
wandering over the hills or by the streams gathering flowers,
and either determining their position from my book, or
coming to the conclusion that they belonged to other orders
of which I knew nothing, and as time went on I found that

there were a very large number of these, including many of
our most beautiful and curious flowers, and I felt that I *must*
get some other book by which I could learn something about
these also. But I knew of no suitable book, I did not even
know that any British floras existed, and having no one to
help me, I was obliged to look among the advertisements of
scientific or educational publications that came in my way.
At length, soon after we came to Neath, David Rees happened
to bring in an old number of the *Gardener's Chronicle*, which
I read with much interest, and as I found in it advertisements and reviews of books, I asked him to bring some more
copies, which he did, and I found in one of them a notice of
the fourth edition of Lindley's " Elements of Botany," which,
as it was said to contain descriptions of all the natural
orders, illustrated by numerous excellent woodcuts, I thought
would be just the thing to help me on. The price, 10s. 6d.,
rather frightened me, as I was always very short of cash; but
happening to have so much in my possession, and feeling
that I *must* have some book to go on with, I ordered it at
Mr. Hayward's shop.

When at length it arrived, I opened it with great expectations, which were, however, largely disappointed, for although
the larger part of the book was devoted to systematic
botany, and all the natural orders were well and clearly
described, yet there was hardly any reference to British
plants—not a single genus was described, it was not even
stated which orders contained any British species and which
were wholly foreign, nor was any indication given of their
general distribution or whether they comprised numerous or
few genera or species. The inclusion of all the natural
orders and the excellent woodcuts illustrating many of
them, and showing the systematic characters by dissections
of the flowers and fruits, were, however, very useful, and
enabled me at once to classify a number of plants which had
hitherto puzzled me. Still, it was most unsatisfactory not to
be able to learn the names of any of the plants I was observing, so one day I asked Mr. Hayward if he knew of any
book that would help me. To my great delight he said he

had Loudon's "Encyclopædia of Plants," which contained all
the British plants, and he would lend it to me, and I could
copy the characters of the British species.

I therefore took it home to Bryn-coch, and for some
weeks spent all my leisure time in first examining it carefully,
finding that I could make out both the genus and the species
of many plants by the very condensed but clear descriptions,
and I therefore copied out the characters of every British
species there given. As Lindley's volume had rather broad
margins, I found room for all the orders which contained
only a moderate number of species, and copied the larger
orders on sheets of thin paper, which I interleaved at the
proper places. Having at length completed this work for
all the flowering plants and ferns, and also the genera of
mosses and the main divisions of the lichens and fungi, I
took back the volume of Loudon, and set to work with
increased ardour to make out all the species of plants I could
find. This was very interesting and quite a new experience
for me, and though in some cases I could not decide to which
of two or three species my plant belonged, yet a considerable
number could be determined without any doubt whatever.

This also gave me a general interest in plants, and a
catalogue published by a great nurseryman in Bristol, which
David Rees got from the gardener, was eagerly read,
especially when I found it contained a number of tropical
orchids, of whose wonderful variety and beauty I had
obtained some idea from the woodcuts in Loudon's Encyclopædia. The first epiphytal orchid I ever saw was at a flowershow
in Swansea, where Mr. J. Dillwyn Llewellyn exhibited a plant
of *Epidendrum fragrans*, one of the less attractive kinds, but
which yet caused in me a thrill of enjoyment which no other
plant in the show produced. My interest in this wonderful
order of plants was further enhanced by reading in the
Gardener's Chronicle an article by Dr. Lindley on one of the
London flower shows, where there was a good display of
orchids, in which, after enumerating a number of the species,
he added, "and *Dendrobium Devonianum*, too delicate and
beautiful for a flower of earth." This and other references to

and descriptions of them gave them, in my mind, a weird and mysterious charm, which was extended even to our native species, and which, I believe, had its share in producing that longing for the tropics which a few years later was satisfied in the equatorial forests of the Amazon.

But I soon found that by merely identifying the plants I found in my walks I lost much time in gathering the same species several times, and even then not being always quite sure that I had found the same plant before. I therefore began to form a herbarium, collecting good specimens and drying them carefully between drying papers and a couple of boards weighted with books or stones. My brother, however, did not approve of my devotion to this study, even though I had absolutely nothing else to do, nor did he suggest any way in which I could employ my leisure more profitably. He said very little to me on the subject beyond a casual remark, but a letter from my mother showed me that he thought I was wasting my time. Neither he nor I could foresee that it would have any effect on my future life, and I myself only looked upon it as an intensely interesting occupation for time that would be otherwise wasted. Even when we were busy I had Sundays perfectly free, and used then to take long walks over the mountains with my collecting box, which I brought home full of treasures. I first named the species as nearly as I could do so, and then laid them out to be pressed and dried. At such times I experienced the joy which every discovery of a new form of life gives to the lover of nature, almost equal to those raptures which I afterwards felt at every capture of new butterflies on the Amazon, or at the constant stream of new species of birds, beetles, and butterflies in Borneo, the Moluccas, and the Aru Islands.

It must be remembered that my ignorance of plants at this time was extreme. I knew the wild rose, bramble, hawthorn, buttercup, poppy, daisy, and foxglove, and a very few others equally common and popular, and this was all. I knew nothing whatever as to genera and species, nor of the large numbers of distinct forms related to each other and grouped into natural orders. My delight, therefore, was great when I was now able to identify the charming little eyebright, the strange-looking cow-wheat and louse-wort, the handsome mullein and the pretty creeping toad-flax, and to find that all of them as well as the lordly foxglove, formed parts of one great natural order, and that under all their superficial diversity of form there was a similarity of structure which, when once clearly understood, enabled me to locate each fresh species with greater ease. The Crucifers, the Pea tribe, the Umbelliferæ, the Compositæ, and the Labiates offered great difficulties, and it was only after repeated efforts that I was able to name with certainty a few of the species, after which each additional discovery became a little less difficult, though the time I gave to the study before I left England was not sufficient for me to acquaint myself with more than a moderate proportion of the names of the species I collected.

Now, I have some reason to believe that this was the turning-point of my life, the tide that carried me on, not to fortune but to whatever reputation I have acquired, and which has certainly been to me a never-failing source of much health of body and supreme mental enjoyment. If my brother had had constant work for me so that I never had an idle day, and if I had continued to be similarly employed after I became of age, I should most probably have become entirely absorbed in my profession, which, in its various departments, I always found extremely interesting, and should therefore not have felt the need of any other occupation or study.

I know now, though I was ignorant of it at the time, that my brother's life was a very anxious one, that the difficulty of finding remunerative work was very great, and that he was often hard pressed to earn enough to keep us both in the very humble way in which we lived. He never alluded to this that I can remember, nor did I ever hear how much our board and lodging cost him, nor ever saw him make the weekly or monthly payments. During the seven years I was with him I hardly ever had more than a few shillings for

personal expenses; but every year or two, when I went home,
what new clothes were absolutely necessary were provided
for me, with perhaps ten shillings or a pound as pocket-
money till my next visit, and this, I think, was partly or
wholly paid out of the small legacy left me by my grandfather. This seemed very hard at the time, but I now see
clearly that even this was useful to me, and was really an
important factor in moulding my character and determining
my work in life. Had my father been a moderately rich
man and had supplied me with a good wardrobe and ample
pocket-money; had my brother obtained a partnership in
some firm in a populous town or city, or had established
himself in his profession, I might never have turned to nature
as the solace and enjoyment of my solitary hours, my whole
life would have been differently shaped, and though I should,
no doubt, have given some attention to science, it seems very
unlikely that I should have ever undertaken what at that
time seemed rather a wild scheme, a journey to the almost
unknown forests of the Amazon in order to observe nature
and make a living by collecting. All this may have been
pure chance, as I long thought it was, but of late years I am
more inclined to Hamlet's belief, when he said—

> " There's a divinity that shapes our ends,
>   Rough-hew them how we will."

Of course, I do not adopt the view that each man's life, in
all its details, is guided by the Deity for His special ends.
That would be, indeed, to make us all conscious automata,
puppets in the hands of an all-powerful destiny. But, as
I shall show later on, I have good reasons for the belief that,
just as our own personal influence and expressed or unseen
guidance is a factor in the life and conduct of our children,
and even of some of our friends and acquaintances, so we are
surrounded by a host of unseen friends and relatives who
have gone before us, and who have certain limited powers of
influencing, and even, in particular cases, almost of determining,
the actions of living persons, and may thus in a great variety
of indirect ways modify the circumstances and character of

any one or more individuals in whom they are specially interested. But a great number of these occurrences in every one's life are apparently what we term chance, and even if all are so, the conclusion I wish to lay stress upon is not affected. It is, that many of the conditions and circumstances that constitute our environment, though at the time they may seem unfortunate or even unjust, yet are often more truly beneficial than those which we should consider more favourable. Sometimes they only aid in the formation of character ; sometimes they also lead to action which gives scope for the use of what might have been dormant or unused faculties (as, I think, has occurred in my own case) ; but much more frequently they seem to us wholly injurious, leading to a life of misery or crime, and turning what in themselves are good faculties to evil purposes. When this occurs in any large number of cases, as it certainly does with us now, we may be sure that it is the system of society that is at fault, and the most strenuous efforts of all who see this should be devoted, not to the mere temporary alleviation of the evils due to it but to the gradual modification of the system itself. This is my present view. At the time of which I am now writing, I had not begun even to think of these matters, although facts which I now see to be of great importance in connection with them were being slowly accumulated for use in after years.

CHAPTER XIV

FIRST LITERARY EFFORTS

IT was during the time that I was most occupied out of doors with the observation and collection of plants that I began to write down, more or less systematically, my ideas on various subjects that interested me. Three of these early attempts have been preserved and are now before me. They all bear dates of the autumn or winter of 1843, when I was between nineteen and twenty years of age.

One of these is a rough sketch of a popular lecture on Botany, addressed to an audience supposed to be as ignorant as I was myself when I began to observe our native flowers. I was led to write it, partly on account of the difficulties I myself had felt in obtaining the kind of information I required, but chiefly on account of a lecture I had attended at Neath by a local botanist of some repute, and which seemed to me so meagre, so uninteresting, and so utterly unlike what such a lecture ought to be, that I wanted to try if I could not do something better. The lecture in question consisted in an enumeration of the whole series of the "Linnæan Classes and Orders," stating their characters and naming a few of the plants comprised in each. It was illustrated by a series of coloured figures on cards about the size of ordinary playing cards, which the lecturer held up one after the other to show what he was talking about. The Linnæan system was upheld as being far the most useful as a means of determining the names of plants, and the natural system was treated as quite useless for beginners, and only suited for experienced botanists.

All this was so entirely opposed to views I had already formed, that I devoted a large portion of my lecture to the question of classification in general, showed that *any* classification, however artificial, was better than none, and that Linnæus made a great advance when he substituted generic and specific names for the short Latin descriptions of species before used, and by classifying all known plants by means of a few well-marked and easily observed characters. I then showed how and why this classification was only occasionally, and as it were accidentally, a natural one; that in a vast number of cases it grouped together plants which were essentially unlike each other ; and that for all purposes, except the naming of species, it was both useless and inconvenient. I then showed what the natural system of classification really was, what it aimed at, and the much greater interest it gave to the study of botany. I explained the principles on which the various natural orders were founded, and showed how often they gave us a clue to the properties of large groups of species, and enabled us to detect real affinities under very diverse external forms.

I concluded by passing in review some of the best marked orders as illustrating these various features. Although crudely written and containing some errors, I still think it would serve as a useful lecture to an audience generally ignorant of the whole subject, such as the young mechanics of a manufacturing town. Its chief interest to me now is, that it shows my early bent towards classification, not the highly elaborate type that seeks to divide and subdivide under different headings with technical names, rendering the whole scheme difficult to comprehend, and being in most cases a hindrance rather than an aid to the learner, but a simple and intelligible classification which recognizes and defines all great natural groups, and does not needlessly multiply them on account of minute technical differences. It has always seemed to me that the natural orders of flowering plants afford one of the best, if not the very best, example of such a classification.

It is this attraction to classification, not as a metaphysically complete system, but as an aid to the comprehension of a subject, which is, I think, one of the chief causes of the success of my books, in almost all of which I have aimed at a simple and intelligible rather than a strictly logical arrangement of the subject-matter.

Another lecture, the draft for which I prepared pretty fully, was on a rather wider subject—" The Advantages of Varied Knowledge "—in opposition to the idea that it was better to learn one subject thoroughly than to know something of many subjects. In the case of a business or profession, something may be said for the latter view, but I treated it as a purely personal matter which led to the cultivation of a variety of faculties, and gave pleasurable occupation throughout life. A few extracts may, perhaps, be permitted from this early attempt. Speaking of a general acquaintance with history, biography, art, and science, I say, " There is an intrinsic value to ourselves in these varied branches of knowledge, so much indescribable pleasure in their possession, so much do they add to the enjoyment of every moment of our existence, that it is impossible to estimate their value, and we would hardly accept boundless wealth, at the cost, if it were possible, of their irrecoverable loss. And if it is thus we feel as to our general store of mental acquirements, still more do we appreciate the value of any particular branch of study we may ardently pursue. What pleasure would remain for the enthusiastic artist were he forbidden to gaze upon the face of nature, and transfer her loveliest scenes to his canvas? or for the poet were the means denied him to rescue from oblivion the passing visions of his imagination? or to the chemist were he snatched from his laboratory ere some novel experiment were concluded, or some ardently pursued theory confirmed? or to any of us were we compelled to forego some intellectual pursuit that was bound up with our every thought? And here we see the advantage possessed by him whose studies have been in various directions, and who at different times has had many different pursuits, for whatever may happen, he will always find something in his surroundings to interest and instruct him."

And further on, as illustrations of the interest in common things conferred by a knowledge of the elementary laws of physical science, I remark—

"Many who marvel at the rolling thunder care not to inquire what causes the sound which is heard when a tightly-fitting cork is quickly drawn from a bottle, or when a whip is cracked, or a pistol fired ; and while they are struck with awe and admiration at the dazzling lightning, look upon the sparks drawn from a cat's back on a frosty evening and the slight crackle that accompanies them as being only fit to amuse a child ; yet in each case the cause of the trifling and of the grand phenomena are the same. He who has extended his inquiries into the varied phenomena of nature learns to despise no fact, however small, and to consider the most apparently insignificant and common occurrences as much in need of explanation as those of a grander and more imposing character. He sees in every dewdrop trembling on the grass causes at work analogous to those which have produced the spherical figure of the earth and planets ; and in the beautiful forms of crystallization on his window-panes on a frosty morning he recognizes the action of laws which may also have a part in the production of the similar forms of plants and of many of the lower animal types. Thus the simplest facts of everyday life have to him an inner meaning, and he sees that they depend upon the same general laws as those that are at work in the grandest phenomena of nature."

I then pass in review the chief arts and sciences, showing their inter-relations and unsolved problems ; and in remarking on the Daguerrotype, then the only mode of photographic portraiture, I make a suggestion that, though very simple, has not yet been carried out. It is as follows :—

"It would be a curious and interesting thing to have a series of portraits taken of a person each successive year. These would show the gradual changes from childhood to old age in a very striking manner ; and if a number of such series from different individuals were obtained, and a brief outline given of their lives during each preceding year, we should have materials not merely for the curious to gaze at, but which might elucidate the problem of how far the mind reacts upon the countenance. We should see the effects of pain or pleasure, of idleness or activity, of dissipation or study, and thus watch the action of the various passions of the mind in modifying the form of the body, and particularly the expression of the features."

Now that photography is so widespread and so greatly improved, it is rather curious that nothing of this kind has been done. Some of our numerous scientific societies might offer to take such photographs of any of their members who would agree to be taken regularly, and would undertake to have one or two of their children similarly taken till they came of age, and also to prepare a very short record each year of the main events or occupations of their lives. If this were widely done in every part of the country, a most interesting and instructive collection of those series which were most complete would be obtained. I have given the concluding passage of the lecture as it appears in the rough draft, which never got rewritten.

"Can we believe that we are fulfilling the purpose of our existence while so many of the wonders and beauties of the creation remain unnoticed around us? While so much of the mystery which man has been able to penetrate, however imperfectly, is still all dark to us? While so many of the laws which govern the universe and which influence our lives are, by us, unknown and uncared for? And this not because we want the power, but the will, to acquaint ourselves with them. Can we think it right that, with the key to so much that we ought to know, and that we should be the better for knowing, in our possession, we seek not to open the door, but allow this great store of mental wealth to lie unused, producing no return to us, while our highest powers and capacities rust for want of use?

"It is true that man is still, as he always has been, subject to error; his judgments are often incorrect, his beliefs false, his opinions changeable from age to age. But experience of error is his best guide to truth, often dearly bought, and, therefore, the more to be relied upon. And what is it but the accumulated experience of past ages that serves us as a beacon light to warn us from error, to guide us in the way of truth. How little should we know had the knowledge acquired by each preceding age died with it ! How blindly should we grope our way in the same obscurity as did our ancestors, pursue the same phantoms, make the same fatal blunders, encounter the same perils, in order to purchase the same truths which had been already acquired by the same process, and lost again and again in bygone ages ! But the wonder-working press prevents this loss ; truths once acquired are treasured up by it for posterity, and each succeeding generation adds something to the stock of acquired knowledge, so that our acquaintance with the works of nature is ever increasing, the range of our inquiries is extended each age, the power of mind over matter becomes, year by year, more complete. Yet our horizon ever widens, the limits to our advance seem more distant than ever, and there seems nothing too noble, too exalted, too marvellous, for the ever-increasing knowledge of future generations to attain to.

"Is it not fitting that, as intellectual beings with such high powers, we should each of us acquire a knowledge of what past generations have taught us, so that, should the opportunity occur, we may be able to add somewhat, however small, to the fund of instruction for posterity? Shall we not then feel the satisfaction of having done all in our power to improve by culture those higher faculties that distinguish us from the brutes, that none of the talents with which we may have been gifted have been suffered to lie altogether idle? And, lastly, can any reflecting mind have a doubt that, by improving to the utmost the nobler faculties of our nature in this world, we shall be the better fitted to enter upon and enjoy whatever new state of being the future may have in store for us?"

These platitudes are of no particular interest, except as showing the bent of my mind at that period, and as indicating a disposition for discursive reading and study, which has been a great advantage to myself, and which has enabled me to write on a variety of subjects without committing any very grievous blunders (so far as my critics have pointed out), and with, I hope, some little profit to my readers.

The only other subject on which I attempted to write at this time was on the manners and customs of the Welsh peasantry as they had come under my personal observation in Brecknockshire and Glamorganshire. I have already described how I came to take some interest in agriculture while surveying in Bedfordshire and the adjacent counties, and this interest was increased by a careful study of Sir Humphry Davy's "Lectures on Agricultural Chemistry," which I met with soon afterwards. I was, therefore, the better able to compare the high-class farming of the home counties with that of the ignorant Welshmen, under all the disadvantages of a poor soil and adverse climate, of distant markets, and the almost entire absence of what the English farmer would consider capital.

Having lived for more than a year on an average Welsh farm at Bryn-coch, while we had often lodged with small farmers and labourers, or at public-houses whose landlords almost always farmed a little land, I got to know a good deal about their ways, and adding to this my own observation of the kind of land they had to farm, and the difficulties under which they laboured, I felt inclined to write a short account of them in the hope that I might perhaps get it accepted by some magazine as being sufficiently interesting for publication. I wrote it out fairly with this intention, and two years afterwards, when in London, I took it to the editor of a magazine (I forget which) who promised to look over it. He returned it in a few days with the remark that it seemed more suited for an agricultural journal than for a popular magazine. I made no other offer of it, and as it was my first serious attempt at writing, though I am afraid it is rather dull, I present it to my readers as one of the landmarks in my literary career. I may add that I have recently visited the Upper Vale of Neath, and renewed my acquaintance with its picturesque scenery. The chief differences that I saw are that some of the smaller farm houses and cottages are in ruins, and that the farms seem to be somewhat larger. Where the ground is fairly level the mowing machine is now used, but in the condition of the farm-yards and the style of the houses I see no advance whatever. Some of the old customs have vanished, for I was unable to obtain any flummery, and on my inquiry for bake-stone bread I found that it was now rarely made. A cake was, however, prepared specially for me, but being made of white American flour it had not the flavour of that which I used so much to enjoy made from the brown flour of home-grown wheat.

THE SOUTH-WALES FARMER

Introductory Remarks

In the following pages I have endeavoured to give a correct idea of the habits, manners, and mode of life of the Welsh hill farmer, a class which, on account of the late Rebecca disturbances, has excited much interest. Having spent some years in Radnorshire, Brecknockshire, Glamorganshire, and other parts of South Wales, and been frequently in the dwellings of the farmers and country people, and had many opportunities of observing their customs and manners, all that I here mention is from my own observation, or obtained by conversation with the parties. I have taken Glamorganshire as the locality of most of what I describe, as I am best acquainted with that part and the borders of Carmarthenshire, where the recent disturbances have been most prominent.

Whenever there is any great difference in neighbouring counties I have noticed it. I may here observe that in Radnorshire the Welsh manners are in a great measure lost with the language, which is entirely English, spoken with more purity than in many parts of England, with the exception of those parts bordering Cardiganshire and Brecknockshire, where the Welsh is still used among the old people, the River Wye, which is the boundary of the latter county and Radnorshire, in its course between Rhayader Gwy and the Hay, also separates the two languages. On the Radnorshire side of the river you will find in nine houses out of ten English commonly spoken, while directly you have crossed the river, there is as great or a still greater preponderance of Welsh. In the country a few miles round the seaport town of Swansea most of the peculiarities I shall mention may be seen to advantage. In the east and south-eastern parts of Glamorganshire, called the Vale of Glamorgan, the appearance of the country and the inhabitants is much more like those of England. The land is very good and fertile, agriculture is much attended to and practised on much better principles. This part, therefore (the neighbourhood of the towns of Cowbridge and Cardiff), is excepted from the following remarks.

The South-Wales Farmer: His Modes of Agriculture, Domestic Life, Customs, and Character.

THE generality of mountain farms in Glamorganshire and most other parts of South Wales are small, though they may appear large when the number of acres only is considered, a large proportion being frequently rough mountain land. On the average they consist of from twenty to fifty acres of arable land in fields of from four to six, and rarely so much as ten acres; the same quantity of rough, boggy, bushy, rushy pasture, and perhaps as much, or twice as much, short-hay meadow, which term will be explained hereafter; and from fifty to five hundred acres of rough mountain pasture, on which sheep and cattle are turned to pick up their living as they can.

Their system of farming is as poor as the land they cultivate. In it we see all the results of carelessness, prejudice, and complete ignorance. We see the principle of doing as well as those who went before them, and no better, in full operation ; the good old system which teaches us not to suppose ourselves capable of improving on the wisdom of our forefathers, and which has made the early polished nations of the East so inferior in every respect to us, whose reclamation from barbarism is ephemeral compared with their long period of almost stationary civilization. The Welshman, when you recommend any improvement in his operations, will tell you, like the Chinaman, that it is an "old custom," and that what did for his forefathers is good enough for him. But let us see if the farmer is so bad as this mode of doing his business may be supposed to make him. In his farmyard we find the buildings with broken and gaping doors, and the floors of the roughest pitching. In one corner is a putrid pond, the overflowings of which empty themselves into the brook below. Into this all the drainings from the dungheaps in the upper part of the yard run, and thus, by evaporation in summer and the running into the brook in winter, full one-half of the small quantity of manure he can obtain (from his cattle spending the greater part of their time on the mountain and in wet bushy pastures) is lost.

The management of his arable land is dreadfully wasteful and injurious.

Of green crops (except potatoes can be so called) he has not the slightest idea, and if he takes no more than three grain crops off the land in succession, he thinks he does very well; five being not uncommon. The first and principal crop is wheat, on which he bestows all the manure he can muster, with a good quantity of lime. He thus gets a pretty good crop. The next year he gets a crop of barley without any manure whatever, and after that a crop of oats, unmanured. He then leaves the field fallow till the others have been treated in the same manner, and then returns to serve it thus cruelly again; first, however, getting his potato crop before his wheat. Some, after the third crop (oats), manure the land as well as they can, and sow barley with clover, which they mow and feed off the second year, and then let it remain as pasture for some time; others, again, have three crops of oats in succession after the wheat and barley, and thus render the land utterly useless for many years.

In this manner the best crops of wheat they can get with abundance of manure, on land above the average quality, is about twenty bushels per acre—ten bushels is, however, more general, and sometimes only seven or eight are obtained.

The rough pastures on which the cattle get their living and waste their manure a great part of the time consist chiefly of various species of rushes and sedges, a few coarse grasses, and gorse and fern on the drier parts. They are frequently, too, covered with brambles, dwarf willows, and alders.

The "short-hay meadows," as they are called, are a class of lands entirely unknown in most parts of England; I shall, therefore, endeavour to describe them.

They consist of large undulating tracts of lands on the lower slopes of the mountains, covered during autumn, winter, and spring with a very short brownish yellow wet turf. In May, June, and July the various plants forming this turf spring up, and at the end of summer are mown, and form "short-hay"; and well it deserves the name, for it is frequently almost impossible to take it up with a hayfork, in which case it is raked up and gathered by armfuls into the cars. The produce varies from two to six hundredweight per acre; four may be about the average, or five acres of land to produce a ton of hay. During the rest of the year it is almost good for nothing. It is astonishing how such stuff can be worth the labour of mowing and making into hay. An English farmer would certainly not do it, but the poor Welshman has no choice; he must either cut his short-hay or have no food for his cattle in the winter; so he sets to, and sweeps away with his scythe a breadth which would astonish an English mower.

The soil which produces these meadows is a poor yellow clay resting on the rock; on the surface of the clay is a stratum of peaty vegetable matter, sometimes of considerable thickness though more generally only a few inches, which collects and retains the moisture in a most remarkable manner, so that though the ground should have a very steep slope the water seems to saturate and cling to it like a sponge, so much so that after a considerable period of dry weather, when, from the burnt appearance of the surface, you would imagine it to be perfectly free from moisture, if you venture to kneel or lie down upon it you will almost instantly be wetted to the skin.

The plants which compose these barren slopes are a few grasses, among which are the sweet vernal grass (*Anthoxanthum odoratum*) and the crested hair grass (*Kœleria cristata*), several Cyperaceæ—species of carex or sedge which form a large proportion, and the feathery cotton grass (*Eriophorum vaginatum*). The toad-rush (*Juncus bufonius*) is frequently very plentiful, and many other plants of the same kind. Several rare or interesting British plants are here found often in great profusion. The Lancashire asphodel (*Narthecium ossifragum*) often covers acres with its delicate yellow and red blossoms. The spotted orchis (*O. maculata*) is almost universally present. The butterwort (*Pinguicula vulgaris*) is also found here, and the beautiful little pimpernel (*Anagallis tenella*). The louseworts (*Pendicularis sylvatica* and *P. palustris*), the melancholy thistle (*Cincus heterophyllus*), and the beautiful blue milkwort (*Polygala vulgaris*), and many others, are generally exceedingly plentiful, and afford much gratification to the botanist and lover of nature.

The number of sheep kept on these farms is about one to each acre of mountain, where they live the greater part of the year, being only brought down to the pastures in the winter, and again turned on the mountain with their lambs in the spring. One hundred acres of pasture and "short-hay meadow" will support from thirty to forty cattle, ten or a dozen calves and oxen being sold each year.

The farmers are almost invariably yearly tenants, consequently little improvement is made even in parts which could be much bettered by draining. The landlord likes to buy more land with his spare capital (if he has any) rather than improve these miserable farms, and the tenant is too poor to lay out money, or if he has it will not risk his being obliged to leave the farm or pay higher rent in return for his permanently improving another person's land.

The hedges and gates are seldom in sufficiently good repair to keep out cattle, and can hardly be made to keep out mountain sheep, who set them completely at defiance, nothing less than a six-foot stone wall, and not always that, serving to confine them. The farmer consequently spends a good deal of his time in driving them out of his young clover (when he has any) or his wheat. He is also constantly engaged in disputes, and not unfrequently litigation, with his neighbours, on account of the mutual trespasses of their stock.

The Welshman is by no means sharp-sighted when his cattle are enjoying themselves in a neighbour's field, especially when the master is from home, otherwise the fear of the "pound" will make him withdraw them after a short time.

On almost every farm water is very plentiful, often far too much so, and it is sometimes run over a meadow, but in such a manner as to lose one-half of the advantage which might be derived from it. The farmer is contented with merely cutting two or three gaps in the watercourse at the top, from which the water flows over the field as it best can, scarcely wetting some parts and making complete pools in others.

Weeding he considers quite an unnecessary refinement, fit only for those who have plenty of money to waste upon their fancies—except now and then, when the weeds have acquired an alarming preponderance over the crop, he perhaps sets feebly to work to extract the more prominent after they have arrived at maturity and the mischief is done. His potatoes are overrun with persicarias, docks, and spurges; his wheat and barley with corn cockle, corn scabious, and knapweed, and his pastures with thistles, elecampine, etc., all in the greatest abundance. If you ask him why he leaves his land in such a disgraceful state, and try to impress upon him how much better crops he would have if he cleared it, he will tell you that he does not think they do much harm, and that if he cleaned them this year, there would be as many as ever next year, and, above all, that he can't afford it, asking you where he is to get money to pay people for doing it.

The poultry, geese, ducks, and fowls are little attended to, being left to pick up their living as well as they can. Geese are fattened by being turned into the corn stubble, the others are generally killed from the yard. The fowls, having no proper places to lay in, are not very profitable with regard to eggs, which have to be hunted for and discovered in all sorts of places. This applies more particularly to Glamorganshire, which is in a great measure supplied with eggs and poultry from Carmarthenshire, or " Sir Gaer" (pronounced there *gar*), as it is called in Welsh, where they manage them much better.

If there happens to be in the neighbourhood any one who farms on the improved English system, has a proper course of crops, with turnips, etc., folds his sheep, and manages things in a tidy manner, it is impossible to make the Welshman believe that such a way of going on pays ; he will persist that the man is losing money by it all the time, and that he only keeps it on because he is ashamed to confess the failure of his new method. Even should the person go on for many years, to all appearance prosperously and in everybody else's eyes be making money by his farm, still the Welshman will declare that he has some other source from which he draws to purchase his dear-bought farming amusement, and that the time will come when he will be obliged to give it up ; and though you tell him that the greater part of the land in England is farmed in that manner, and ask him whether he thinks they can all be foolish enough to go on losing money year after year, he is still incredulous, says that he knows " the nature of farming," and that such work as that can never pay. While the ignorance which causes this incredulity exists, it is evidently a difficult task to improve him.

Domestic Life, Customs, etc.

The house is a tiled, white-washed edifice, in the crevices of which wall rue, common spleenwort, and yarrow manage generally to vegetate, notwithstanding their (at the very least) annual coat of lime. It consists on the ground floor of a rather large and very dark room, which serves as kitchen and dining-room for the family, and a rather better one used as a parlour on high days or when visitors call ; this latter frequently serves as the bedroom of the master and mistress. The kitchen, which is the theatre of the Welsh farmer's domestic life, has either a clay floor or one of very uneven stone paving, and the ceiling is in many cases composed of merely the floor boards of the room above, through the chinks of which everything going on aloft can be very conveniently heard and much seen. The single window is a small and low one, and this is rendered almost useless by the dirtiness of the glass, some window drapery, a Bible, hymn book and some old newspapers on the sill, and a sickly-looking geranium or myrtle, which seems a miracle of vital tenacity in that dark and smoky atmosphere. On one side may be discerned an oak sideboard brilliantly polished, on the upper part of which are rows of willow pattern plates and dishes, in one corner an open cupboard filled with common gaudily-coloured china, and in the other a tall clock with a handsome oak case. Suspended from the ceiling is a serious impediment to upright walking in the shape of a bacon rack, on which is, perhaps, a small supply of that article and some dried beef, also some dried herbs in paper, a large collection of walking sticks, and an old gun. In the chimney opening a coal fire in an iron grate takes the place of the open hearth and smoky peat of Radnorshire and other parts. A long substantial oak table, extending along the room under the window, an old armchair or two, a form or bench and two or three stools, complete the furniture of the apartment. From the rack before mentioned is generally suspended a piece of rennet for making cheese, and over the mantelpiece is probably a toasting-fork, one brass and two tin candlesticks, and a milk strainer with a hole in the bottom of it ; on the dresser, too, will be perceived a brush and comb which serve for the use of the whole family, and which you may apply to your own head (if you feel so inclined) without any fear of giving offence.

Upstairs the furniture is simple enough : two or three plain beds in each room with straw mattresses and home-made blankets, sheets being entirely unknown or despised ; a huge oak chest full of oatmeal, dried beef, etc., with perhaps a chest of drawers to contain the wardrobe ; a small looking-glass which distorts the gazer's face into a mockery of humanity ; and a plentiful supply of fleas, are all worth noticing. Though the pigs are not introduced into the family quite so familiarly as in Ireland, the fowls seem to take their place. It is nothing uncommon for them to penetrate even upstairs ; for we were once ourselves much puzzled to account for the singular phenomenon of finding an egg upon the bed, which happened twice or we might have thought it put there by accident. It was subsequently explained to us that some persons thought it lucky for the fowls to lay there : the abundance of fleas was no longer a mystery. The bed in the parlour before mentioned serves, besides its ostensible use, as a secret cupboard, where delicacies may be secured from the junior members of the family. I have been informed by an acquaintance whose veracity I can rely on (and indeed I should otherwise find no difficulty in believing it) that one day, being asked to take some bread and cheese in a respectable farmhouse, the wheat bread (a luxury) was procured from some mysterious part of the bed, either between the blankets or under the mattress, which my informant could not exactly ascertain. The only assistants in the labours of the farm, besides the sons and daughters, is generally a female servant, whose duties are multifarious and laborious, including driving the horses while ploughing and in haytime, and much other out-of-door work. If you enter the house in the morning, you will probably see a huge brass pan on the fire filled with curdled milk for making cheese. Into this the mistress dips her red and not particularly clean arm up to the elbow, stirring it round most vigorously. Meals seem to be prepared solely for the men, as you seldom see the women sit down to table with them. They will either wait till the others have done or take their dinner on their laps by the fire. The breakfast consists of hasty-pudding or oatmeal porridge, or cheese with thin oatmeal cakes or barley bread, which are plentifully supplied at all meals, and a basin of milk for each person ; for dinner there is perhaps the same, with the addition of a huge dish of potatoes, which they frequently break into their basin of milk or eat with their cheese ; and for supper, often milk with flummery or "siccan" (pronounced *shiccan*). As this is a peculiar and favourite Welsh dish, I will describe its composition. The oat bran with some of the meal left in it is soaked for several days in water till the acetous fermentation commences; it is then strained off, producing a thin, starchy liquid. When wanted for use this is boiled, and soon becomes nearly of the consistence and texture of blancmange, of a fine light brown colour and a peculiar acid taste which, though at first disagreeable to most persons, becomes quite pleasant with use. This is a dish in high repute with all real Welshman. Each person is provided with a basin of new milk, cold, and a spoon, and a large dish of hot flummery is set on the table, each person helping himself to as much as he likes (and that is often a great deal), putting it in his basin of milk ; and it is, I have no doubt, very wholesome and nourishing food. I must mention that the women, both in the morning and evening (and frequently at dinner too), treat themselves to a cup of tea, which is as universal a necessary among the fair sex here as in other parts of the kingdom. They prefer it, too, without milk, which they say takes away the taste, and as it is generally made very weak, that may be the case. Once or twice a week a piece of bacon or dry beef is added at dinner or supper, more as a relish to get down the potatoes than as being any food in itself. The beef in particular is so very high-dried and hard as almost to defy the carver's most strenuous efforts. The flavour is, nevertheless, at times very fine when the palate gets used to it, though the appearance is far from inviting, being about the colour and not far from the hardness of the black oak table. They generally keep it in a large chest in oatmeal (which was before mentioned). Often, when lodging at a little country inn, have we, when just awake in the morning, seen one of the children come stealthily into the room, open the lid of the huge chest, climb over the edge of it, and, diving down, almost disappear in its recesses, whence, after sundry efforts and strainings, he has reappeared, dragging forth a piece of the aforesaid black beef, which is obtained thus early that it may be soaked a few hours before boiling, to render it more submissive to the knife.

From the foregoing particulars it will be seen that these people live almost entirely on vegetable food. When a cow or a pig is killed, for a day or two they luxuriate on fresh meat; but that is the exception, not the rule. Herrings, too, they are fond of as a relish, as well as cockles and other indigestible food; but neither these nor the beef and bacon can be considered to be the staple food of the peasantry, which is, in one form or another, potatoes, oatmeal, bread, cheese, and milk.

The great consumption of oatmeal produces, as might be expected, cutaneous diseases, though, generally speaking, the people are tolerably healthy. They have a great horror of the doctor, whom they never send for but when they think there is some great danger. So long as the patient is free from pain they think all is right. They have not the slightest idea of what an invalid ought to eat. If gruel is ordered, they make a lumpy oatmeal pudding, to which, however, the sick man will frequently prefer bread and cheese. When they have gone on in this way till the unhappy individual is in the greatest danger and the medical attendant insists upon his directions being attended to, they unwillingly submit; and if the patient dies, they then impute it entirely to the doctor, and vow they will never call him in to kill people again.

As in most rural districts, by constant inter-marriages every family has a host of relations in the surrounding country. All consider it their duty to attend a funeral, and almost every person acquainted with the deceased attends as a mark of respect. Consequently the funerals are very large, often two or three hundred persons, and when the corpse has to be carried a distance, most of them come on horseback, which, with the varied colours of the women's dresses and the solemn sounds of a hymn from a hundred voices, as they wend their way along some lonely mountain road, has a most picturesque and interesting effect. This large company generally meet at the house, where provisions are ready for all who choose to partake of them. The well-known beautiful custom of adorning the graves with flowers and evergreens is much practised.

When a birth takes place in a family all the neighbours and relations call within a few days to inquire after the health of the mother and child, and take a cup of tea or bread and cheese, and every one brings some present, either a pound of sugar, quarter pound of tea, or a shilling or more in money, as they think best. This is expected to be returned when the givers are in a similar situation.

The " bidding," which is a somewhat similar custom at a marriage, is not quite so general, though it is still much used in Carmarthenshire. When a young couple are married they send notice to all their friends, that "on a day named they intend to have a ' bidding,' at which they request their company, with any donation they may think proper, which will be punctually returned when they are called upon on a similar occasion." At such biddings £20 or £30 are frequently collected, and sometimes much more, and as from various causes they are not called upon to return more than one-half, they get half the sum clear, and a loan without interest of the other half to commence life with.

The national dress or costume of the men (if ever they had any) is not now in use ; that of the women, however, is still very peculiar. Both use principally home-made articles, spinning their own wool and sending it to the factory to be made into flannel or cloth. They also dye the wool black themselves, using in the operation the contents of certain well-known domestic utensils, which is kept stewing over the fire some days, emitting a most unsavoury odour, which, however, they assert to be very wholesome. The men generally wear a square-cut coat of home-made pepper-and-salt coloured cloth, waistcoat and breeches or trousers of the same, and a round low crowned hat ; or occasionally fustian trousers and gay flannel waistcoat with bright metal buttons, coloured neckerchief, home-knit stockings of black sheep's wool, and lace-up boots. Shirts of checked coarse flannel—cotton shirts and sheets being considered equally luxurious. One of the most striking parts of the women's dress is the black beaver hat, which is almost universally worn and is both picturesque and becoming. It is made with a very high crown, narrowing towards the top, and a broad, perfectly flat brim, thus differing entirely from any man's hat. They frequently give thirty shillings for one of these hats, and make them last the greater part of their lives. The body dress consists of what they call a bedgown, or *betcown*, as it is pronounced, which is a dress made quite plain, entirely open in front (like a gentleman's dressing gown), with sleeves a little short of the elbow. A necessary accompaniment to this is an apron, which ties it up round the waist. The bedgown is invariably formed of what they call flannel, which is a stuff formed by a mixture of wool, cotton, and sometimes a little silk. It is often striped black or dark blue, or brown and white, with alternate broad and narrow stripes, or red and black, but more frequently a plaid of several colours, the red and black being wool, the white or blue cotton, and often a narrow yellow stripe of silk, made in plaid patterns of every variety of size and colour. The apron is almost always black-and-white plaid, the only variety being in the form and size of the pattern, and has a pretty effect by relieving the gay colours of the other part of the dress. They in general wear no stays, and this, with the constant habit of carrying burdens on the head, produces almost invariably an upright carriage and good figure, though rather inclined to the corpulency of Dutch beauties. On their necks they usually wear a gay silk kerchief or flannel shawl, a neat white cap under the hat ; laced boots and black worsted stockings complete their attire. In Carmarthenshire a jacket with sleeves is frequently worn by the women, in other respects their dress does not much differ from what I have described.

The women and girls carry (as before mentioned) great loads upon their heads, fifty or sixty pounds weight, and often much more. Large pitchers (like Grecian urns) of water or milk are often carried for long distances on uneven roads, with both hands full at the same time. They may be often seen turning round their heads to speak to an acquaintance and tripping along with the greatest unconcern, but never upsetting the pitcher. The women are almost invariably stout and healthy looking, notwithstanding their hard work and poor living. These circumstances, however, make them look much older than they really are. The girls are often exceedingly pretty when about fifteen to twenty, but after that, hard work and exposure make their features coarse, so that a girl of five-and-twenty would often be taken for nearer forty.

All, but more especially the young ones, ride most fearlessly, and at fairs they may be seen by dozens racing like steeple-chasers.

Many of these farmers are freeholders, cultivating their own land and living on the produce ; but they are generally little, if any, better off than the tenants, leaving the land in the same manner, thus showing that it is not altogether want of leases and good landlords that makes them so, but the complete ignorance in which they pass their lives.

All that I have hitherto said refers solely to the poorer class, known as hill farmers. In the valleys and near the town where the land is better, there are frequently better educated farmers, who assimilate more to the English in their agricultural operations, mode of living, and dress.

In all the mining districts, too, there is another class—the colliers and furnacemen, smiths, etc., who are as different from the farmers in everything as one set of men can be from another. When times are good their wages are such as to afford them many luxuries, which the poor farmer considers far too extravagant. Instead of living on vege-table diet with cheese and buttermilk, they luxuriate on flesh and fowl, and often on game too, of their own procuring. But in their dress is the greatest difference. The farmer is almost always dressed the same, except that on Sundays and market-day it is newer. But the difference between the collier or furnaceman at his work—when he is half naked, begrimed from head to foot, labouring either in the bowels of the earth or among roaring fires, and looking more like demon than man—and on holidays dressed in a suit of clothes that would not disgrace an English gentleman, is most remarkable. It is nothing uncommon to see these men dressed in coat and trousers of *fine black cloth*, elegant waistcoat, fine shirt, beaver hat, Wellington boots, and a fine silk handkerchief in his pocket ; and instead of being ridiculous, as the clumsy farmer would be in such a dress, wearing it with a quiet, unconcerned, and gentlemanly air. The men at the large works, such as Merthyr Tydfil, are more gaudy in their dress, and betray themselves much more quickly than the colliers of many other districts.

It is an undoubted fact, too, that the persons engaged in the collieries and ironworks are far more intellectual than the farmers, and pay more attention to their own and their children's education. Many of them indeed are well informed on most subjects, and in every respect much more highly civilized than the farmer.

The wages which these men get—in good times £2 or £3 per week—prevents them, with moderate care, from being ever in any great distress. They likewise always live well, which the poor farmer does not, and though many of them have a bit of land and all a potato ground, the turnpike grievances, poor-rates, and tithes do not affect them as compared with the farmers, to whom they are a grievous burden, making the scanty living with which they are contented hard to be obtained.

The rents, too, continue the same as when their produce sold for much more and the above-mentioned taxes were not near so heavy. The consequence is that the poor farmer works from morning to night after his own fashion, lives in a manner which the poorest English labourer would grumble at, and as his reward, perhaps, has his goods and stock sold by his landlord to pay the exorbitant rent, averaging 8s. or 10s. per acre for such land as I have described.

LANGUAGE, CHARACTER, ETC.

The Welsh farmer is a veritable Welshman. He can speak English but very imperfectly, and has an abhorrence of all Saxon manners and innovations. He is frequently unable to read or write, but can sometimes con over his Welsh Bible, and make out an unintelligible bill; and if in addition he can read a little English and knows the four first rules of arithmetic, he may be considered a well-educated man. The women almost invariably neither read nor write, and can scarcely ever understand two words of English. They fully make up for this, however, by a double share of volubility and animation in the use of their own language, and their shrill clear voices are indications of good health, and are not unpleasant. The choleric disposition usually ascribed to the Welsh is, I think, not quite correct. Words do not often lead to blows, as they take a joke or a satirical expression very good humouredly, and return it very readily. Fighting is much more rarely resorted to than in England, and it is, perhaps, the energy and excitement with which they discuss even common topics of conversation that has given rise to the misconception. They have a ready and peculiar wit, something akin to the Irish, but more frequently expressed so distantly and allegorically as to be unintelligible to one who does not understand their modes of thought and peculiarities of idioms, which latter no less than the former they retain even when they converse in English. They are very proud of their language, on the beauty and expression of which they will sometimes dilate with much animation, concluding with a triumphant assertion that theirs *is* a language, while the English is none, but merely a way of speaking.

The language, though at times guttural, is, when well spoken, both melodious and impressive. There are many changes in the first letters of words, for the sake of euphony, depending on what happens to precede them ; *m* and *b*, for instance, are often changed into *f* (pronounced *v*), as *melin* or *felin*, a mill ; *mel* or *fel*, honey. The gender is often changed in the same manner, as *bach* (masculine), *fach* (feminine), small ; *mawr* (m.), *fawr* (f.), great. The modes of making the plural is to an Englishman rather *singular*, a syllable being taken off instead of being added, as is usually the case with us, as *plentyn*, a child ; *plent*, children : *mochyn*, a pig ; *moch*, pigs. But in other cases a syllable or letter is added.

Their preachers or public speakers have much influence over them. During a discourse there is the most breathless attention, and at the pauses a universal thrill of approbation. Allegory is their chief speciality, and seems to give the hearers the greatest pleasure, and the language appears well fitted for giving it its full effect.

As might be expected from their ignorance, they are exceedingly superstitious, which is rather increased than diminished in those who are able to read by their confining their studies almost wholly to the Bible. The forms their superstitions take are in general much the same as in Scotland, Ireland, and other remote parts of the kingdom. Witches and wizards and white witches, as they are called, are firmly believed in, and their powers much dreaded. There is a witch within a mile of where I am now writing who, according to report, has performed many wonders. One man who had offended her she witched so that he could not rise from his bed for several years, but he was at last cured by inviting the witch to tea and making friends with her. Another case was of a man driving his pig to market when the witch passed by. The pig instantly refused to move, sat up on its hind legs against the hedge in such a manner as no pig was ever seen to do before, and, as it could not be persuaded to walk, was carried home, where it soon died. These and dozens of other similar stories are vouched for by eye-witnesses, one of whom told me this. A still more extraordinary instance of the woman's supernatural powers must be mentioned. She is supposed to have the power of changing herself into different shapes at pleasure, that of a hare seeming to be with her, as with many other witches, the favourite one, as if they delighted in the persecution that harmless animal generally meets with. It is related that one day, being pursued by men and dogs in this shape, the pursuers came to a coal mine the steam-engine of which was in full work, bringing up coal. The witch-hare jumped on to the woodwork which supports the chains, when immediately they refused to move, the engine stopped, pumps, everything remained motionless, and amid the general surprise the witch escaped. But the pit could never be worked again, the pumps and the engine were taken away, and the ruins of the engine house and parts of the other machinery are now pointed out as an undoubted and visible proof of the witch's power.

The witch, being aware of her power over the minds of the people, makes use of it for her own advantage, borrowing her neighbours' horses and farming implements, which they dare not refuse her.

But the most characteristic and general superstition of this part of the country is the "corpse candle." This is seen in various shapes and heard in various sounds ; the normal form, from which it takes its name, being, however, a lighted candle, which is supposed to foretell death, by going from the house in which the person dies along the road where the coffin will be carried to the place of burial. It is only a few of the most hardy and best educated who dare to call in question the reality of this fearful omen, and the evidence in support of it is of such a startling and voluminous character, that did we not remember the trials and burnings and tortures for witchcraft and demonianism, and all the other forms of superstition in England but a few years ago, it would almost overpower our common sense.

I will mention a few cases which have been told me by the persons who were witnesses of them, leaving out the hundreds of more marvellous ones which are everywhere to be heard secondhand.

A respectable woman, in a house where we lodged, assured us that on the evening before one of her children died, she saw a lighted candle moving along about three feet from the ground from the foot of the stairs, across the room towards her, that it came close up to her apron and then vanished, and that it was as distinct and plainly visible as the other candles which were in the room.

Another case is of a collier who, going one morning into the pit before any of the other men were at work, heard the coal waggons coming along, although he knew there could be no one then at work. He stood still at the side of the passage, the waggons came along drawn by horses as usual, a man he knew walking in front and another at the side, and the dead body of one of his fellow workmen was in one of the waggons. In the course of the day he related what he had seen to some of the workmen (one of whom told me the story), declaring his belief that the man whose body he had seen would meet with an accident before long. About a year afterwards the man *was* killed by an accident in the pit. The two men seen were near him, and brought him out in the waggon, and their being obliged to stop at the particular place and every other circumstance happened exactly as had been described. This is as the story was told me by a man who declares he heard the prophecy and saw the fulfilment a year afterwards. When such stories are told and believed, it is, of course, useless arguing against the absurdity of it. They naturally say they must believe their own senses, and they are not sufficiently educated to appreciate any general argument you may put to them. There seems to be no fixed time within which the death should follow the "candle" (as all these appearances are called), and therefore when a person sees or thinks he sees anything at night, he sets it down as corpse candle, and by the time he gets home the fright has enlarged it into something marvellously supernatural, and the first corpse that happens to be carried that way is considered to be the fulfilment of it.

There is a general belief that if the person who meets a candle immediately lies down on his back, he will see the funeral procession with every person that will be present, and the corpse with the candle in his hand. There are many strongly authenticated instances of this. One man, on lying down in this manner, saw that it was himself who carried the candle in his hand. He went home, went to bed, never rose from it, but died in a week. These and numberless other stories of a similar character foster the belief in these uneducated people ; indeed, it is so general that you can hardly meet a person but can tell you of several marvellous things he has seen himself, besides hundreds vouched for by his neighbours.

They have an account of the origin of this warning in the story of an ancient Welsh bishop, who, while being burnt to death by the Catholics, declared that if his religion was true, a candle should precede every death in the Diocese of St. David's, going along the exact road the coffin would be carried. They are very incredulous when you tell them that these corpse candles are in great repute in Radnorshire, which is not in the Diocese of St. David's, and that there are the same appearances under a different name in Ireland.

A celebrated astrologer or conjurer, as he is called in Carmarthenshire, is a living proof of the superstition of the Welsh. This man has printed cards, openly professing to cast nativities, etc., of one of which the following is a literal copy :—

" *Nativities Calculated,*

" In which are given the general transactions of the native through life, viz. Description (without seeing the person), temper, disposition, fortunate or unfortunate in their general pursuits, Honour, Riches, Journeys and Voyages, success therein, and what places best to travel to or reside in ; Friends and Enemies, Trade or Profession best to follow and whether fortunate in speculations, viz. Lottery, dealing in foreign markets, &c., &c., &c.

" Of Marriage, if to marry :—The description, temper and disposition of the person ; from whence, rich or poor, happy or unhappy in marriage, &c., &c., &c. Of children, whether fortunate or not, &c., &c., &c.

" Deduced from the influence of the Sun and Moon with the Planetary Orbs at the time of birth.

" Also judgment and general issue in sickness, disease, &c. By HENRY HARRIES.

" All letters addressed to him or his father, Mr. JOHN HARRIES, Surgeon, Cwrtycadno, must be post paid or will not be received."

He is, however, most generally consulted when money, horses, sheep, etc., are stolen. He then, without inquiring the time of birth or any other particulars, and without consulting the stars, pretends to know who they are and what they come for. He is, however, generally not at home, and his wife then treats them well, and holds them in conversation till he returns, when he immediately gives them some particulars of the neighbourhood they live in, and pretends to describe the person who stole the goods and the house he lives in, etc., and endeavours to frighten the thief by giving out that he will mark him so that everybody shall know him. In some few cases this succeeds, the person, fearful of the great conjurer's power, returns the goods, and the conjurer then gets great credit. In other cases he manages to tell them something which they cannot tell how he became aware of, and then, even if nothing more is heard of the goods, he still keeps up his fame. Two cases have come under my own observation, in which the parties have gone, in one case forty the other sixty miles, to consult this man about some stolen money ; and though in neither case was the desired end obtained, they were told so much about themselves that they felt sure he must have obtained his knowledge by supernatural means. They accordingly spread his name abroad as a wonderful man, who knew a great deal more than other people. The name of his house, " Cwrt y cadno," is very appropriate, as it means in English " The Fox's Court."

Besides these and numberless other instances of almost universal belief in supernatural agency, their superstition as well as their ignorance is further shown by their ascribing to our most harmless reptiles powers of inflicting deadly injury. The toad, newt, lizard, and snake are, they imagine, virulently poisonous, and they look on with horror, and will hardly trust their eyes, should they see them handled with impunity. The barking of dogs at night, hooting of owls, or any unusual noise, dreams, etc., etc., are here, as in many parts of England, regarded as dark omens of our future destiny, mysterious warnings sent to draw aside the veil of futurity and reveal to us, though obscurely, impending danger, disease or death.

Reckoned by the usual standards on these subjects, the religion of the lower orders of Welshmen may be said to be high in the scale, while their morality is decidedly low. This may appear a contradiction to some persons, but those who are at all acquainted with mankind well know that, however luxuriantly religion in its outward forms and influence on the tongue may flourish in an uncultivated soil, it is by no means necessarily accompanied by an equal growth of morality. The former, like the flower of the field, springs spontaneously, or with but little care ; the latter, like the useful grain, only by laborious cultivation and the careful eradication of useless or noxious weeds.

If the number of chapels and prayer-meetings, the constant attendance on them, and the fervour of the congregation can be accounted as signs of religion, it is here. Besides the regular services on the Sabbath and on other days, prayer meetings are held early in the morning and late at night in different cottages by turns, where the uneducated agriculturist or collier breathes forth an extemporary prayer. The Established Church is very rarely well attended. There is not enough of an exciting character or of originality in the service to allure them, and the preacher is too frequently an Englishman who speaks the native tongue, but as a foreigner.

Their preachers, while they should teach their congregation moral duties, boldly decry their vices, and inculcate the commandments and the duty of doing to others as we would they should do unto us, here, as is too frequently the case throughout the kingdom, dwell almost entirely on the mystical doctrine of the atonement—a doctrine certainly not intelligible to persons in a state of complete ignorance, and which, by teaching them that they are not to rely on their own good deeds, has the effect of entirely breaking away the connection between their religion and the duties of their everyday life, and of causing them to imagine that the animal excitement which makes them groan and shriek and leap like madmen in the place of worship, is the true religion which will conduce to their happiness here, and lead them to heavenly joys in a world to come.

Among the youth of both sexes, however, the chapel and prayer meeting is considered more in the light of a " trysting" place than as a place of worship, and this is one reason of the full attendance, especially at the evening services. And as the meetings are necessarily in a thinly populated country, often distant, the journey, generally performed on horseback, affords opportunities for converse not to be neglected.

Thus it will not be wondered at, even by those who affirm the connection between religion and morality, that the latter is, as I said before, at a very low ebb. Cheating of all kinds, when it can be done without being found out, and all the lesser crimes are plentiful enough. The notoriety which Welsh juries and Welsh witnesses have obtained (not unjustly) shows how little they scruple to break their word or their oath. Having to give their evidence through the medium of an interpreter gives them an advantage in court, as the counsel's voice and manner have not so much effect upon them. They are, many of them, very good witnesses as far as sticking firmly to the story they have been instructed in goes, and returning the witticisms of the learned counsel so as often to afford much mirth. To an honest jury a Welsh case is often very puzzling, on account of its being hardly possible to get at a single fact but what is sworn against by an equal number of witnesses on the opposite side ; but to a Welsh jury, who have generally decided on their verdict before the trial commences, it does not present any serious difficulty.

The morals and manners of the females, as might be expected from entire ignorance, are very loose, and perhaps in the majority of cases a child is born before the marriage takes place.

But let us not hide the poor Welshman's virtues while we expose his faults. Many of the latter arise from his desire to defend his fellow countrymen from what he considers unfair or unjust persecution, and many others from what he cannot himself prevent—his ignorance. He is hospitable even to the Saxon. His fire, a jug of milk, and bread and cheese being always at your service. He works hard and lives poorly. He bears misfortune and injury long before he complains. The late Rebecca disturbances, however, show that he may be roused, and his ignorance of other effectual measures should be his excuse for the illegal and forcible means he took to obtain redress—means which, moreover, have been justified by success. It is to be hoped that he will not have again to resort to such outrages as the only way to compel his rulers to do him justice.

A broader system of education is much needed in the Principality. Almost all the schools, it is true, teach the English language, but the child finds the difficulty of acquiring even the first rudiments of education much increased by his being taught them in an unfamiliar tongue of which he has perhaps only picked up a few common-place expressions. In arithmetic, the new language presents a greater difficulty, the method of enumerating being different from their own ; in fact, many Welsh children who have been to school cannot answer a simple question in arithmetic till they have first translated it into Welsh. Unless, therefore, they happen to be thrown among English people or are more than usually well instructed, they get on but little with anything more than speaking English, which those who have been to school generally do very well. Whatever else they have learnt is soon lost for want of practice. It would be very useful to translate some of the more useful elementary works in the different branches of knowledge into Welsh, and sell them as cheaply as possible. The few little Welsh books to be had (and they are very few) are eagerly purchased and read with great pleasure, showing that if the means of acquiring knowledge are offered him, the Welshman will not refuse them.

I will now conclude this brief account of the inhabitants of so interesting a part of our island, a part which will well repay the trouble of a visit, as much for its lovely vales, noble mountains, and foaming cascades, as for the old customs and still older language of the inhabitants of the little white-washed cottages which enliven its sunny vales and barren mountain slopes.

CHAPTER XV

In April, 1843, my father died at Hoddesdon, at the age of seventy-two, and was buried in the family vault in St. Andrew's churchyard, Hertford. As my sister's school was not paying very well, and it was necessary to economize as much as possible, the house was given up early the following year, my mother took an engagement as housekeeper in a gentleman's family at Isleworth, and my sister obtained a post as teacher at an episcopal college, then just founded by the Bishop of Georgia (Dr. Elliott), at Montpelier Springs, seventeen miles from Macon, and left England in August, 1844. In the following year, at the invitation of the parents of some of the pupils, she removed to Robinson, near Montgomery, Alabama, as mistress of a private school much needed in the district; and she remained there till she returned to England in 1846.

Shortly before I came of age in January, 1844, my brother told me that as he had no work in prospect it was necessary that I should leave him and look out for myself; so I determined to go up to London and endeavour to obtain some employment.

As the period of my home and school life and subsequent tutelage under my brother now came to an end, and I had for the future to make my own way in the world, this affords a suitable occasion for a brief review of the chief points in my character, which may now be considered to have been fairly determined, although some portions of it had not yet had opportunity for full development. I do not think that at this time I could be said to have shown special superiority in any of the higher mental faculties, but I possessed a strong desire to know the causes of things, a great love of beauty in form and colour, and a considerable but not excessive desire for order and arrangement in whatever I had to do. If I had one distinct mental faculty more prominent than another, it was the power of correct reasoning from a review of the known facts in any case to the causes or laws which produced them, and also in detecting fallacies in the reasoning of other persons. This power has greatly helped me in all my writings, especially those on natural history and sociology. The determination of the direction in which I should use these powers was due to my possession in a high degree of the two mental qualities usually termed emotional or moral, an intense appreciation of the beauty, harmony and variety in nature and in all natural phenomena, and an equally strong passion for justice as between man and man—an abhorrence of all tyranny, all compulsion, all unnecessary interference with the liberty of others. These characteristics, combined with certain favourable conditions, some of which have already been referred to, have determined the direction of the pursuits and inquiries in which I have spent a large portion of my life.

It will be well to state here certain marked deficiencies in my mental equipment which have also had a share in determining the direction of my special activities. My greatest, though not perhaps most important, defect is my inability to perceive the niceties of melody and harmony in music ; in common language, I have no ear for music. But as I have a fair appreciation of time, expression, and general effect, I am deeply affected by grand, pathetic, or religious music, and can at once tell when the heart and soul of the musician is in his performance, though any number of technical errors, false notes, or disharmonies would pass unnoticed. Another and more serious defect is in verbal memory, which, combined with the inability to reproduce vocal sounds, has rendered the acquirement of all foreign languages very difficult and distasteful. This, with my very imperfect school training, added to my shyness and want of confidence, must have caused me to appear a very dull, ignorant, and uneducated person to numbers of chance acquaintances. This deficiency has also put me at a great disadvantage as a public speaker. I can rarely find the right word or expression to enforce or illustrate my argument, and constantly feel the same difficulty in private conversation. In writing it is not so injurious, for when I have time for deliberate thought I can generally express myself with tolerable clearness and accuracy. I think, too, that the absence of the flow of words which so many writers possess has caused me to avoid that extreme diffuseness and verbosity which is so great a fault in many scientific and philosophical works.

Another important defect is in the power of rapidly seeing analogies or hidden resemblances and incongruities, a deficiency which, in combination with that of language, has produced the total absence of wit or humour, paradox or brilliancy, in my writings, although no one can enjoy and admire these qualities more than I do. The rhythm and pathos, as well as the inimitable puns of Hood, were the delight of my youth, as are the more recondite and fantastic humour of Mark Twain and Lewis Carroll in my old age. The faculty which gives to its possessor wit or humour is also essential to the high mathematician, who is almost always witty or poetical as well; and I was therefore debarred from any hope of success in this direction; while my very limited power of drawing or perception of the intricacies of form were equally antagonistic to much progress as an artist or a geometrician.

Other deficiencies of great influence in my life have been my want of assertiveness and of physical courage, which, combined with delicacy of the nervous system and of bodily constitution, and a general disinclination to much exertion, physical or mental, have caused that shyness, reticence, and love of solitude which, though often misunderstood and leading to unpleasant results, have, perhaps, on the whole, been beneficial to me. They have helped to give me those long periods, both at home and abroad, when, alone and surrounded only by wild nature and uncultured man, I could ponder at leisure on the various matters that interested me. Thus was induced a receptiveness of mind which enabled me at different times to utilize what appeared to me as sudden intuitions— flashes of light leading to a solution of some problem which was then before me ; and these flashes would often come to me when, pen in hand, I was engaged in writing on a subject on which I had no intention or expectation of saying anything new.

There is one other point in which most of my scientific friends and readers will hold that I am deficient, but which in a popular writer on science may be considered to be an advantage. It is, that though fond of order and systematic arrangement of all the parts of a subject, and especially of an argument, I am yet, through my want of the language-faculty, very much disinclined to use technical terms wherever they can be avoided. This is especially the case when a subject is elaborately divided up under various subordinate groups and sub-groups, each with a quite new technical name. This often seems to me more confusing than enlightening, and when other writers introduce different terms of their own, or use them in a somewhat different sense, or still further sub-divide the groups, the complication becomes too great for the non-specialist to follow.

Before leaving the sketch of my mental nature at the threshold of my uncontrolled life, I may properly say a few words on the position I had arrived at in regard to the great question of religious belief. I have already shown that my early home training was in a thoroughly religious but by no means rigid family, where, however, no religious doubts were ever expressed, and where the word "atheist" was used with bated breath as pertaining to a being too debased almost for human society. The only regular teaching I received was to say or hear a formal prayer before going to bed, hearing grace before and after dinner, and learning a collect every Sunday morning, the latter certainly one of the most stupid ways of inculcating religion ever conceived. On Sunday evenings, if we did not go to church or chapel, my father would read some old sermon, and when we did go we were asked on our return what was the text. The only books allowed to be read on Sundays were the " Pilgrim's Progress " or " Paradise Lost," or some religious tracts or moral tales, or the more interesting parts of the Bible were read by my mother, or we read ourselves about Esther and Mordecai or Bel and the Dragon, which were as good as any story book. But all this made little impression upon me, as it never dealt sufficiently with the mystery, the greatness, the ideal and emotional aspects of religion, which only appealed to me occasionally in some of the grander psalms and hymns, or through the words of some preacher more impassioned than usual.

As might have been expected, therefore, what little religious belief I had very quickly vanished under the influence of philosophical or scientific scepticism. This came first upon me when I spent a month or two in London with my brother John, as already related in my sixth chapter ; and during the seven years I lived with my brother William, though the subject of religion was not often mentioned, there was a pervading spirit of scepticism, or free-thought as it was then called, which strengthened and confirmed my doubts as to the truth or value of all ordinary religious teaching.

He occasionally borrowed interesting books which I usually read. One of these was an old edition of Rabelais' works, which both interested and greatly amused me ; but that which bears most upon the present subject was a reprint of lectures on Strauss' "Life of Jesus," which had not then been translated into English. These lectures were, I think, delivered by some Unitarian minister or writer, and they gave an admirable and most interesting summary of the whole work. The now well-known argument, that all the miracles related in the Gospels were mere myths, which in periods of ignorance and credulity always grow up around all great men, and especially around all great moral teachers when the actual witnesses of his career are gone and his disciples begin to write about him, was set forth with great skill. This argument appeared conclusive to my brother and some of his friends with whom he discussed it, and, of course, in my then frame of mind it seemed equally conclusive to me, and helped to complete the destruction of whatever religious beliefs still lingered in my mind. It was not till many years afterwards that I saw reason to doubt this whole argument, and to perceive that it was based upon pure assumptions which were not in accordance with admitted historical facts.

My brother never went to church himself, but for the first few years I was with him he sent me once every Sunday ; but, of course, the only effect of this was to deepen my spirit of scepticism, as I found no attempt in any of the clergymen to reason on any of the fundamental questions at the root of the Christian and every other religion. Many of our acquaintances were either church- or chapel-goers, but usually as a matter of form and convention, and, on the whole, religion seemed to have no influence whatever on their conduct or conversation. The majority, especially of the younger men, were either professors of religion who thought or cared nothing about it, or were open sceptics and scorners.

In addition to these influences my growing taste for various branches of physical science and my increasing love of nature disinclined me more and more for either the observances or the doctrines of orthodox religion, so that by the time I came of age I was absolutely non-religious, I cared and thought nothing about it, and could be best described by the modern term " agnostic."

The next four years of my life were also of great import- ance both in determining the direction of my activity, and in laying the foundation for my study of the special subjects through which I have obtained most admiration or notoriety. This period will be dealt with in another chapter, as it proved to be that which, through a series of what may be termed happy accidents, laid the foundation for everything of import- ance that succeeded them.

CHAPTER XVI

As I came of age in January, 1844, and there was nothing doing at Neath, I left my brother about the middle of December so as to spend the Christmas with my mother and sister at Hoddesdon, after which I returned to London, sharing my brother John's lodging till I could find some employment. At that time the tithe-commutation surveys were nearly all completed, and the rush of railway work had not begun : surveying was consequently very slack. As my brother William, who had a large acquaintance among surveyors and engineers all over the south of England, could not find employment, except some very small local business, I felt it to be quite useless for me to seek for similar work. I therefore determined to try for some post in a school to teach English, surveying, elementary drawing, etc. Through some school agency I heard of two vacancies that might possibly suit. The first required, in addition to English, junior Latin and algebra. Though I had not looked at a Latin book since I left school, I thought I might possibly manage ; and as to algebra, I could do simple equations, and had once been able to do quadratics, and felt sure I could keep ahead of beginners. So with some trepidation I went to interview the master, a rather grave but kindly clergyman. I told him my position, and what I had been doing since I left school. He asked me if I could translate Virgil, at which I hesitated, but told him I had been through most of it at school. So he brought out the book and gave

me a passage to translate, which, of couse, I was quite unable to do properly. Then he set me a simple equation,
which I worked easily. Then a quadratic, at which I stuck.
So he politely remarked that I required a few months' hard
work to be fitted for his school, and wished me good
morning.

My next attempt was more hopeful, as drawing, surveying, and mapping were required. Here, again, I met a
clergyman, but a younger man, and more easy and friendly
in his manner. I had taken with me a small coloured map
I had made at Neath to serve as a specimen, and also one or
two pencil sketches. These seemed to satisfy him, and as I
was only wanted to take the junior classes in English reading, writing, and arithmetic, teach a very few boys surveying, and beginners in drawing, he agreed to engage me. I
was to live in the house, preside over the evening preparation
of the boarders (about twenty in number), and to have, I
think, thirty or forty pounds a year, with which I was quite
satisfied. I was to begin work in about a fortnight. My
employer was the Rev. Abraham Hill, headmaster of the
Collegiate School at Leicester.

I stayed at the school a little more than a year, and
should probably have remained some years longer, and
perhaps even have been a junior school assistant all my
life, but for a quite unexpected event—the death of my
brother William. I was very comfortable at the school,
owing to the kindness of Mr. and Mrs. Hill, and of the
opportunities afforded me for reading, study, and the observation of nature. In my duties I got on fairly well, as the
boys were mostly well-behaved, though, of course, my ignorance and shyness led to some unpleasantness. The first
evening I sat with the boys at their work, one of the older
ones came to me to ask me to explain a difficult passage to
him in some classic—I forget which—evidently to test my
knowledge or ignorance. So I declined even to look at it,
and told him that I taught English only, and that for all
other information they were to go to Mr. Hill himself. On
another occasion the classical assistant master asked me to

take the lowest class in Greek for him, and I was obliged to tell him I did not even know the Greek alphabet. But these little unpleasantnesses once got over did not recur. There were two assistant masters in the school, both pleasant men, but as they did not live in the house I did not see a great deal of them. In drawing, I had only beginners ; but I soon found I had to improve myself, so I sketched a good deal, but could never acquire the freedom of touch of my brother William, and before I left, one of my scholars drew very nearly, if not quite, as well as I did.

I had a very comfortable bedroom, where a fire was lit every afternoon in winter, so that with the exception of one hour with the boys and half an hour at supper with Mr. and Mrs. Hill, my time after four or five in the afternoon was my own. After a few weeks, finding I knew a little Latin, I had to take the very lowest class, and even that required some preparation in the evening. Mr. Hill was a good mathematician, having been a rather high Cambridge wrangler, and finding I was desirous of learning a little more algebra, offered to assist me. He lent me Hind's algebra, which I worked all through successfully, and this was followed by the same author's trigonometry, which I also went through, with occasional struggles. Then I attacked the Differential Calculus, and worked through that; but I could never fully grasp the essential principle of it. Finally, I began the Integral Calculus, and here I found myself at the end of my tether. I learnt some of the simpler processes, but very soon got baffled, and felt that I wanted some faculty necessary for seeing my way through what seemed to me an almost trackless labyrinth. Whether, under Mr. Hill's instruction, I should ultimately have been able to overcome these difficulties I cannot positively say, but I have good reason to believe that I never should have done so. Briefly stated, just as no amount of teaching or practice would ever have made me a good musician, so, however much time and study I gave to the subject, I could never have become a good mathematician. Whether all this work did me any good or not, I am rather doubtful. My after-life being directed to

altogether different studies, I never had occasion to use my
newly acquired knowledge, and soon forgot most of the
processes. But it gave me an interest in mathematics which
I have never lost ; and I rarely come across a mathematical
investigation without looking through it and trying to follow
the reasoning, though I soon get lost in the formulæ. Still,
the ever-growing complexity of the higher mathematics
has a kind of fascination for me as exhibiting powers of the
human mind so very far above my own.

There was in Leicester a very good town library, to which
I had access on paying a small subscription, and as I had
time for several hours' reading daily, I took full advantage of
it. Among the works I read here, which influenced my
future, were Humboldt's " Personal Narrative of Travels in
South America," which was, I think, the first book that gave
me a desire to visit the tropics. I also read here Prescott's
" History of the Conquests of Mexico and Peru," Robertson's
" History of Charles V. " and his " History of America," and
a number of other standard works. But perhaps the most
important book I read was Malthus's " Principles of Population," which I greatly admired for its masterly summary of
facts and logical induction to conclusions. It was the first
work I had yet read treating of any of the problems of philosophical biology, and its main principles remained with me
as a permanent possession, and twenty years later gave me
the long-sought clue to the effective agent in the evolution of
organic species.

It was at Leicester that I was first introduced to a subject
which I had at that time never heard of, but which has played
an important part in my mental growth—psychical research,
as it is now termed. Some time in 1844 Mr. Spencer Hall
gave some lectures on mesmerism illustrated by experiments,
which I, as well as a few of the older boys, attended. I was
greatly interested and astonished at the phenomena exhibited,
in some cases with persons who volunteered from the audience ; and I was also impressed by the manner of the lecturer,

which was not at all that of the showman or the conjurer. At the conclusion of the course he assured us that most persons possessed in some degree the power of mesmerising others, and that by trying with a few of our younger friends or acquaintances, and simply doing what we had seen him do, we should probably succeed. He also showed us how to distinguish between the genuine mesmeric trance, and any attempt to imitate it.

In consequence of this statement, one or two of the elder boys tried to mesmerise some of the younger ones, and in a short time succeeded ; and they asked me to see their experiments. I found that they could produce the trance state, which had all the appearance of being genuine, and also a cataleptic rigidity of the limbs by passes and by suggestion, both in the trance and afterwards in the normal waking state. This led me to try myself in the privacy of my own room, and I succeeded after one or two attempts in mesmerising three boys from twelve to sixteen years of age, while on others within the same ages I could produce no effect, or an exceedingly slight one. During the trance they seemed in a state of semi-torpor, with apparently no volition. They would remain perfectly quiescent so long as I did not notice them, but would at once answer any questions or do anything I told them. On the two boys with whom I continued to experiment for some time, I could produce catalepsy of any limb or of the whole body, and in this state they could do things which they could not, and certainly would not have done in their normal state. For example, on the rigid out-stretched arm I would hang an ordinary chair at the wrist, and the boy would hold it there for several minutes, while I sat down and wrote a short letter for instance, without any complaint, or making any remark when I took it off. I never left it more than five minutes because I was afraid that some injury might be caused by it. I soon found that this rigidity could be produced in those who had been mesmerised by suggestion only, and in this way often fixed them in any position, notwithstanding their efforts to change it. One experiment was to place a shilling on the table in front of a

boy, and then say to him, "Now, you can't touch that shilling."
He would at once move his hand towards it, but when halfway it would seem to stick fast, and all his efforts could not
bring it nearer, though he was promised the shilling if he
could take it.

Every phenomenon of suggestion I had seen at the lecture,
and many others, I could produce with this boy. Giving him
a glass of water and telling him it was wine or brandy, he
would drink it, and soon show all the signs of intoxication,
while if I told him his shirt was on fire he would instantly
strip himself naked to get it off. I also found that he had
community of sensation with myself when in the trance. If
I held his hand he tasted whatever I put in my mouth, and
the same thing occurred if one or two persons intervened
between him and myself; and if another person put substances at random into my mouth, or pinched or pricked me
in various parts of the body, however secretly, he instantly
felt the same sensation, would describe it, and put his hand
to the spot where he felt the pain.

In like manner any sense could be temporarily paralyzed
so that a light could be flashed on his eyes or a pistol fired
behind his head without his showing the slightest sign of
having seen or heard anything. More curious still was the
taking away the memory so completely that he could not tell
his own name, and would adopt any name that was suggested
to him, and perhaps remark how stupid he was to have forgotten it; and this might be repeated several times with
different names, all of which he would implicitly accept.
Then, on saying to him, "Now you remember your own
name again; what is it?" an inimitable look of relief would
pass over his countenance, and he would say, 'Why, P——,
of course," in a way that carried complete conviction.

But perhaps the most interesting group of phenomena to
me were those termed phreno-mesmerism. I had read, when
with my brother, George Combe's "Constitution of Man," with
which I had been greatly interested, and afterwards one of
the writer's works on Phrenology, and at the lecture I had

seen some of the effects of exciting the phrenological organs
by touching the corresponding parts of the patient's head.
But as I had no book containing a chart of the organs, I
bought a small phrenological bust to help me in determining
the positions.

Having my patient in the trance, and standing close to
him, with the bust on my table behind him, I touched succes-
sively several of the organs, the position of which it was easy
to determine. After a few seconds he would change his
attitude and the expression of his face in correspondence
with the organ excited. In most cases the effect was unmistakable, and superior to that which the most finished actor could
give to a character exhibiting the same passion or emotion.

At this very time the excitement caused by painless
surgical operations during the mesmeric trance was at its full
height, as I have described it in my "Wonderful Century"
(chapter xxi.), and I had read a good deal about these, and
also about the supposed excitement of the phrenological
organs, and the theory that these latter were caused by
mental suggestion from the operator to the patient, or what
is now termed telepathy. But as the manifestations often
occurred in a different form from what I expected, I felt
sure that this theory was not correct. One day I intended
to touch a particular organ, and the effect on the patient
was quite different from what I expected, and looking
at the bust while my finger was still on the boy's head, I
found that I was not touching the part I supposed, but an
adjacent part, and that the effect exactly corresponded to the
organ touched and not to the organ I *thought* I had touched,
completely disproving the theory of suggestion. I then tried
several experiments by looking away from the boy's head
while I put my finger on it at random, when I always found
that the effect produced corresponded to that indicated by
the bust. I thus established, to my own satisfaction, the
fact that a real effect was produced on the actions and speech
of a mesmeric patient by the operator touching various parts
of the head; that the effect corresponded with the natural
expression of the emotion due to the phrenological organ

situated at that part—as combativeness, acquisitiveness, fear, veneration, wonder, tune, and many others ; and that it was in no way caused by the will or suggestion of the operator.

As soon as I found that these experiments were successful I informed Mr. Hill, who made no objection to my continuing them, and several times came to see them. He was so much interested that one evening he invited two or three friends who were interested in the subject, and with my best patient I showed most of the phenomena. At the suggestion of one of the visitors I told the boy he was a jockey, and was to get on his horse and be sure to win the race. Without another word from me he went through the motions of getting on horseback, of riding at a gallop, and after a minute or two he got excited, spoke to his horse, appeared to use his spurs, shake the reins, then suddenly remain quiet, as if he had passed the winning-post ; and the gentleman who had suggested the experiment declared that his whole motions, expressions, and attitudes were those of a jockey riding a race. At that time I myself had never seen a race. The importance of these experiments to me was that they convinced me, once for all, that the antecedently incredible may nevertheless be true ; and, further, that the accusations of imposture by scientific men should have no weight whatever against the detailed observations and statements of other men, presumably as sane and sensible as their opponents, who had witnessed and tested the phenomena, as I had done myself in the case of some of them. At that time lectures on this subject were frequent, and during the holidays, which I generally spent in London with my brother, we took every opportunity of attending these lectures and witnessing as many experiments as possible. Knowing by my own experience that it is quite unnecessary to resort to trickery to produce the phenomena, I was relieved from that haunting idea of imposture which possesses most people who first see them, and which seems to blind most medical and scientific men to such an extent as to render them unable to investigate the subject fairly, or to arrive at any trustworthy conclusions in regard to it.

How I was introduced to Henry Walter Bates I do not exactly remember, but I rather think I heard him mentioned as an enthusiastic entomologist, and met him at the library. I found that his specialty was beetle collecting, though he also had a good set of British butterflies. Of the former I had scarcely heard, but as I already knew the fascinations of plant life I was quite prepared to take an interest in any other department of nature. He asked me to see his collection, and I was amazed to find the great number and variety of beetles, their many strange forms and often beautiful markings or colouring, and was even more surprised when I found that almost all I saw had been collected around Leicester, and that there were still many more to be discovered. If I had been asked before how many different kinds of beetles were to be found in any small district near a town, I should probably have guessed fifty or at the outside a hundred, and thought that a very liberal allowance. But I now learnt that many hundreds could easily be collected, and that there were probably a thousand different kind within ten miles of the town. He also showed me a thick volume containing descriptions of more than three thousand species inhabiting the British Isles. I also learnt from him in what an infinite variety of places beetles may be found, while some may be collected all the year round, so I at once determined to begin collecting, as I did not find a great many new plants about Leicester. I therefore obtained a collecting bottle, pins, and a store-box; and in order to learn their names and classification I obtained, at wholesale price through Mr. Hill's bookseller, Stephen's " Manual of British Coleoptera," which henceforth for some years gave me almost as much pleasure as Lindley's Botany, with my MSS. descriptions, had already done.

This new pursuit gave a fresh interest to my Wednesday and Saturday afternoon walks into the country, when two or three of the boys often accompanied me. The most delight-ful of all our walks was to Bradgate Park, about five miles from the town, a wild, neglected park with the ruins of a mansion, and many fine trees and woods and ferny or bushy slopes. Sometimes the whole school went for a picnic, the park at that time being quite open, and we hardly ever met any one. After we got out of the town there was a wide grassy lane that led to it, which itself was a delightful walk and was a good collecting ground for both plants and insects. For variety we had the meadows along the course of the little river Soar, which were very pleasant in spring and summer. Twice during the summer the whole of the boarders were taken for a long day's excursion. The first time we went to Kenilworth Castle, about thirty miles distant, driving in coaches by pleasant country roads, and passing through Coventry. Towards the autumn we had a much longer excursion, partly by coach and partly by canal boat, to a very picturesque country with wooded hills and limestone cliffs, rural villages, and an isolated hill, from the top of which we had a very fine and extensive view. I think it must have been in Derbyshire, near Wirksworth, as there is a long canal tunnel on the way there. One of the rough out-of-door sketches made on this occasion is reproduced here on a reduced scale, as well as a more finished drawing of some village, perhaps near Leicester, as they may possibly enable some reader to recognize the localities, and also serve to show the limits of my power as an artist.

At midsummer there was the usual prize-giving, accompanied by recitation; and to introduce a little variety I wrote a prologue, in somewhat boyish style, to be spoken by a chubby boy about twelve years old; and it took me a good deal of trouble to drill him into appropriate emphasis and action. It went off very well, and as it was to some extent a programme as well as a prologue, I give here as much of it as I can recollect.

PROLOGUE.

With Greek and Latin, French, and other stuff,
And Euclid too, and Algebra enough,
For this half-year I'm glad to say we've done,
And the long looked-for hour at length is come,
That brings before us this superb array
Of company to grace our holiday.

IN DERBYSHIRE.

(*From pencil sketch by A. R. Wallace.* 1844.)

[*To face p.* 238, VOL I.

A VILLAGE NEAR LEICESTER.

(*Pencil drawing by A. R. Wallace.* 1844.)

[*To face p.* 238, VOL. I.

> We bid you welcome ! and hope each may find
> Something we've chosen suited to his mind ;
> Our bill of fare contains some curious dishes
> To satisfy your various tastes and wishes.
> And first, to show our classic lore, we'll speak
> What Sophocles composed in sounding Greek,
> Repeat the words his olden heroes said,
> And from their graves call back the mighty dead.
> Then in Rome's Senate we will bid you stand,
> The Conscript Fathers ranged on either hand,
> When Cicero th' expectant silence broke,
> And cruel Verres trembled while he spoke.
> In modern Rome's soft language we'll rehearse
> Immortal Tasso's never-dying verse :
> In German we've a name you all know well,
> The brave, the free, the patriot, William Tell ;
> And then, for fear all this dry stuff they'll tire on,
> To please the ladies we've a piece from Byron.
> Next, we've the one-legged goose—that *rara avis*,
> Whose history will be told by Master Davis,
> And Monsieur Tonson's griefs we're sure will call
> A little hearty laughter from you all.

With a few concluding lines which I cannot remember.

Just before the Christmas holidays (or perhaps on the fifth of November) I wrote a slight serio-comic play, the subject being "Guy Faux." While following history pretty closely as to the chief characters and events, I purposely introduced a number of anachronisms, as umbrellas, macintoshes, lucifer matches, half-farthings then just issued. I also made use of some modern slang, and concluded with a somewhat mock-heroic speech by the judge when sentencing the criminal. The boys acted their parts very well, and the performance was quite a success.

Early in the following year (February, 1846) I received the totally unexpected news of the death of my brother William at Neath. He had been in London to give evidence before a committee on the South Wales Railway Bill, and returning at night caught a severe cold by being chilled in a wretched third-class carriage, succeeded by a damp bed at Bristol. This brought on congestion of the lungs, to which he speedily succumbed. I and my brother John went down to Neath to the funeral, and as William had died without a will, we had to take out letters of administration. Finding from my brother's papers that he had obtained a small local business, and that there was railway work in prospect, I determined to take his place, and at once asked permission of Mr. Hill to be allowed to leave at Easter.

My year spent at Leicester had been in many ways useful to me, and had also a determining influence on my whole future life. It satisfied me that I had no vocation for teaching, for though I performed my duties I believe quite to Mr. Hill's satisfaction, I felt myself out of place, partly because I knew no subject—with the one exception of surveying—sufficiently well to be able to teach it properly, but mainly because a completely subordinate position was distasteful to me, although I could not have had a more considerate employer than Mr. Hill. The time and opportunity I had for reading was a great advantage to me, and gave me an enduring love of good literature. I also had the opportunity of hearing almost every Sunday one of the most impressive and eloquent preachers I have ever met with—Dr. John Brown, I think, was his name. He was one of the few Church of England clergymen who preached extempore, and he did it admirably so that it was a continual pleasure to listen to him. But I was too firmly convinced of the incredibility of large portions of the Bible, and of the absence of sense or reason in many of the doctrines of orthodox religion to be influenced by any such preaching, however eloquent. My return to some form of religious belief was to come much later, and from a quite different source.

But, as already stated, the events which formed a turning-point in my life were, first, my acquaintance with Bates, and through him deriving a taste for the wonders of insect-life, opening to me a new aspect of nature, and later on finding in him a companion without whom I might never have ventured on my journey to the Amazon. The other and equally important circumstance was my reading Malthus, without which work I should probably not have hit upon the theory of natural selection and obtained full credit for its independent discovery. My year spent at Leicester must, therefore, be considered as perhaps the most important in my early life.

CHAPTER XVII

RESIDENCE AT NEATH

AT Easter I bade farewell to Leicester and went to Neath with my brother John, in order to wind up our brother William's affairs. We found from his books that a considerable amount was owing to him for work done during the past year or two, and we duly made out accounts of all these and sent them in to the respective parties. Some were paid at once, others we had to write again for and had some trouble to get paid. Others, again, were disputed as being an extravagant charge for the work done, and we had to put them in a lawyer's hands to get settled. One gentleman, whose account was a few pounds, declared he had paid it, and asked us to call on him. We did so, and, instead of producing the receipt as we expected, he was jocose about it, asked us what kind of business men we were to want him to pay twice ; and when we explained that it was not shown so in my brother's books, and asked to look at the receipt, he coolly replied, " Oh, I never keep receipts ; never kept a receipt in my life, and never was asked to pay a bill twice till now ! " In vain we urged that we were bound as trustees for the rest of the family to collect all debts shown by my brother's books to be due to him, and that if he did not pay it, we should have to lose the amount ourselves. He still maintained that he had paid it, that he remembered it distinctly, and that he was not going to pay it twice. At last we were obliged to tell him that if he did not pay it we *must* put it in the hands of a lawyer to take what steps he thought necessary: then he gave way, and said, " Oh, if you are going to law about such a trifle, I suppose I must pay it again!" and, counting out the money, added, " There it is; but I paid it before, so give me a receipt *this* time," apparently considering himself a very injured man.   This little experience annoyed me much, and, with others of the same nature later on, so disgusted me with business as to form one of the reasons which induced me to go abroad.

When we had wound up William's affairs as well as we could, my brother John returned to London, and I was left to see if any work was to be had, and in the mean time devoted myself to collecting butterflies and beetles.   While at Leicester I had been altogether out of the business world, and do not remember even looking at a newspaper, or I might have heard something of the great railway mania which that year reached its culmination.   I now first heard rumours of it, and some one told me of a civil engineer in Swansea who wanted all the surveyors he could get, and that they all had two guineas a day, and often more.   This I could hardly credit, but I wrote to the gentleman, who soon after called on me, and asked me if I could do levelling.   I told him I could, and had a very good level and levelling staves.   After some little conversation he told me he wanted a line of levels up the Vale of Neath to Merthyr Tydfil for a proposed railway, with cross levels at frequent intervals, and that he would give me two guineas a day, and all expenses of chain and staff men, hotels, etc.   He gave me all necessary instructions, and said he would send a surveyor to map the route at the same time.   This was, I think, about mid-summer, and I was hard at work till the autumn, and enjoyed myself immensely.   It took me up the south-east side of the valley, of which I knew very little, along pleasant lanes and paths through woods and by streams, and up one of the wildest and most picturesque little glens I have ever ex-plored.   Here we had to climb over huge rocks as big as houses, ascend cascades, and take cross-levels up steep banks and precipices all densely wooded.   It was surveying under difficulties, and excessively interesting.   After the first rough

levels were taken and the survey made, the engineers were
able to mark out the line provisionally, and I then went over
the actual line to enable the sections to be drawn as required
by the Parliamentary Standing Orders.

In the autumn I had to go to London to help finish the
plans and reference books for Parliament. There were about
a dozen surveyors, draughtsmen, and clerks in a big hotel in
the Haymarket, where we had a large room upstairs for work,
and each of us ordered what we pleased for our meals in the
coffee-room. Towards the end of November we had to work
very late, often till past midnight, and for the last few days of
the month we literally worked all night to get everything
completed.

In this year of wild speculation it is said that plans and
sections for 1263 new railways were duly deposited, having a
proposed capital of £563,000,000, and the sum required to be
deposited at the Board of Trade was so much larger than the
total amount of gold in the Bank of England and notes in
circulation at the time, that the public got frightened, a panic
ensued, shares in the new lines which had been at a high
premium fell almost to nothing, and even the established lines
were greatly depreciated. Many of the lines were proposed
merely for speculation, or to be bought off by opposing lines
which had a better chance of success. The line we were at
work on was a branch of the Great Western and South Wales
Railway then making, and was for the purpose of bringing the
coal and iron of Merthyr Tydfil and the surrounding district
to Swansea, then the chief port of South Wales. But we had
a competitor along the whole of our route in a great line
from Swansea to Yarmouth, by way of Merthyr, Hereford,
Worcester, and across the midland agricultural counties,
called, I think, the East and West Junction Railway, which
sounded grand, but which had no chance of passing. It
competed, however, with several other lines, and I heard that
many of these agreed to make up a sum to buy off its opposition. Not one-tenth of the lines proposed that year were
ever made, and the money wasted upon surveyors, engineers,
and law expenses must have amounted to millions.

Finding it rather dull at Neath living by myself, I persuaded my brother to give up his work in London as a journeyman carpenter and join me, thinking that, with his practical experience and my general knowledge, we might be able to do architectural, building, and engineering work, as well as surveying, and in time get up a profitable business. We returned together early in January, and continued to board and lodge with Mr. Sims in the main street, where I had been very comfortable, till the autumn, when, hearing that my sister would probably be home from America the following summer, and my mother wishing to live with us, we took a small cottage close to Llantwit Church, and less than a mile from the middle of the town. It had a nice little garden and yard, with fowl-house, shed, etc., going down to the Neath Canal, immediately beyond which was the river Neath, with a pretty view across the valley to Cadoxton and the fine Drumau mountain.

Having the canal close at hand and the river beyond, and then another canal to Swansea, made us long for a small boat, and not having much to do, my brother determined to build one, so light that it could be easily drawn or carried from the canal to the river, and so give access to Swansea. It was made as small and light as possible to carry two or, at most, three persons. When finished, we tried it with much anxiety, and found it rather unstable, but with a little ballast at the bottom and care in moving, it did very well, and was very easy to row. One day I persuaded my mother to let me row her to Swansea, where we made a few purchases ; and then came back quite safely till within about a mile of home, when, passing under a bridge, my mother put her hand out to keep the boat from touching, and leaning over a little too much, the side went under water, and upset us both. As the water was only about two or three feet deep we escaped with a thorough wetting. The boat was soon bailed dry, and then I rowed on to Neath Bridge, where my mother got out and walked home, and did not trust herself in our boat again, though I and my brother had many pleasant excursions.

Our chief work in 1846 was the survey of the parish of
Llantwit-juxta-Neath, in which we lived. The agent of the
Gnoll Estate had undertaken the valuation for the tithe commutation, and arranged with me to do the survey and make
the map and the necessary copies. When all was finished and
the valuation made, I was told that I must collect the payment from the various farmers in the parish, who would afterwards deduct it from their rent. This was a disagreeable
business, as many of the farmers were very poor ; some could
not speak English, and could not be made to understand what
it was all about ; others positively refused to pay ; and the
separate amounts were often so small that it was not worth
going to law about them, so that several were never paid at
all, and others not for a year afterwards. This was another
of the things that disgusted me with business, and made me
more than ever disposed to give it all up if I could but get
anything else to do.

We also had a little building and architectural work. A
lady wanted us to design a cottage for her, with six or seven
rooms, I think, for £200. Building with the native stone
was cheap in the country, but still, what she wanted was
impossible, and at last she agreed to go to £250, and with
some difficulty we managed to get one built for her for this
amount. We also sent in a design for a new Town Hall
for Swansea, which was beyond our powers, both of design
and draughtsmanship ; and as there were several established
architects among the competitors, our very plain building and
poor drawings had no chance. But shortly afterwards a
building was required at Neath for a Mechanics' Institute,
for which £600 was available. It was to be in a narrow side
street, and to consist of two rooms only, a reading room and
library below, and a room above for classes and lectures. We
were asked to draw the plans and supervise the execution,
which we did, and I think the total cost did not exceed the
sum named by more than £50. It was, of course, very plain,
but the whole was of local stone, with door and window-
quoins, cornice, etc., hammer-dressed ; and the pediments over
the door and windows, arched doorway, and base of squared

blocks gave the whole a decidedly architectural appearance. It is now used as a free library, and through the kindness of Miss Florence Neale, of Penarth, I am enabled to give a photographic reproduction of it.

This reminds me that the Mechanics' Institution was, I think, established by Mr. William Jevons, a retired merchant or manufacturer of Liverpool, and the uncle of William Stanley Jevons, the well-known writer on Logic and Political Economy. Mr. Jevons was the author of a work on "Systematic Morality," very systematic and very correct, but as dry as its title. He had a good library, and was supposed in Neath to be a man of almost universal knowledge. I think my brother William had become acquainted with him after I left Neath, as he attended the funeral, and I and John spent the evening with him. When I came to live in Neath after my brother's death, I often saw him and occasionally visited him, and I think borrowed books, and the following winter, finding I was interested in science generally, he asked me to give some familiar lectures or lessons to the mechanics of Neath, who then met, I think, in one of the schoolrooms. I was quite afraid of undertaking this, and tried all I could to escape, but Mr. Jevons was very persistent, assured me that they knew actually nothing of science, and that the very simplest things, with a few diagrams and experiments, would be sure to interest them. At last I reluctantly consented, and began with very short and simple talks on the facts and laws of mechanics, the principle of the lever, pulley, screw, etc., falling bodies and projectiles, the pendulum, etc.

I got on fairly well at first, but on the second or third occasion I was trying to explain something which required a rather complex argument which I thought I knew perfectly, when, in the middle of it, I seemed to lose myself and could not think of the next step. After a minute's dead silence, Mr. Jevons, who sat by me, said gently—"Never mind that now. Go on to the next subject." I did so, and after a few minutes, what I had forgotten became clear to me, and I returned to it, and went over it with success. I gave these lessons for two winters, going through the

FREE LIBRARY, NEATH.
(*Designed by A. R. Wallace.* 1847.)

[*To face p.* 246, VOL. I.

elementary portions of physics; and after a week in Paris in 1847, I gave to the same audience a general account of the city, with special reference to its architecture, museums, and gardens, showing that it was often true that "they did these things better in France."[1]

There was also in Neath a Philosophical Society with a small library and reading room, in connection with which occasional lectures were given. Sir G. B. Airy, the Astronomer Royal, gave a lecture there on the return of Halley's Comet shortly before we came to Neath. He recommended them to purchase a good telescope of moderate size and have it properly mounted, so as to be able to observe all the more remarkable astronomical phenomena. A telescope was actually obtained with, I think, a four- or five- inch object glass, and as there was no good position for it available, a kind of square tower was built attached to the library, high enough to obtain a clear view, on the top of which it was proposed to use the telescope. But the funds for a proper mounting and observatory roof not being forthcoming, the telescope was hardly ever used, owing to the time and trouble always required to carry upstairs and prepare for observation any astronomical telescope above the very smallest size.

During the two summers that I and my brother John lived at Neath we spent a good deal of our leisure time in wandering about this beautiful district, on my part in search of insects, while my brother always had his eyes open for any uncommon bird or reptile. One day when I was insect hunting on Crymlyn Burrows, a stretch of very interesting sandhills, rock, and bog near the sea, and very rich in curious plants, he came upon several young vipers basking on a rock.

___

[1] In 1895 I received a letter from Cardiff, from one of the workmen who attended the Neath Mechanics' Institution, asking if the author of " Island Life," the " Malay Archipelago," and other books is the same Mr. Alfred Wallace who taught in the evening science classes to the Neath Abbey artificers. He writes— "I have often had a desire to know, as I benefited more while in your class—if you are the same Mr. A. Wallace—than I ever was taught at school. I have often wished I knew how to thank you for the good I and others received from your teaching.—(Signed) MATTHEW JONES."

They were about eight or nine inches long. As they were quite still, he thought he could catch one by the neck, and endeavoured to do so, but the little creature turned round suddenly, bit his finger, and escaped. He immediately sucked out the poison, but his whole hand swelled considerably, and was very painful. Owing, however, to the small size of the animal the swelling soon passed off, and left no bad effects. Another day, towards the autumn, we found the rather uncommon black viper in a wood a few miles from Neath. This he caught with a forked stick, to which he then tied it firmly by the neck, and put it in his coat pocket. Meeting a labourer on the way, he pulled it out of his pocket, wriggling and twisting round the stick and his hand, and asked the man if he knew what it was, holding it towards him. The man's alarm was ludicrous. Of course, he declared it to be deadly, and for once was right, and he added that he would not carry such a thing in his pocket for anything we could give him.

Though I have by no means a very wide acquaintance with the mountain districts of Britain, yet I know Wales pretty well; have visited the best parts of the lake district; in Scotland have been to Loch Lomond, Loch Katrine, and Loch Tay; have climbed Ben Lawers, and roamed through Glen Clova in search for rare plants;—but I cannot call to mind a single valley that in the same extent of country comprises so much beautiful and picturesque scenery, and so many interesting special features, as the Vale of Neath. The town itself is beautifully situated, with the fine wooded and rock-girt Drumau Mountain to the west, while immediately to the east are well-wooded heights crowned by Gnoll House, and to the south-east, three miles away, a high rounded hill, up which a chimney has been carried from the Cwm Avon copper-works in the valley beyond, the smoke from which gives the hill much the appearance of an active volcano. To the south-west the view extends down the valley to Swansea Bay, while to the north-east stretches the Vale of Neath itself, nearly straight for twelve miles, the river winding in a level fertile valley about a quarter to half a mile wide, bounded on each side by abrupt hills, whose lower slopes are finely wooded,

YSGWD GLADYS, VALE OF NEATH.

[*To face p.* 249, VOL. I.

and backed by mountains from fifteen hundred to eighteen hundred feet high. The view up this valley is delightful, its sides being varied with a few houses peeping out from the woods, abundance of lateral valleys and ravines, with here and there the glint of falling water, while its generally straight direction affords fine perspective effects, sometimes fading in the distance into a warm yellow haze, at others affording a view of the distant mountain ranges beyond.

At twelve miles from the town we come to the little village of Pont-nedd-fychan (the bridge of the little Neath river), where we enter upon a quite distinct type of scenery, dependent on our passing out of the South Wales coal basin, crossing the hard rock-belt of the millstone grit, succeeded by the picturesque crags of the mountain limestone, and then entering on the extensive formation of the Old Red Sandstone. The river here divides first into two, and a little further on into four branches, each in a deep ravine with wooded slopes or precipices, above which is an undulating hilly and rocky country backed by the range of the great forest of Brecon, with its series of isolated summits or vans, more than two thousand feet high, and culminating in the remarkable twin summits of the Brecknock Beacons, which reach over twenty-nine hundred feet. Within a four-mile walk of Pont-nedd-fychan there are six or eight picturesque waterfalls or cascades, one of the most interesting, named Ysgwd Gladys, being a miniature of Niagara, inasmuch as it falls over an overhanging rock, so that it is easy to walk across behind it. A photograph of this fall is given here. Another, Ysgwd Einon Gam, is much higher, while five miles to the west, near Capel Coelbren, is one of the finest waterfalls in Wales, being surpassed only, so far as I know, by the celebrated falls above Llanrhaiadr in the Berwyn Mountains. From the open moor it drops suddenly about ninety feet into a deep ravine, with vertical precipices wooded at the top all round. In summer the stream is small, but after heavy rains it must be a very fine sight, as it falls unbroken into a deep pool below, and then flows away down a thickly wooded glen to the river Tawe.

Within a mile of Pont-nedd-fychan is the Dinas rock, a tongue of mountain limestone jutting out across the mill-stone grit, and forming fine precipices, one of which was called the Bwa-maen or bow rock, from its being apparently bent double.   Lower down there are also some curious waving lines of apparent stratification, but on a recent examination I am inclined to think that these are really glacial groovings caused by the ice coming down from Hirwain, right against these ravines and precipices, and being thus heaped up and obliged to flow away at right angles to its former course.

But the most remarkable and interesting of the natural phenomena of the upper valley is Porth-yr-Ogof (the gate-way of the cavern), where the river Mellte runs for a quarter of a mile underground.   The entrance is under a fine arch of limestone rock overhung with trees, as shown in the accompanying photograph.   The outlet is more irregular and less lofty, and is also less easily accessible ; but the valley just below has wooded banks, open glades, and fantastic rocks near the cave, forming one of the most charmingly picturesque spots imaginable.   It is also very interesting to walk over the underground river along a hollow strewn with masses of rock, and with here and there irregular funnels, where the water can be heard and in one place seen.   The whole place is very instructive, as showing us how many of the narrow limestone gorges, bounded by irregular perpendicular rocks with no sign of water-wear, have been formed.   Caves abound in all limestone regions, owing to the dissolving power of rain-water penetrating the fissures of the rock, and finding outlets often at a distance of many miles and then gushing forth in a copious spring.   Where a range of such caverns lies along an ancient valley, and are not very far below the surface, they in time fall in, and, partially blocking up the drainage, cause the caverns to be filled up and still further enlarged.   In time the fallen portion is dissolved and worn away, other portions fall in, and in course of ages an open valley is formed, bounded by precipices with fractured surfaces, and giving the idea of their being rent open by some tremendous convulsion of nature—a favourite expression of the old geologists.

PORTH-YR-OGOF, VALE OF NEATH.

[*To face p.* 250, VOL. I.

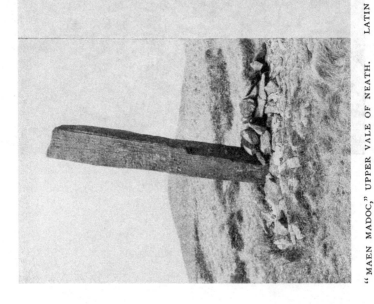

"MAEN MADOC," UPPER VALE OF NEATH.　　LATIN INSCRIPTION ON "MAEN MADOC."

I have already (in chap. xi.) described one of the curious "standing stones" near the source of the Llia river, but there is a still more interesting example about a mile and a half north-west of Ystrad-fellte, where the old Roman road—the Sarn Helen—crosses over the ridge between the Nedd and the Llia valleys. This is a tall, narrow stone, roughly quadrilateral, on one of the faces of which there is a rudely inscribed Latin inscription, as seen in the photograph, and in copy of the letters given opposite. It reads as follows :—

DERVACI FILIUS JUSTI IC IACIT

meaning [The body] of Dervacus the son of Justus lies here. It will be seen that the letters D, A, and I in Dervaci, and the T and I in Justi are inverted or reversed, probably indicating that the cutting was done by an illiterate workman, who placed them as most convenient when working on an erect stone. The stone itself is probably British, and was utilized as a memorial of a Roman soldier who died near the place.

One of our most memorable excursions was in June, 1846, when I and my brother spent the night in this water-cave. I wanted to go again to the top of the Beacons to see if I could find any rare beetles there, and also to show my brother the waterfalls and other beauties of the upper valley. Starting after an early breakfast we walked to Pont-nedd-fychan, and then turned up the western branch to the Rocking Stone, a large boulder of millstone-grit resting on a nearly level surface, but which by a succession of pushes with one hand can be made to rock considerably. It was here I obtained one of the most beautiful British beetles, *Trichius fasciatus*, the only time I ever captured it. We then went on to the Gladys and Einon Gam falls; then, turning back followed up the river Nedd for some miles, crossed over to the cavern, and then on to Ystrad-fellte, where we had supper and spent the night, having walked leisurely about eighteen or twenty miles.

The next morning early we proceeded up the valley to the highest farm on the Dringarth, then struck across the mountain to the road from Hirwain to Brecon, which we

followed to the bridge over the Taff, and then turned off
towards the Beacons, the weather being perfect. It was a
delightful walk, on a gradual slope of fifteen hundred feet in
a mile and a half, with a little steeper bit at the end, and the
small overhanging cap of peat at the summit, as already
described in chapter xi. I searched over it for beetles,
which were, however, very scarce, and we then walked along
the ridge to the second and higher triangular summit, peeped
with nervous dread on my part over the almost perpendicular
precipice towards Brecon, noted the exact correspondence
in slope of the two peat summits, and then back to the ridge
and a little way down the southern slope to where a tiny spring
trickles out—the highest source of the river Taff—and there,
lying on the soft mountain turf, enjoyed our lunch and the
distant view over valley and mountain to the faint haze
of the British Channel. We then returned to the western
summit, took a final view of the grand panorama around us,
and bade farewell to the beautiful mountain, the summit of
which neither of us visited again, though I have since been
very near it. We took nearly the same route back, had a
substantial tea at the little inn at Ystrad-fellte, and then,
about seven o'clock, walked down to the cave to prepare our
quarters for the night. I think we had both of us at this
time determined, if possible, to go abroad into more or less
wild countries, and we wanted for once to try sleeping
out-of-doors, with no shelter or bed but what nature provided.

Just inside the entrance of the cave there are slopes of
water-worn rock and quantities of large pebbles and boulders,
and here it was quite dry, while farther in, where there were
patches of smaller stones and sand, it was much colder and
quite damp, so our choice of a bed was limited to rock or
boulders. We first chose a place for a fire, and then searched
for sufficient dead or dry wood to last us the night. This
took us a good while, and it was getting dusk before we lit
our fire. We then sat down, enjoying the flicker of the flame
on the roof of the cavern, the glimmer of the stars through
the trees outside, and the gentle murmur of the little river
beside us. After a scanty supper we tried to find a place

where we could sleep with the minimum of discomfort, but with very little success. We had only our usual thin summer clothing, and had nothing whatever with us but each a small satchel, which served as a pillow. As the cave faces north the rocky floor had not been warmed by the sun, and struck cold through our thin clothing, and we turned about in vain for places where we could fit ourselves into hollows without feeling the harsh contact of our bones with the rock or pebbles. I found it almost impossible to lie still for half an hour without seeking a more comfortable position, but the change brought little relief. Being midsummer, there were no dead leaves to be had, and we had no tool with which to cut sufficient branches to make a bed. But I think we had determined purposely to make no preparation, but to camp out just as if we had come accidentally to the place in an unknown country, and had been compelled to sleep there. But very little sleep was to be had, and while in health I have never passed a more uncomfortable night. Luckily it was not a long one, and before sunrise we left our gloomy bedroom, walked up to the main road to get into the sunshine, descended into the Nedd valley and strolled along, enjoying the fresh morning air and warm sun till we neared Pont-nedd-fychan, when, finding a suitable pool, we took a delightful and refreshing bath, dried our bodies in the sun, and then walked on to the little inn, where we enjoyed our ample dish of eggs and bacon, with tea, and brown bread-and-butter. We then walked slowly on, collecting and exploring by paths and lanes and through shady woods on the south bank of the river, till we reached our lodgings at Neath, having thoroughly enjoyed our little excursion.

A few months later one of our walks had a rather serious sequel. We started after breakfast one fine Sunday morning for a walk up the Dulais valley, returning by Pont-ar-dawe, and about four in the afternoon found ourselves near my old lodgings at Bryn-coch. We accordingly went in and, of course, were asked to stay to tea, which was just being got ready. The Misses Rees, with their usual hospitality, made a huge plate of buttered toast with their home-made bread which was very substantial, and, being very hungry after our long walk, we made a hearty meal of it. My brother felt no ill effects from this, but in my case it brought on a severe attack of inflammation of the stomach and bowels, which kept me in bed some weeks, and taught me not to overtax my usually good digestion.

During my residence at Neath I kept up some correspondence with H. W. Bates, chiefly on insect collecting. We exchanged specimens, and, I think in the summer of 1847, he came on a week's visit, which we spent chiefly in beetle-collecting and in discussing various matters, and it must have been at this time that we talked over a proposed collecting journey to the tropics, but had not then decided where to go. Mr. Bates' widow having kindly returned to me such of my letters as he had preserved, I find in them some references to the subjects in which I was then interested. I will, therefore, here give a few extracts from them.

In a letter written November 9, I finish by asking: "Have you read 'Vestiges of the Natural History of Creation,' or is it out of your line?" And in my next letter (December 28), having had Bates' reply to the question, I say: "I have rather a more favourable opinion of the 'Vestiges' than you appear to have. I do not consider it a hasty generalization, but rather as an ingenious hypothesis strongly supported by some striking facts and analogies, but which remains to be proved by more facts and the additional light which more research may throw upon the problem. It furnishes a subject for every observer of nature to attend to; every fact he observes will make either for or against it, and it thus serves both as an incitement to the collection of facts, and an object to which they can be applied when collected. Many eminent writers support the theory of the progressive development of animals and plants. There is a very philosophical work bearing directly on the question—Lawrence's 'Lectures on Man'—delivered before the Royal College of Surgeons, now published in a cheap form. The great object of these 'Lectures' is to illustrate the different races of mankind, and the manner in which they probably originated, and he arrives at the conclusion (as also does Pritchard in his work on the 'Physical History of Man') that the varieties of the human race have not been produced by any external causes, but are due to the development of certain distinctive peculiarities in some individuals which have thereafter become propagated through an entire race. Now, I should say that a permanent peculiarity not produced by external causes is a characteristic of 'species' and not of mere 'variety,' and thus, if the theory of the 'Vestiges' is accepted, the Negro, the Red Indian, and the European are distinct species of the genus Homo.

"An animal which differs from another by some decided and permanent character, however slight, which difference is undiminished by propagation and unchanged by climate and external circumstances, is universally held to be a distinct *species ;* while one which is not regularly transmitted so as to form a distinct race, but is occasionally reproduced from the parent stock (like Albinoes), is generally, if the difference is not very considerable, classed as a *variety.* But I would class both these as distinct *species,* and I would only consider those to be *varieties* whose differences are produced by external causes, and which, therefore, are not propagated as distinct races. . . . As a further support to the 'Vestiges,' I have heard that in his 'Cosmos' the venerable Humboldt supports its views in almost every particular, not excepting those relating to animal and vegetable life. This work I have a great desire to read, but fear I shall not have an opportunity at present. Read Lawrence's work ; it is well worth it."

This long quotation, containing some very crude ideas, would not have been worth giving except for showing that at this early period, only about four years after I had begun to take any interest in natural history, I was already speculating upon the origin of species, and taking note of everything bearing upon it that came in my way. It also serves to show the books I was reading about this time, as well as my appreciation of the "Vestiges," a book which, in my opinion, has always been undervalued, and which when it first appeared was almost as much abused, and for very much the same reasons, as was Darwin's "Origin of Species," fifteen years later.

In a letter dated April 11, 1846, there occur the following remarks on two books about which there has been little difference of opinion, and whose authors I had at that time no expectation of ever calling my friends. "I was much pleased to find that you so well appreciated Lyell. I first read Darwin's 'Journal' three or four years ago, and have lately re-read it. As the Journal of a scientific traveller, it is second only to Humboldt's 'Personal Narrative'—as a work of general interest, perhaps superior to it. He is an ardent admirer and most able supporter of Mr. Lyell's views. His style of writing I very much admire, so free from all labour, affectation, or egotism, and yet so full of interest and original thought. . . . I quite envy you, who have friends near you attached to the same pursuits. I know not a single person in this little town who studies any one branch of natural history, so that I am quite alone in this respect." My reference to Darwin's "Journal" and to Humboldt's "Personal Narrative" indicate, I believe, the two works to whose inspiration I owe my determination to visit the tropics as a collector.

In September, 1847, my sister returned home from Alabama, and from that time till I left for Para, in the following year, we lived together at Llantwit Cottage. To commemorate her return she invited my brother and me to go to Paris for a week, partly induced by the fact that everywhere in America she was asked about it, while we were very glad to have her as an interpreter. The last letter to Bates before our South American voyage is occupied chiefly with an account of this visit, a comparison of Paris with London, and especially an account of the museums at the Jardin des Plantes as compared with the British Museum. Towards the end of this long letter the following passages are the only ones that relate to the development of my views. After referring to a day spent in the insect-room at the British Museum on my way home, and the overwhelming numbers of the beetles and butterflies I was able to look over, I add: "I begin to feel rather dissatisfied with a mere local collection; little is to be learnt by it. I should like to take some one family to study thoroughly,

principally with a view to the theory of the origin of species.
By that means I am strongly of opinion that some definite
results might be arrived at." And at the very end of the
letter I say : "There is a work published by the Ray Society
I should much like to see, Oken's 'Elements of Physio-
philosophy.' There is a review of it in the *Athenæum*. It
contains some remarkable views on my favourite subject—the
variations, arrangements, distribution, etc., of species."

These extracts from my early letters to Bates suffice to
show that the great problem of the origin of species was
already distinctly formulated in my mind ; that I was not
satisfied with the more or less vague solutions at that time
offered ; that I believed the conception of evolution through
natural law so clearly formulated in the "Vestiges" to be, so
far as it went, a true one ; and that I firmly believed that a
full and careful study of the facts of nature would ultimately
lead to a solution of the mystery.

There is one other subject on which I obtained conclusive
evidence while living at Neath, which may here be briefly
noticed. I have already described how at Leicester I became
convinced of the genuineness of the phenomena of mesmerism, and was able thoroughly to test them myself. I also
was able to make experiments which satisfied me of the truth
of phrenology, and had read sufficient to enable me to understand its general principles. But during my early residence
at Neath after my brother's death, I heard two lectures on
the subject, and in both cases I had my character delineated
with such accuracy as to render it certain that the positions
of all the mental organs had been very precisely determined.
It must be understood that the lecturers were both strangers,
and that they each gave only a single lecture on their way
to more important centres. In each case I received a large
printed sheet, with the organs and their functions stated, and
a number placed opposite to each to indicate its comparative
size. In addition to this, there was a written delineation of
character, but in each case it only professed to be a sketch,
as I could not then afford the higher fee for a full written
development of character. As these two documents have

fortunately been preserved and are now before me, it will be interesting to see how closely the main features of my character were stated by these two itinerant lecturers about sixty years ago.

I will take, first, that of Mr. Edwin Thomas Hicks, who called himself " Professor of Phrenology," and whose delineation was the less detailed of the two. It is as follows :—

" The intellectual faculties are very well combined in your head, you will manifest a good deal of perception, and will pay great attention to facts, but as soon as facts are presented you begin to reason and theorize upon them ; you will be constantly searching for causes, and will form your judgment from the analogy which one fact bears to another. You have a good development of number and order, will therefore be a good calculator, will excel in mathematics, and will be very systematic in your arrangements and plans. You possess a good deal of firmness in what you consider to be right, but you want self-confidence. You are cautious in acting and speaking, quick in temper, but kind and good in disposition."

The above estimate, although partial, and dealing almost entirely with the intellectual faculties, is yet wonderfully accurate, if we consider that it is founded upon a necessarily hasty examination, and a comparison of the proportionate development of the thirty-seven distinct organs which the examiner recognized. It is not generally known that even when the size or development of each organ is accurately given the determination of the resulting character is not a simple matter, as it depends upon a very careful study of the infinitely varied *combinations* of the organs, the result of which is sometimes very different from what might be anticipated. A good phrenologist has to make, first, a very accurate determination of the comparative as well as the absolute size of all the organs, and then a careful estimate of the probable result of the special combination of organs in each case ; and in both there will be a certain amount of difference even between equally well-trained observers, while in special details there may be a considerable difference in the final estimate, especially when the two observers are not equal in knowledge and experience.

The first sentence in the estimate is wonderfully accurate and comprehensive, since it gives in very few words the exact combination of faculties which have been the effective agents in all the work I have done, and which have given me whatever reputation in science, literature, and thought which I possess. It is the result of the organs of comparison, causality, and order, with firmness, acquisitiveness, concentrativeness, constructiveness, and wonder, all above the average, but none of them excessively developed, combined with a moderate faculty of language, which enables me to express my ideas and conclusions in writing, though but imperfectly in speech. I feel, myself, how curiously and persistently these faculties have acted in various combinations to determine my tastes, disposition, and actions. Thus, my organ of order is large enough to make me *wish* to have everything around me in its place, but not sufficient to enable me to keep them so, among the multiplicity of interests and occupations which my more active intellectual faculties lead me to indulge in.

The next sentence is also fairly accurate, as at school I always found arithmetic easy, but Mr. Hicks did not, perhaps, know that my rather small organ of wit would prevent my ever " excelling " in mathematics. That I am "systematic in my arrangements and plans " is, however, quite correct. My want of self-confidence has already been stated in my own estimate of my character ; and the last sentence is also fairly precise and accurate.

Among the other organs not referred to in the written character, there are a few worth noting. Inhabitiveness, giving attachment to place, is among my smaller faculties, while Locality, giving power of remembering places and the desire to travel, is noted as being one of the largest. Individuality, giving power of remembering names and dates, is rather small, while Time is given as the smallest of all, in both cases strictly corresponding with the amount of each faculty I possess. Again, Veneration is among the smallest indicated, and is shown in my character by my disregard for mere authority or rank, its place being taken by Ideality and Wonder, both marked as well developed, and which lead to my intense delight in the grand, the beautiful, or the mysterious in nature or in art.

Coming now to the estimate of the other lecturer, Mr. James Quilter Rumball, an M.R.C.S. and author of some medical works, we have a more detailed and careful "Phrenological Development," founded on the comparative sizes of thirty-nine organs. It is as follows, only omitting a few words at the end, which are of a purely private and personal nature.

"(*a*) There is some delicacy in the nervous system, and consequent sensitiveness which unfits it for any very long-continued exertion; but this may be overcome by a strong will. There is some tendency to indigestion; this requires air and exercise.

"(*b*) The power of fixing the attention is very good indeed, and there is very considerable perceptive power, so that this gentleman should learn easily and remember well, notwithstanding verbal memory is but moderate. Concentrativeness is the chief organ upon which all the memories depend, and this is undoubtedly large.

"(*c*) He has some vanity, and more ambition. He may occasionally exhibit a want of self-confidence; but general opinion ascribes to him too much. In this, opinion is wrong: he knows that he has not enough; he may assume it, but it will sit ill.

"(*d*) If Wit were larger he would be a good Mathematician; but without it, however clear and analytical the mind may be, it wants breadth and depth, and so I do not put down his mathematical talents as first-rate, although Number is good. The same must be said of his classical abilities—good, but not first-rate.

"(*e*) He has some love for music from his Ideality, but I do not find a good ear, or sufficient time; he has, however, mechanical ability sufficient to produce enough of both, especially for the flute, if he so choose.

"(*f*) As an artist, he would excel if his vision were perfect : he has every necessary faculty, even to Imitation.

"(*g*) He is fond of argument, and not easily convinced ; he would exhibit physical courage if called upon ; and although he loves money—as who does not ?—so far from there being any evidence of greediness, he is benevolent and liberal, but probably not extravagant. This part of his disposition is, however, so evenly balanced that there is not likely to be much peculiarity.

"(*h*) His domestic affections are his best. Conscientiousness ought to be one more, but I do not see what will try it.

"J. Q. Rumball."

I will make a few remarks on this estimate, referring to the lettered paragraphs : (*a*) This is more medical than phrenological, but it is strikingly accurate. So long as I was at school I suffered from indigestion; but my after life, largely spent in the open air, has almost entirely removed this slight constitutional failing. (*b*) A very accurate statement. (*c*) This is strikingly correct. (*d*) I have already shown how my experience at Leicester exactly accorded with this estimate. (*e*) This also is an exact statement of my relation to music. (*f*) Here I think Mr. Rumball has gone somewhat beyond his own detailed estimate of the development of my organs of Weight, Form, and Size, which are put at only a little above the average. The position of these organs over the frontal sinus renders their estimate very difficult, and I am inclined to think they are really a little below rather than above the average. At the same time I did draw a little without any teaching worth the name, and I have a high appreciation of good design, and especially of the artistic touch, so that if my attention had been wholly devoted to the study and practice of art, I may possibly have succeeded. But my occupations and tastes led me in other directions, while the progress of photography rendered sketching less and less necessary.

(*g*) The first statement here is not only correct, but it is really the main feature of my intellectual character. I can

hardly write with ease, unless I am seeking to prove something. Mere narrative is distasteful to me. The remainder
of the section calls for no special observation.

(*h*) I will only remark that the defect here pointed out
does undoubtedly exist, and it has been of some use to me to
know it.

On the whole, it appears to me that these two expositions
of my character, the result of a very rapid examination of the
form of my head by two perfect strangers, made in public
among, perhaps, a dozen others, all waiting at the end of
an evening lecture, are so curiously exact in so many distinct points as to demonstrate a large amount of truth—
both in the principle and in the details—of the method by
which they were produced. A short account of the evidence
in support of Phrenology is given in my "Wonderful
Century" (Chapter xx.), and those who are interested in the
subject will there see that the supposed "localization of
motor areas," by Professor Ferrier and others, which are
usually stated to be a disproof of the science, are really one
of its supports, the movements produced being merely those
which express the emotions due to the excitation of the
phrenological organ excited. When I touched the organ of
Veneration in one of my boy patients at Leicester he fell
upon his knees, closed his palms together, and gazed upwards,
with the facial expression of a saint in the ecstasy of adora-
tion. Here are very definite movements of a great number
of the muscles of the whole body, and some of the movements
observed by Professor Ferrier were almost as complex, and
almost as clearly due to the physical expression of a familiar
and powerful emotion.

I will here briefly record a few family events which succeeded my departure from England early in 1848. My
brother, not having enough surveying or other work to live
upon, took a small house and a few acres of good pasture land
near the town, in order to keep cows and supply milk. This
he tried for a year, my mother and sister living with him,
doing the house work, while he carried the milk daily into

the town in a small pony-cart. But the rent was too high, and it did not pay ; so in the spring of 1849, he gave it up and sailed for California in April, soon after the discoveries of gold there and when San Francisco was a city of huts and tents, and he lived there till his death in 1895, having only once visited England, in the winter of 1850–51, in order to marry the only daughter of his former employer, Mr. Webster.

Shortly after this my sister married Mr. Thomas Sims, eldest son of the Mr. Sims with whom I and my brother had lodged in Neath. He had taught himself the then undeveloped art of photography, and he and his wife settled first in Weston-super-Mare, and afterwards came to London, where I lived with them in Upper Albany Street, after my return from the Amazon.

CHAPTER XVIII

THE JOURNEY TO THE AMAZON

WHAT decided our going to Para and the Amazon rather than to any other part of the tropics was the publication in 1847, in Murray's Home and Colonial Library, of " A Voyage up the Amazon," by Mr. W. H. Edwards. This little book was so clearly and brightly written, described so well the beauty and the grandeur of tropical vegetation, and gave such a pleasing account of the people, their kindness and hospitality to strangers, and especially of the English and American merchants in Para, while expenses of living and of travelling were both very moderate, that Bates and myself at once agreed that this was the very place for us to go to if there was any chance of paying our expenses by the sale of our duplicate collections. I think we read the book in the latter part of the year (or very early in 1848), and we immediately communicated with Mr. Edward Doubleday, who had charge of the butterflies at the British Museum, for his advice upon the matter. He assured us that the whole of northern Brazil was very little known, that some small collections they had recently had from Para and Pernambuco contained many rarities and some new species, and that if we collected all orders of insects, as well as land-shells, birds, and mammals, there was no doubt we could easily pay our expenses. Thus encouraged, we determined to go to Para, and began to make all the necessary arrangements. We found that by sailing in early spring we should reach Para at the beginning of the dry season, which is both

ALFRED R. WALLACE.   1848.
(*From a daguerrotype.*)

[*To face p.* 264, VOL. I.

the most agreeable for new-comers and the best for making collections. We arranged, therefore, to meet in London towards the end of March to study the collections at the British Museum, make purchases of books, collecting apparatus, and outfit, arrange with an agent to receive and dispose of our collections, and make inquiries as to our passage.

By a curious coincidence we found that Mr. Edwards, whose book had determined us to go to the Amazon, was in London exhibiting a very fine ivory crucifix of Italian workmanship. We called upon him in a street out of Regent Street, and we had an interesting talk about the country. He kindly gave us letters of introduction to some of his American friends in Para, among others, to Mr. Leavens at the Saw Mills, with whom we went on our short expedition up the Tocantins river. We also saw the crucifix, which was certainly a very fine work of art, carved out of an unusually large mass of ivory. Mr. Edwards, who, though a little older than myself, is still alive, writes to me (October 23, 1904) that the crucifix was the work of a monk of St. Nicholas, Genoa, and was purchased by Mr. C. Edwards Lester, United States consul in that city. A brother of our Mr. Edwards purchased it for ten thousand dollars, and exhibited it successfully in many American cities. He died, however, in 1847, and as it was necessary to sell it, our Mr. Edwards, who was his executor, brought it to London, and was exhibiting it with the object of finding a purchaser. But the Louis Philippe revolution in France occurred just at the time he arrived in London, and caused such disturbances and excitement throughout Europe as to be very unfavourable for the disposal of works of art, and he was obliged to take it back to America. In a year or two it was sold to the Catholics, and he thinks it is now in one of their churches at Cleveland, Ohio. Nearly forty years later I had the pleasure of visiting Mr. Edwards at his residence in Coalburgh, West Virginia, as will be referred to in its proper place.

Among the interesting visits we paid while in London was one to Dr. Horsfield at the India Museum, who showed

us the cases in which he had brought home his large collection of butterflies from Java. These were stout, oblong
boxes, about three feet long by two feet wide and two feet
deep. Inside these were vertical grooves, about two inches
apart, to hold the boards corked on both sides, on which the
insects were pinned. The advantages were that a large number
of specimens were packed in a small space, and at much
less cost than in store boxes, while any insects which should
accidentally get loose would fall to the bottom, where a small
vacant space was left, and do no injury to other specimens.
It seemed such an excellent plan that we had a case made
like it, and sent home our first collections in it ; but though
it answered its purpose it was very inconvenient, and quite
unsuited to a travelling collector. We therefore returned to
the old style of store box, which we got made in the country,
while a very good substitute for cork was found in some of
the very soft woods, or in slices of the midribs of palms.

We were fortunate in finding an excellent and trustworthy
agent in Mr. Samuel Stevens, an enthusiastic collector of
British Coleoptera and Lepidoptera, and brother of Mr. J. C.
Stevens, the well-known natural history auctioneer, of King
Street, Covent Garden. He continued to act as my agent
during my whole residence abroad, sparing no pains to dispose
of my duplicates to the best advantage, taking charge of my
private collections, insuring each collection as its despatch was
advised, keeping me supplied with cash, and with such stores
as I required, and, above all, writing me fully as to the progress of the sale of each collection, what striking novelties it
contained, and giving me general information on the progress
of other collectors and on matters of general scientific interest.
During the whole period of our business relations, extending
over more than fifteen years, I cannot remember that we ever
had the least disagreement about any matter whatever.

Mr. Bates' parents having kindly invited me to spend a
week with them before we sailed, we left London early in
April for Leicester, where I was very hospitably entertained,
and had an opportunity of visiting some of my old friends. I
also practised shooting and skinning birds ; and as the ship

we were to sail in was somewhat delayed, I spent some days in the wild district of Charnwood Forest, which I had often wished to visit. At length, everything being ready, and our date of sailing being fixed for April 20, we left Leicester by coach a few days before that date, and stayed, I think, at Bakewell, in order to visit Chatsworth and see the palm and orchid houses, then the finest in England. The next day we went on to Liverpool, where we arrived late, after a cold and rather miserable journey outside a stage-coach.

The next morning we called upon Mr. J. G. Smith, the gentleman who had collected butterflies at Pernambuco and Para, at his office, and he invited us to dine with him in the evening, when he showed us his collection, and gave us much information about the country, the people, and the beauties of nature. During the day we got our luggage on board, saw our berths, and other accommodation, which was of the scantiest, and heard that the ship was to sail the next day. In the morning, after breakfast at our inn, we made a few final purchases, received a letter of introduction to the con-signee of the vessels, and bade farewell to our native land.

At that time there were very few steamships, and most of the ocean trade was still carried on in sailing vessels. Ours was one of the smallest, being a barque of 192 tons, named the *Mischief*, and said to be a very fast sailer. We were told that she was ranked A 1 at Lloyds, and that we might there-fore be quite sure that she was thoroughly seaworthy. We were the only passengers, and were to have our meals with the captain and mate, both youngish men, but of whom, owing to my deficient individuality, I have not the slightest recol-lection. Soon after we got out to sea the wind rose and increased to a gale in the Bay of Biscay, with waves that flooded our decks, washed away part of our bulwarks, and was very near swamping us altogether. All this time I was in my berth prostrate with sea-sickness, and it was only, I think, on the sixth day, when the weather had become fine and the sea smooth, that I was able to go on deck just as we had a distant sight of Maderia. Shortly afterwards we got into the region of the trade-wind, and had fine, bright weather all the rest of the voyage. We passed through part of the celebrated Sargasso Sea, where the surface is covered with long stretches of floating sea-weed, not brought there by storms from the distant shore, but living and growing where it is found, and supporting great numbers of small fish, crabs, mollusca, and innumerable low forms of marine life. And when we left this behind us, the exquisite blue of the water by day and the vivid phosphorescence often seen at night were a constant delight, while our little barque, with every sail set, and going steadily along day and night about ten knots an hour, was itself a thing of beauty and a perpetual enjoyment.

At length the water began to lose its blue colour, becoming first greenish, then olive, and finally olive-yellow, and one morning we saw on the horizon the long, low line of the land, and on the next, when we came on deck before sunrise, found ourselves anchored opposite the city of Para, twenty-nine days after leaving Liverpool. From this date till I landed at Deal, in October 1852, my adventures and experiences are given in my book, "A Narrative of Travels on the Amazon and Rio Negro," a cheap edition of which is comprised in "The Minerva Library of Famous Books."

In order that no large gap may occur in these memories of my life, I will give here a general outline of my travels, with such incidental remarks or recollections as may occur to me. To begin with, I will give a short description of my impressions written to my old friend and schoolfellow, Mr. George Silk, about a fortnight after our arrival, to supplement the more detailed but less impulsive account in my published narrative.

"We have been staying for near a fortnight at the country house (called here Rosinha) of Mr. Miller, the consignee of the vessel and the captain's brother, about half a mile out of the city. We have just taken a house ourselves rather nearer the woods, and to-morrow expect to be in it. We have an old nigger who cooks for us. The city of Para is a curious, outlandish looking place, the best part of it very like

Boulogne, the streets narrow and horribly rough—no pave-
ment. The public buildings handsome, but out of repair or
even ruinous. The squares and public places covered with
grass and weeds like an English common. Palm trees of
many different kinds, bananas and plantains abundant in all
the gardens, and orange trees innumerable, most of the roads
out of the city being bordered on each side with them.
Bananas and oranges are delicious. I eat them at almost
every meal. Beef is the only meat to be constantly had, not
very good, but cheap—2¾d. a pound. Coffee grows wild all
about the city, yet it is imported for use, the people are so
lazy. Every shade of colour is seen here in the people
from white to yellow, brown, and black—negroes, Indians,
Brazilians, and Europeans, with every intermediate mixture.
The Brazilians and Portuguese are very polite, and have all
the appearance of civilization. Naked nigger children abound
in the streets.

"Within a mile of the city all around is the forest, extending uninterruptedly many hundreds and even, in some
directions, thousands of miles into the interior. The climate
is beautiful. We are now at the commencement of the dry
season. It rains generally for an hour or two every evening,
though not always. Before sunrise the thermometer is about
75°, in the afternoon 85° to 87°, the highest I have yet noted.
This is hot, but by no means oppressive. I enjoy it as much
as the finest summer weather in England. We have been
principally collecting insects at present. The variety is
immense ; we have already got about four hundred distinct
kinds."

In fulfilment of a promise I made before I left Neath, I
wrote a letter to the members of the Mechanics' Institution,
after I had been nine months in the country, and as my
mother preserved a copy of it, I will give the more important
parts of it here. After a few preliminary observations, I
proceed thus :—

" Previous to leaving England I had read many books of
travels in hot countries, I had dwelt so much on the enthusiastic

descriptions most naturalists give of the surpassing beauty of tropical vegetation, and of the strange forms and brilliant colours of the animal world, that I had wrought myself up to a fever-heat of expectation, and it is not to be wondered at that my early impressions were those of disappointment. On my first walk into the forest I looked about, expecting to see monkeys as plentiful as at the Zoological Gardens, with humming-birds and parrots in profusion. But for several days I did not see a single monkey, and hardly a bird of any kind, and I began to think that these and other productions of the South American forests are much scarcer than they are represented to be by travellers. But I soon found that these creatures were plentiful enough when I knew where and how to look for them, and that the number of different kinds of all the groups of animals is wonderfully great. The special interest of this country to the naturalist is, that while there appears at first to be so few of the higher forms of life, there is in reality an inexhaustible variety of almost all animals. I almost think that in a single walk you may sometimes see more quadrupeds, birds, and even some groups of insects in England than here. But when seeking after them day after day, the immense variety of strange forms and beautiful colours is really astonishing. There are, for instance, few places in England where during one summer more than thirty different kinds of butterflies can be collected; but here, in about two months, we obtained more than four hundred distinct species, many of extraordinary size, or of the most brilliant colours.

"There is, however, one natural feature of this country, the interest and grandeur of which may be fully appreciated in a single walk: it is the "virgin forest." Here no one who has any feeling of the magnificent and the sublime can be disappointed; the sombre shade, scarce illumined by a single direct ray even of the tropical sun, the enormous size and height of the trees, most of which rise like huge columns a hundred feet or more without throwing out a single branch, the strange buttresses around the base of some, the spiny or furrowed stems of others, the curious and even extraordinary creepers and climbers which wind around them, hanging in long festoons from branch to branch, sometimes curling and twisting on the ground like great serpents, then mounting to the very tops of the trees, thence throwing down roots and fibres which hang waving in the air, or twisting round each other form ropes and cables of every variety of size and often of the most perfect regularity. These, and many other novel features—the parasitic plants growing on the trunks and branches, the wonderful variety of the foliage, the strange fruits and seeds that lie rotting on the ground—taken altogether surpass description, and produce feelings in the beholder of admiration and awe. It is here, too, that the rarest birds, the most lovely insects, and the most interesting mammals and reptiles are to be found. Here lurk the jaguar and the boa-constrictor, and here amid the densest shade the bell-bird tolls his peal. But I must leave these details and return to some more general description.

"The whole country for some hundreds of miles around Para is almost level, and seems to be elevated on the average about thirty or forty feet above the river, the only slopes being where streams occur, which flow in very shallow and often scarcely perceptible valleys. The great island of Marajó, opposite Para, is equally flat, and the smaller island of Mexiana (pronounced Mishiána), which is about forty miles long, is even more so, there not being, I believe, a rise or fall of ten feet over the whole of it. Up the river Tocantins, however, about one hundred and fifty miles south-west of Para, the land begins to rise. At about a hundred miles from its mouth, the bed of the river becomes rocky and the country undulating, with hills four or five hundred feet high, entirely covered with forest except at a few places on the banks where some patches of open grass land occur, probably the site of old cultivation and kept open by the grazing of cattle.

"The whole of the Para district is wonderfully intersected by streams, and the country being so flat, there are frequently cross-channels connecting them together. Up all these the tide flows, and on their banks all the villages, estates, and

native huts are situated. There is probably no country in
the world that affords such facilities for internal communication by water.

"The climate of Para cannot be spoken of too highly.
The temperature is wonderfully uniform, the average daily
variation of the thermometer being only 12° F. The lowest
temperature at night is about 74°, the highest in the day
about 86°, but with occasional extremes of 70° and 90°
Though I have been constantly out at all times of the day,
and often exposed to the vertical sun, I have never suffered
any ill effects from the heat, or even experienced so much inconvenience from it as I have often done during a hot summer
at home. There are two principal divisions of the year
into the wet and dry seasons, called here winter and summer.
The wet season is from January to June, during which
time it rains more or less every day, but seldom the whole
day, the mornings usually being fine. The dry season is by
no means what it is in some parts of the world; it still rains
every two or three days, and it is a rare thing for more than
a week to pass without a shower, so that vegetation is never
dried up, and a constant succession of fruits and flowers and
luxuriant foliage prevails throughout the year. Notwithstanding the amount of water everywhere, Para is very
healthy. The English and Americans who have lived here
the longest look the healthiest. As for myself, I have enjoyed
the most perfect health and spirits without the necessity for
nearly so many precautions as are required at home.

"The vegetable productions of the country around Para
are very numerous and interesting. There are upwards of
thirty different kinds of palms, and in almost every case the
leaves, stems, or fruits are useful to man. One elegant species,
the stem of which, though not thicker than a man's arm,
rises to a height of sixty or eighty feet, produces a small
blackish fruit, from which a creamy preparation is made, of
which everybody becomes very fond, and which forms a large
part of the subsistence of the natives. From the fibres of one
kind ropes are made, which are in general use for the cables
of native vessels as they are almost indestructible in water.

The houses of the Indians are often entirely built of various parts of palm-trees, the stems forming posts and rafters, while the leaf-stalks, often twenty feet long, placed side by side and pegged together, make walls and partitions. Not a particle of iron is needed, the various parts of the roofs being fastened together with the lianas or forest-ropes already described, while, as both stem and leaf-stalks split perfectly straight no tools whatever are needed besides the heavy bush-knife which every countryman carries.

" The calabash tree supplies excellent basins, while gourds of various sizes and shapes are formed into spoons, cups, and bottles ; and cooking-pots of rough earthenware are made everywhere. Almost every kind of food, and almost all the necessaries of life, can be here grown with ease, such as coffee and cocoa, sugar, cotton, farinha from the mandioca plant (the universal bread of the country), with vegetables and fruits in inexhaustible variety. The chief articles of export from Para are india-rubber, brazil-nuts, and piassaba (the coarse stiff fibre of a palm, used for making brooms for street-sweeping), as well as sarsaparilla, balsam-capivi, and a few other drugs. Oranges, bananas, pine-apples, and water-melons are very plentiful, while custard-apples, mangoes, cashews, and several other fruits abound in their season. All are very cheap, as may be judged by the fact that a bushel basket of delicious oranges may be purchased for sixpence or a shilling.

" Coming to the animal world, a forest country is often disappointing because so few of the larger animals can be seen, though some of them may be often heard, especially at night. The monkeys are in every way the most interesting, and are the most frequently to be met with. A large proportion of American monkeys have prehensile tails, which are so powerful in some of the species that they can hang their whole weight upon it and swing about in the air with only a few inches of the tip twisted round a branch. If disturbed in such a position they swing themselves off, catching hold of boughs hand over hand, and rapidly disappear. They live entirely in the tree-tops, hardly ever descending to the ground,

and in this region of forests they can travel hundreds of
miles without requiring to do so, so that they are almost as
independent of the earth as are the swifts and the humming-
birds. They vary in size from the little marmosets, not so
large as a squirrel, up to the howling monkeys the size of a
large shepherd's dog. Of what are commonly termed wild
beasts the jaguar or onça (somewhat similar to a leopard, but
stouter) is the most powerful and dangerous, and is very
destructive to horses and cattle. The puma (often called the
American lion), though equally large, is much less dangerous.
Tapirs, agoutis, armadillos, and sloths are not uncommon, but
are very rarely seen. Birds are very abundant, and many
are exceedingly beautiful. Macaws, parrots, toucans, trogons,
chatterers, and tanagers, are all common, and often of the
most gorgeous colours, while the lovely little humming-birds,
though not so numerous as in the mountain districts, are to
be seen in every garden. In the islands of Mexiana and
Marajo, those splendid birds the scarlet ibis and the roseate
spoonbill abound, together with great numbers of storks,
herons, ducks, divers, and other aquatic birds; while in the
forests of the mainland the fine crested curassows and the
elegant trumpeters are among the larger ground-feeders.

"Lizards swarm everywhere in a variety of strange forms
—the curious geckos, which can walk about the ceilings by
means of suckers on their toes; the large iguanas, which cling
to branches by their prehensile tails, and whose flesh is a
delicacy; and the large ground lizards, three or four feet long.
Frogs of all kinds abound, and some of the little tree frogs
are so gaily coloured as to be quite pretty. The rivers are
full of turtles of many kinds, one of the largest being very
plentiful and as delicate eating as the well-known marine
turtle of City feasts. Snakes, though not often seen, are
really very numerous, but comparatively few are poisonous.

"Fish abound in all the rivers, and many of them are of the
very finest quality. One very large fish, called the pirarucú,
is three or four feet long, and when slightly salted and dried
in the sun can be kept for any time, and takes the place of
salt cod, kippered haddocks, and red herrings in Europe.

" The inhabitants of Para, as of all Brazil, consist of three distinct races : the Portuguese and their descendants with a few other Europeans, the native Indians, and the Negroes together with a considerable number of mixed descent. The Indians in and near Para are all 'tame Indians,' being Roman Catholics in religion and speaking Portuguese, though many speak also the Lingoa-Geral or common Indian language. They are the chief boatmen, fishermen, hunters, and cultivators in the country, while many of them work as labourers or mechanics in the towns. The negroes were originally all slaves, but a large number are now free, some having purchased their freedom, while others have been freed by their owners by gift or by will. Most of the sugar and cocoa plantations are worked partly by slave and partly by hired labour. The negroes, here as elsewhere, are an exceedingly talkative and contented race, as honest as can be expected under the circumstances, and when well treated exceedingly faithful and trustworthy. Generally they are not hard-worked, and are treated with comparative kindness and lenity.

" The people of all races are universally polite, and are generally temperate and peaceful. The streets of Para are more free from drunkenness and quarrels than any town of like size in England or Wales ; yet in the time of Portuguese rule there were some fearful insurrections, brought on by oppressive government. But now, foreigners of all sorts can live in perfect safety, and on excellent terms with the native residents and officials, though, of course, they have to conform to the customs of the country, and obey all the laws and regulations, which latter are sometimes inconvenient and troublesome."

*   *   *   *   *   *

Shortly after writing this letter I went on a collecting expedition up the river Guamá, and soon after my return, in July, 1849, my younger brother Herbert came out to join me in order to see if he had sufficient taste for natural history to become a good collector. I had decided to start up the Amazon as soon as I could find an opportunity, and after a month in the suburbs of Para we left in a small empty boat

returning to Santarem, where we intended to stay for some
time. Dr. Richard Spruce, the now well-known traveller and
botanist, came out in the same ship with my brother, and was
accompanied by a young Englishman, Mr. King, as an
assistant and pupil in botany ; and as Dr. Spruce was a well-
educated men, a most ardent botanist, and of very pleasing
manners and witty conversation, we very much enjoyed the
short time we were together. My brother was the only one
of our family who had some natural capacity as a verse-writer,
and I will therefore supplement my rather dry descriptions
by some bright verses he sent home, giving his impressions of
Para and the voyage to Santarem, which occupied twenty-
eight days, the distance being about seven hundred miles.

" FROM PARA TO SANTAREM.

" Well ! here we are at anchor
    In the river of Pará ;
We have left the rolling ocean
    Behind us and afar ;
Our weary voyage is over,
    Sea-sickness is no more,
The boat has come to fetch us
    So let us go on shore.
How strange to us the aspect
    This southern city wears !
The ebon niggers grinning,
    The Indians selling wares ;
The lasses darkly delicate,
    With eyes that ever kill,—
All breathe to us in whispers
    That we are in Brazil.

" The streets are green and pleasant,
    The natives clad in white ;
We miss the noise of coaches,
    But miss it with delight.
The hairy sheep is biting
    The grass between the stones,
And many a pig is grunting
    In half familiar tones ;
And through the green *janellas* *
    (Which we should like to raise)
Dark eyes of the senhoras
    Upon the strangers gaze.

* Venetian shutters in place of sashes.

The many foreign faces,
  The lingo stranger still,—
All breathe to us in whispers
  That we are in Brazil.

" We stroll about the suburbs,
  Beneath the mango groves,
Where friends appoint their meetings
  And lovers seek their loves ;
Where fruit and docé vendors,
  With many a varied cry,
Invite the evening stroller
  Their luxuries to buy.
Here soars the lofty coco,
  Here feathery palm-trees rise,
And the green broad-leaved banana
  Swells forth 'neath sunny skies.
The cooling water-melon,
  The wild pine by the rill,—
All breathe to us in whispers
  That we are in Brazil.

*   *   *   *   *

" Once more upon the waters,
  Adieu to thee, Para,
Adieu, kind friends, whose latticed homes
  Are fading now afar.
We sail 'mid lovely islands,
  Where man has seldon trod,
Where the wild deer and the onça
  Are owners of the sod ;
By forests high and gloomy,
  Where never a ray of sun
Can pierce its way to enter
  Those shades so thick and dun.
The cry of parrots overhead,
  The toucan with his bill,—
All breathe to us in whispers
  That we are in Brazil.

" And now upon the Amazon,
  The waters rush and roar—
The noble river that flows between
  A league from shore to shore ;
Our little bark speeds gallantly,
  The porpoise, rising, blows,
The gull darts downward rapidly
  At a fish beneath our bows,

The far-off roar of the onça,
    The cry of the whip-poor-will—
All breathe to us in whispers
    That we are in Brazil.

" By many an Indian cottage,
    By many a village green,
Where naked little urchins
    Are fishing in the stream,
With days of sunny pleasure,
    But, oh, with weary nights,
For here upon the Amazon
    The dread mosquito bites—
Inflames the blood with fever,
    And murders gentle sleep,
Till, weary grown and peevish,
    We've half a mind to weep !
But still, although they torture,
    We know they cannot kill,—
All breathe to us in whispers
    That we are in Brazil.

" And now the wave around us
    Has changed its muddy hue,
For we are on the Tapajoz,
    And Santarem 's in view;
Fair Santarem, whose sandy beach
    Slopes down into the wave,
Where mothers wash their garments,
    And their happy children lave.
Now comes the welcome greeting,
    The warm embrace of friends,
And here, then, for a season,
    The toil of voyaging ends.
The silent Indian sentry,
    The mud fort on the hill,—
All breathe to us in whispers
    That we are in Brazil."

We remained at Santarem about three months, including
a visit to Monte Alegre, a village on the opposite or north
side of the river, where we had heard there were some very
interesting caves, and where we found the great water-lily, the
Victoria regia, growing abundantly in a backwater of the
Amazon. Santarem and Monte Alegre both differ from
almost all the rest of the places on the banks of the Amazon
in being open country, with rocky hills dotted all over with

low trees and shrubs, and with only isolated patches of forest for many miles round. This peculiarity of vegetation was accompanied by an equal peculiarity of insect life, especially in the butterflies, which were almost all different from any I had found at Para, and many of them wonderfully beautiful. Here I first obtained evidence of the great river limiting the range of species. At Santarem I found a lovely butterfly about the size of our largest peacocks or red-admirals, but entirely of different shades of the most exquisite sky-blue of a velvety texture (*Callithea sapphirina*), while on the opposite side of the river was a closely allied species of an almost indigo-blue colour, and with different markings underneath. Dr. Spruce assured me that, though he had studied all the known plants of the Amazon before leaving England, he felt quite puzzled when collecting at Santarem, because almost every shrub and tree he found there proved to be a new species.

We greatly enjoyed our short residence at Santarem, both on account of the delightful climate, the abundance of good milk, which we could get nowhere else after leaving Para, and for the pleasant friends we met there. The following descriptive verses by my brother may therefore appropriately follow here :—

"A DESCRIPTION OF SANTAREM.

" I stand within a city,
   A city strangely small ;
'Tis not at all like Liverpool,
   Like London, not at all.
The blue waves of the Tapajoz
   Are rippling at its feet,
Where anchored lie the light canoes—
   A Lilliputian fleet.
The scream of parrots overhead,
   The cry of the whip-poor-will,
All tell me you're in England,
   And I am in Brazil.

" I wander through the city,
   Where everything is new :
The grinning, white-toothed negroes,
   The pigs of varied hue ;
The naked little children,
   With skins of every dye,

> Some black, some brown, some lighter,
>   Some white as you or I.
> A dozen such in family,
>   With bellies all to fill,
> Would be no joke in England ;
>   'Tis nothing in Brazil ! "

Then follows his farewell verses, well expressing the regret we both felt at leaving it. I may just note here that his reference to "blue pig" is not imagination only. Among the quantities of pigs that roamed about the city and suburbs (really little more than a large straggling village) was one whose nearly black skin was seen in certain lights to be distinctly blue ; and to have found the real "blue pig," which under the name of the "Blue Boar" is a not uncommon inn-sign at home, greatly delighted my brother.

> " FAREWELL TO SANTAREM.
>
> " My skiff is waiting on the shore,
>   And on the wave is my canoe ;
> Ye citizens of Santarem,
>   To each and all, adieu !
> The hour has come to bid, with grief,
> Adieu to milk and tender beef.
>
> " Adieu, the fort upon the hill,
>   And yon cathedral's domes,
> Like guardian giants gazing down
>   Upon thy lowly homes ;
> Ye naked children, all adieu,
> And thou strange pig with skin of blue !
>
> " Farewell, the forest's deep recess,
>   Where Sol can never come ;
> Farewell, the campo's sandy plain,
>   The lizards in the sun.
> To water-melons cool, adieu ;
> And farewell, old black cook, to you.
>
> " Adieu, thy shores, broad Tapajoz,
>   Within thy heaven-dyed wave,
> At noonday's silent, sultry hour
>   I've joy'd to plunge and lave.
> Adieu ! to-morrow's noonday sun,
> I'll bathe in yellow Amazon."

On reaching the city of Barra at the mouth of the Rio Negro we found a strange and even now unaccountable poverty both in insects and birds, although there was fine virgin forest within a walk, with roads and paths and fine rocky streams. All seemed barren and lifeless as compared with the wonderful productiveness of Para. It was, therefore, necessary to seek other localities in search of rarities. I accordingly went a three days' journey up the Rio Negro to obtain specimens of the umbrella-bird, one of the most remarkable birds of these regions, my brother going in another direction to see what he could discover.

After a month I returned to Barra, and after some months of almost constant wet weather went to a plantation in the Amazon above Barra for two months, where I made a tolerable collection, while my brother went to Serpa, lower down on the Amazon ; and on returning I prepared for my long intended voyage to the Upper Rio Negro in hopes of getting into a new and more productive country. As soon as a much overdue vessel had arrived, bringing letters and remittances from England, I was ready to start for a journey of unknown duration. After a year's experience it was now clear that my brother was not fitted to become a good natural-history collector, as he took little interest in birds or insects, and without enthusiasm in the pursuit he would not have been likely to succeed. We therefore arranged that he should stay at or near Barra for a few months of the dry season, make what collections he could, then return to Para on his way home. I left him what money I could spare, and as he was now well acquainted with the country, and could, if absolutely necessary, get an advance from our agents at Para, I had little doubt that he would get home without difficulty. But I never saw him again. When he reached Para, towards the end of May, 1851, he at once took a passage to England in a ship to leave early in June, but before it sailed he was seized with yellow fever, then prevalent in the town, and though at first seeming to get better, died a few days afterwards. Mr. Bates was at Para at the time, preparing for his second long journey up the Amazon. He was with him when he was taken ill, and did all he could in getting medical assistance and helping to nurse him. But just when my brother was at his worst, two days before his death, he was himself attacked with the same disease, which rendered him absolutely helpless for ten days, though, being of a stronger and more hardened constitution, he finally recovered. Mr. Miller, the Vice-consul, with whom I and Bates had stayed when we arrived at Para, was with my brother when he died. This gentleman had severe brain-fever not long afterwards, and also died ; but he told Mr. Bates that a few hours before my brother's death he had said that "it was sad to die so young." In one of his last letters home he had spoken quite cheerfully, saying, "When I arrive in England I have my plans, which I can better tell than write." And then referring to his brother John's emigration to California, and some idea that he, Herbert, might go there too, he says, "I do not like the Californian scheme for many reasons. I should like to have seen John's first letter. No doubt *he* is sure to get on. I wish I was a little less poetical ; but, as I am what I am, I must try and do the best for myself I can." I rather think he had the idea of getting some literary work to do, perhaps on a country newspaper or magazine, and it is not unlikely that *that* was what he was best fitted for.

I may here briefly explain why he had no regular employ-ment to fall back upon. Owing to the fact that I left home when I was fourteen (he being then only seven and a half), and that when I happened to be at home afterwards he was often away at school, I really knew very little of him till he came to me at Para. Until I left school he had been taught at home by my father, and afterwards went for a year or two to a cheap boarding school in Essex. As it was necessary for him to learn something, he was placed with a portmanteau and bag-maker in Regent Street, where he was at first a mere shop-boy, and as he showed little aptitude for learning the trade, and was not treated very kindly by his master, he was rather miserable, and was taken away after a year. My brother William then got him into the pattern-shop at the Neath Abbey Iron Works soon after I had gone to Leicester.

There he remained, lodging near the works, and when we went to live at Neath, spending his Sundays with us. At this time he took to writing verses, and especially enigmas in the style of W. Mackworth Praed, and these appeared almost weekly in some of the local papers. But he evidently had no inclination or taste for mechanical work, and though he spent, I think, about four years in the pattern-shops he never became a good workman ; and as he saw no prospect of ever earning more than a bare subsistence as a mechanic, and perhaps not even that, he gladly came out to me, when he had just completed his twentieth year. His misfortune was that he had no thorough school training, no faculty for or love of mechanical work, and was not possessed of sufficient energy to overcome these deficiencies of nature and nurture.

The remainder of my South American travels consisted of two voyages up the Rio Negro. On the first I went beyond the boundaries of Brazil, and crossed by a road in the forest to one of the tributaries of the Orinoko. Returning thence I visited a village up a small branch of the Rio Negro, where there is an isolated rocky mountain, the haunt of the beautiful Cock of the Rock; afterwards going up the Uaupés as far as the second cataract at Juaurité. I then returned with my collections to Barra, having determined to go much farther up the Uaupés in order to obtain, if possible, the white umbrella bird which I had been positively assured was found there ; and also in the hopes of finding some new and better collecting ground near the Andes. These journeys were made, but the second was cut short by delays and the wet season. My health also had suffered so much by a succession of fevers and dysentery that I did not consider it prudent to stay longer in the country.

Although during the last two journeys in the Rio Negro and Orinoko districts I had made rather large miscellaneous collections, and especially of articles of native workmanship, I never found any locality at all comparable with Para as a collecting ground. The numerous places I visited along more than a thousand miles of river, all alike had that poverty of insect and bird-life which characterized Barra itself, a poverty which is not altogether explicable. The enormous difficulties and delays of travel made it impossible to be at the right place at the right season; while the excessive wetness of the climate rendered the loss of the only month or two of fine weather irreparable for the whole year. The comparative scantiness of native population at all the towns of the Rio Negro, the small amount of cultivation, the scarcity of roads through the forest, and the want of any guide from the experience of previous collectors, combined to render my numerous journeys in this almost totally unknown region comparatively unproductive in birds and insects. As it happened (owing to Custom House formalities at Barra), the whole of my collections during the last two voyages were with me on the ship that was burnt, and were thus totally lost. On the whole, I am inclined to think that the best places now available for a collector in the country I visited are at the San Jeronym and Juarité falls on the River Uaupés, and at Javita, on a tributary of the Orinoko, if the whole of the dryest months could be spent there. So far as I have heard, no English traveller has to this day ascended the Uaupés river so far as I did, and no collector has stayed any time at Javita, or has even passed through it. There is, therefore, an almost unknown district still waiting for exploration by some competent naturalist.

One letter I wrote from Guia on the Upper Rio Negro, three months after my arrival there, has been preserved, and from it I extract the following passage :—

"I have been spending a month with some Indians three days' journey up a narrow stream (called the Cobati River). From there we went half a day's journey through the forest to a rocky mountain where the celebrated 'Gallos de Serra' (Cocks of the Rock) breed. But we were very unfortunate, for though I had with me ten hunters and we remained nine days at the Serra, suffering many inconveniences (having only taken farinha and salt with us), I only got a dozen gallos, whereas I had expected in less time to have secured

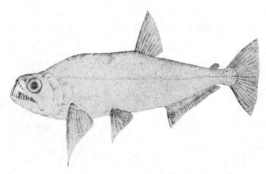

1. CYNODON SCOMBROIDES. FAM. CHARACINIDÆ.

(One-fourth natural size.)

2. XIPHOSTOMA LATERISTRIGA. FAM. CHARACINIDÆ.

(One-third natural size.)

[*To face p.* 285, VOL. I.

fifty. Insects, there were none at all; and other good birds excessively rare.

"My canoe is now getting ready for a further journey up to near the sources of the Rio Negro in Venezuela, where I have reason to believe I shall find insects more plentiful, and at least as many birds as here. On my return from there I shall take a voyage up the great river Uaupés, and another up the Isanna, not so much for my collections, which I do not expect to be very profitable there, but because I am so much interested in the country and the people that I am determined to see and know more of it and them than any other European traveller. If I do not get profit, I hope at least to get some credit as an industrious and persevering traveller."

I then go on to describe the materials I was collecting for books on the palms and on the fishes of these regions, and also for a book on the physical history of the Amazon valley. Only the "Palms" were published, but I give here a few copies of the drawings I made of about two hundred species of Rio Negro fishes, which I had hoped to increase to double that number had I remained in the country.

The two first figures (*Cynodon scombroides* and *Xiphostoma lateristriga*) belong to the family Characinidæ, a group which abounds in the fresh waters of tropical America and Africa, where it replaces the carps (Cyprinidæ) of Europe and the Old World generally, though not very closely allied to them. Many of the species are very like some of our commonest river-fish, such as gudgeons, dace, roach, tench, and bream, and I have drawings of no less than sixty-five species of the family. They are all, I believe, eatable, but are not held to be fishes of the best quality.

The next figure (*Pimelodus holomelas*) is an example of the family Siluridæ, which is found in the fresh waters of all parts of the world. The cat-fishes of North America and the sturgeons of Eastern Europe belong to it. I obtained thirty-four species on the Rio Negro, many being of a large size. They are generally bottom-feeding fishes and are

greatly esteemed, the flesh being very fat and rich, quite
beyond any of our English fishes.

The next figure (*Plecostomus guacari*) is one of the
Loricariidæ, which are allied to the Siluridæ, but characterized
by hard bony scales or plates, and dangerous bony spines to
the dorsal and pectoral fins. Many are of very strange and
repulsive forms, and though eatable are not esteemed. I
obtained seven species of these curious fishes.

The remaining two figures serve to illustrate the family
Cichlidæ, one of the most abundant and characteristic groups
of South American fishes. All are of moderate size, and feed
partially or entirely on vegetable substances, especially fruits
which grow on the river-banks and when ripe fall into the
water. They are caught with fruits as a bait, and the fisherman gently lashes the water with his rod so as to imitate
the sound of falling fruit, thus attracting the fish. Some of
these are the most delicious fish in the world, both delicate
and fat, to such an extent that the water they are boiled in
is always served at table in basins, and is a very delicious
broth, quite different to any meat broth and equal to the
best. It is more like a very rich chicken broth than anything else. I obtained twenty-two species of this family of
fishes, the little *Pterophyllum scalaris*, called the butterfly
fish, being one of the most fantastic of fresh-water fishes.
The other, *Cichlosoma severum*, is one of the best for the table.

I have presented my collection of fish drawings to the
British Museum of Natural History, and I am indebted to
Mr. C. Tate Regan, who has charge of this department, for
giving me the names of the species represented. In a paper
read before the Zoological Society in August, 1905, he states
that he has named about a hundred species, and that a large
portion of the remainder are probably new species, showing
how incomplete is our knowledge of the fishes of the Amazon
and its tributaries.

Looking back over my four years' wanderings in the
Amazon valley, there seem to me to be three great features
which especially impressed me, and which fully equalled or

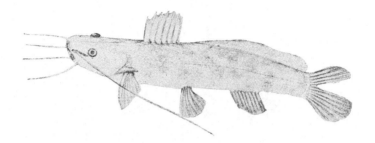

3. PIMELODUS HOLOMELAS. FAM. SILURIDÆ.

(One-third natural size.)

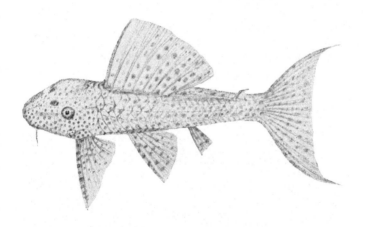

4. PLECOSTOMUS GUACARI. FAM. LORICARIIDÆ.

(One-third natural size.)

[*To face p.* 286, VOL. I.

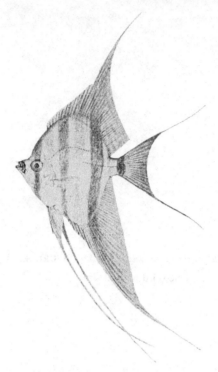

5. PTEROPHYLLUM SCALARA. FAM. CICHLIDÆ.

(One-third natural size.)

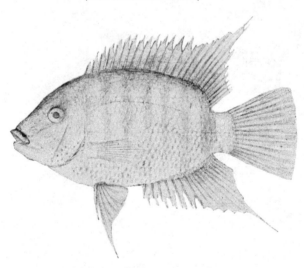

6. CICHLOSOMA SEVERUM. FAM. CICHLIDÆ.

(One-third natural size.)

[*To face p.* 286, VOL. I.

even surpassed my expectations of them. The first was the virgin forest, everywhere grand, often beautiful and even sublime. Its wonderful variety with a more general uniformity never palled. Standing under one of its great buttressed trees—itself a marvel of nature—and looking carefully around, noting the various columnar trunks rising like lofty pillars, one soon perceives that hardly two of these are alike. The shape of the trunks, their colour and texture, the nature of their bark, their mode of branching and the character of the foliage far overhead, or of the fruits or flowers lying on the ground, have an individuality which shows that they are all distinct species differing from one another as our oak, elm, beech, ash, lime, and sycamore differ. This extraordinary variety of the species is a general though not universal characteristic of tropical forests, but seems to be nowhere so marked a feature as in the great forest regions which encircle the globe for a few degrees on each side of the equator. An equatorial forest is a kind of natural arboretum where specimens of an immense number of species are brought together by nature. The western half of the island of Java affords an example of such a forest-region which has been well explored, botanically ; and although almost all the fertile plains have been cleared for cultivation, and the forests cover only a small proportion of the country, the number of distinct species of forest-trees is said to be over fifteen hundred. Now the whole island is only about as large as Ireland, and has a population of over twenty millions ; and as the eastern half of the island has a much drier climate, where there are forests of teak and much more open country, it is certain that this enormous variety of species is found in a wonderfully small area, probably little larger than Wales. I have no doubt that the forests of the Amazon valley are equally rich, while there are not improbably certain portions of their vast extent which are still richer.

The second feature, that I can never think of without delight, is the wonderful variety and exquisite beauty of the butterflies and birds, a variety and charm which grow upon one month after month and year after year, as ever new and

beautiful, strange and even mysterious, forms are continually
met with.   Even now I can hardly recall them without a
thrill of admiration and wonder.

The third and most unexpected sensation of surprise and
delight was my first meeting and living with man in a state
of nature—with absolute uncontaminated savages! This was
on the Uuapés river, and the surprise of it was that I did not
in the least expect to be so surprised. I had already been
two years in the country, always among Indians of many
tribes ; but these were all what are called tame Indians, they
wore at least trousers and shirt ; they had been (nominally)
converted to Christianity, and were under the government of
the nearest authorities ; and all of them spoke either Portuguese or the common language, called "Lingoa-Geral."

But these true wild Indians of the Uaupés were at once
seen to be something totally different. They had nothing
that we call clothes ; they had peculiar ornaments, tribal
marks, etc. ; they all carried weapons or tools of their own
manufacture ; they were living in a large house, many
families together, quite unlike the hut of the tame Indians ;
but, more than all, their whole aspect and manner were
different—they were all going about their own work or
pleasure which had nothing to do with white men or their
ways ; they walked with the free step of the independent
forest-dweller, and, except the few that were known to my
companion, paid no attention whatever to us, mere strangers
of an alien race. In every detail they were original and
self-sustaining as are the wild animals of the forests, absolutely
independent of civilization, and who could and did live their
own lives in their own way, as they had done for countless
generations before America was discovered. I could not
have believed that there would be so much difference in the
aspect of the same people in their native state and when
living under European supervision. The true denizen of the
Amazonian forests, like the forest itself, is unique and not
to be forgotten.

H. E. WALLACE.   AGE 8.

(*From a pencil sketch by Miss Townsend.*)

HERBERT EDWARD WALLACE.　AGE 20.

(*From a silhouette.*)

CHAPTER XIX

"IN MEMORIAM"

In memory of
HERBERT EDWARD WALLACE,
who died of yellow fever at Para, June 8, 1851,
Age 22 years.

DURING the three or four years my brother lived at Neath he contributed a considerable number of verses and enigmas to the local newspapers, while some of his old notebooks contain many others in an unfinished state. While on the Amazon he wrote several more, and I will here give a few samples of these, which may perhaps be thought worth preserving, and as a memento of a young life prematurely closed in a distant land. He was a great admirer of Hood and of Longfellow, and several of his little poems are reflections of their writings, while the enigmas were inspired by those of William Mackworth Praed.

The only two likenesses of my brother we possess are copied here. The first is from a pencil sketch by an old friend of the family (Miss Townsend), taken at Hoddesdon when he was about eight years old, which was always considered a striking likeness. The other is a copy of a black silhouette taken before he came out to the Amazon in 1849, when he was just twenty years old.

My lamented friend Dr. Spruce kindly sent me two letters he received from my brother in the interval between our parting at Santarem and his return to Para, and as they are probably the last he ever wrote I give them here (omitting one or two personal matters) in order to show his usual good spirits and random style of writing.

" Barra, March 15, 1850.

"DEAR SIR,

"A lodge is gained at last.   Here we are in a Barra !

> ' Here we work with Net and Trigger
> By the famous river Nigger,'

on whose midnight waters never is heard the hum of the sanguinary carapaná,[1] where 'sleep, which knits up the ravelled sleeve of care,' hath no intruder.   By-the-by, talking of sleep reminds me of redés.[2]   All the redés in Barra possess a title.   Why?   Because they are Barra-nets. This you may think far-fetched.   Well!   I will own 'tis rather distant; perhaps you would like one a little nearer? Good.   As we left Obydos, remarking the woody declivity on our right, the following sublime comparative similitude burst forth spontaneously.   Why is this hill like a dead body running?   Because, says I—but no! you must really try to guess it ; however, I will enclose the answer to refer to in case of failure.   [See p. 291.]

*   *   *   *   *   *

"With best wishes for your health and success, and kind remembrances to Mr. King and Santarem friends.

"I remain, yours respectfully
"EDWARD WALLACE."

" Serpa, December 29, 1850.

"DEAR SIR,

"I have just returned from a month's excursion among the lakes and byways of the mighty Amazon, and whilst reposing my weary limbs amid the luxurious folds of a redé, drinking a fragrant cup of the sober beverage, and

---

[1] Carapaná is the native name of the mosquito.
[2] Redé or net, the local name for "hammock."

meditating (but cheerfully) upon the miseries of human nature, I received notice of your arrival in the Barra.

"So you have at last gained that 'lodge' so long pictured in the vista of imagination. You are at last in that Promised Land—a land flowing with caxáça and farinha;[1] a land where a man may literally, and safely, sleep without breeches—a luxury which must be enjoyed to be appreciated.

"I am now waiting for a passage to Para, from thence to return to England. There is a vessel caulking here I expect will go in two or three weeks. I have a small collection of birds and butterflies, but new species of the latter are very scarce.

*   *   *   *   *   *

"The Christmas festa is now over, and this little village has resumed its wonted tranquillity. I suppose you intend soon to proceed up the Rio Negro; no doubt my brother is now glorying in ornithological rarities, and revelling amid the sweets of lepidopterous loveliness. But enough! A little while and the wintry sea is roaring around my pillow; then shall I envy you in your snug redés far from the restless billow; then, whilst vainly endeavouring to swallow preserved salmon or other ship luxury, I shall long for my Amazonian appetite and roasted pirarucú; then—— But I will not anticipate hours which are inevitable. I hope yourself and Mr. King are in good health. In this respect I have no cause to complain. Wishing you both a prosperous and a pleasant time, I must now remain,

"Yours sincerely,

"EDWARD WALLACE."

It is evident from this letter that the usual dilatoriness and difficulties of Amazonian travel delayed his arrival at Para about four months beyond the time he calculated on. The answer to the enigma in the first letter, which he says he has enclosed, I did not receive; but I have no doubt it is as follows: "Because it is a corpse (copse) sloping away from

---

[1] Native rum and mandioca meal.

the town." "Slope," "sloping," were at that time slang
words for escaping or running away, "understanded by the
people," which perhaps they may not be now. I may add
here that he did not like the name Herbert (his first name),
and so took to his second—Edward.

The friends of temperance often complain of the want of
a good song. I think the following, written by my brother
about 1848, may perhaps be considered suitable till a better
one is written :—

"THE CUP OF TEA.

I.

"Some love to sip their Burgundy,
    And some prefer Champagne ;
Some like the wines of sunny France,
    And some the grape of Spain.
There's some will take their brandy neat,
    While others mix with water ;
There's some drink only Indian ale,
    And others London porter.
Away with poisons such as these,
    No Alcohol for me !
Oh, fill me up the sober cup,
    The social cup of Tea.

II.

"Some love to sing of ancient times,
    And drinking customs preach ;
Such customs are—as Shakespeare saith—
    More honoured in the breach ;
For we can sing a joyous song
    Without the aid of wine,
And court the muse without a glass
    To spur the lagging rhyme.
Then take the pledge, be one of us,
    And join our melody—
' Oh, fill me up the sober cup,
    The social cup of Tea.'

III.

"We pray for that long-wished-for hour
    When Bacchus shall be slain,
John Barleycorn be trodden down
    And ne'er rise up again ;

> When man, begun to know himself,
>   Shall maddening bowls resign,
> And Temperance, with a mighty hand,
>   ' Dash down the Samian wine.'
> Here's to the death of Alcohol !
>   And still our song shall be,
> ' Oh, fill me up the sober cup,
>   The social cup of Tea.' "

The next verses, suggested by a well-known old song, show his early love of humanity and aspirations for an improved social state. It was probably written at Neath about 1847 or 1848.

> " THE LIGHT OF DAYS TO COME.

> " The light of other days is faded,
>   But we will not repine,
> Nor waste the precious hours as they did,
>   The dwellers in that time.
> We will not sign in gloom and sadness
>   O'er what can ne'er return,
> But rather share the mirth and gladness
>   In the light we now discern.

> " The past brought luxury and pleasure
>   To few beneath the sun,
> But equal all shall share the treasure
>   Of the light of days to come.
> Knowledge shall strengthen each endeavour
>   To set the future right,
> And Justice with her sword shall sever
>   The iron hand of Might.

> " The fields where warriors have commanded,
>   And men have fought for fame,
> Shall in a future age be branded
>   With an inglorious name.
> Bright souls who perish unassuming,
>   Your work is not yet done,
> Like scattered seed your deeds shall bloom in
>   The Light of days to come."

I preserve the following fantastic little poem because it so well describes the mode of house-building of the dwellers in the grand equatorial forests which supply so many of man's wants in a way unknown in the colder climes.

"THE INDIAN'S HUT.

" 'Twas on the mighty Amazon,
  We floated with the tide,
While steep and flowery were the banks
  That rose on either side,
And where the green bananas grow,
  An Indian's cot I spied.

" Like to the halls of Solomon,
  Yon humble dwelling rose,
Without the grating of the saw
  Or echoing hammers blows ;
For all its parts are bound with rope,
  Which in the forest grows.

" Those wild fantastic slender cords
  Which hang from branches high,
The place of staple, screw, and nail,
  With equal strength supply,
And pole and rafter firm and fast
  All silently they tie.

" All silently, for stake and pole
  Were sharpened where they grew ;
And where the house was built, no axe
  Was lifted up to hew,
But slow and still the Indian worked,
  His wife and children too.

" ' Oh, for a lodge ! ' thus Cowper cried ;
  And here's a peaceful home,
A quiet spot, a calm retreat,
  Where care can seldom come.
Adieu ! thou silent Indian cot,
  My fate it is to roam."

I give the following verses on the Cayman or Alligator of the Amazon because I remember how pleased my brother was with the quotation from Macbeth, which so aptly applies to this dangerous reptile.

" SONG OF THE CAYMAN.

(Written, 1850.)

" 'Thy bones are marrowless, thy blood is cold :
Thou hast no speculation in those eyes
Which thou dost glare with.'

" I bask in the waveless waters
    When the sun is shining on high,
Watching the Indian children
    With a grim and greedy eye ;
Woe to the careless bather
    Who ventures where I lie.

" I float on the midnight waters
    With my deathly demon head ;
My skin is an iron armour
    Which flattens the hunter's lead ;
And my eyes are a living terror,
    Glassy as those of the dead.

" I hear the house-dog prowling,
    And without a ripple sink ;
Down to the stream he cometh
    And enters the water to drink,
I rise again as noiseless
    And seize him on the brink.

" I dwell not in rushing waters,
    But in woodland pool and lake,
Where the cowfish and the turtle
    Lie sleeping 'neath the brake ;
I seize the senseless dreamers,
    And a merry meal I make.

" Midnight deeds have I witness'd,
    But never shudder'd to see.
Tremble not, thou murderer pale !
    Go ! leave the corpse to me,
And not a hair or a whiten'd bone
    I'll leave to speak of thee."

I preserve the next little poem because I feel sure that the first three verses were inspired by the memories of his childhood, while the conclusion indicates those deeper feelings still more dominant in that which follows it.

"Voices.

"I remember voices
    In my early home,
Pleasant and familiar,
    Breathed in sweetest tone—

"Little manly voices,
    Brothers then were near,
Soft and kindly voices ;
    Of my sisters dear.

"Grave and tender voices,
    Voices now no more,
In the ear of childhood
    Whispered golden lore.

"I remember voices,
    Tones of later years,
Passionate and tearful,
    Full of hopes and fears.

"Eloquent and earnest,
    Seeming firm and true,
Trusting to these voices
    I've had cause to rue.

"Friendship's voice deceived me,
    And the maid I loved,
Vain of wealth and beauty,
    False and fickle proved.

"I remember voices,
    Now I hear but one,
The silent voice within me
    Speaks to me alone—

"' Calm amid the tempests,
    Live in peace with me,
Thou shalt learn Earth's wisdom
    And Heaven's mystery.'"

The following poem is probably the last written by my brother. There is no draft or note of it in his rough note-book, and it is written out carefully on a sheet of thin letter-paper which he probably obtained in Para. It was there-fore almost certainly written during the two weeks before his fatal illness.

"Our Better Moments.

" Uncalled they come across the mind,
  We know not why or how,
And with instinctive reverence
  Ignoble feelings bow :
A power strange, yet holy too,
  Breathes through our every sense ;
Each atom of our being feels
  Its subtle influence.
High visions, noble thinkings, flash
  Like meteors through the brain,
If Paradise was lost to us,
  'Tis surely come again !
Better moments ! Better moments ! Ye are sunny angels' wings,
Sent to shed a holier radiance o'er all dim and worldly things.

" Perchance we love to watch awhile,
  In simple child-like mood,
The waving of the summer grass,
  The ebbing of the flood,
Or lie upon a mossy bank
  In some secluded shade,
When sudden, from before our gaze,
  The grass—the waters fade ;
And giving up our being's rein
  To unknown guiding hands,
We float in passive confidence
  To voiceless spirit lands.
Better moments ! Better moments ! Ye are sunny angels' wings,
Sent to shed a holier radiance o'er all dim and worldly things.

" Or sitting in a leafy wood,
  Some still and breathless hour,
The joyous twitter of a bird
  Has strange unconscious power ;
The power to send through ev'ry nerve
  A thrill of soft delight ;
A better moment, like the dawn,
  Steals in with ambient light ;
The soul expands, and lovingly
  Takes in its pure embrace,
All life ! all nature ! high or mean,
  Of colour, tongue, or race.
Better moments ! Better moments ! Ye are sunny angels' wings,
Sent to shed a holier radiance o'er all dim and worldly things.

" A thousand various scenes and tones
        Awake the better thought,
By which our duller years of life
        Become inspired and taught.
In olden times there rudely came
        Handwriting on the wall,
And prostrate souls fell horror-struck
        At that wild spirit-call ;
But now God's momentary gleam
        Is sent into the soul
To guide uncertain wavering feet
        To Life's high solemn goal.
Better moments !   Better moments !   Ye are sunny angels' wings,
Sent to shed a holier radiance o'er all dim and worldly things."

Of the numerous versified enigmas he wrote, I print four
of the best.   They may interest some of my younger readers.
They are not difficult to guess, but I give the solutions at the
end.

ENIGMAS.

I.

" There was a Spanish gentleman
        Of high and noble mien,
Who riding into Seville's town
        One summer's eve was seen ;
He came among us suddenly,
        And vanished as he came ;
We only knew him as my First,
        But never knew his name.

" We saw him at the opera,
        We met him at the ball,
The very point of chivalry
        A pattern for us all ;
And oft upon my Second seen
        Where Seville's beauties came,
But still we knew him as my First,
        And did not know his name.

" 'Twas *I* who brought that gentleman
        From out another clime,
'Twas I upon my Second stood
        With skins of smuggled wine ;

And ye were duller far than me,
Proud gentlemen of Spain,
To only know him as my First,
And never know his name."

II.

(Written in 1847.)

" Know ye my Second, the green and the beautiful,
Sitting alone by the sea,
Weeping in sadness o'er children undutiful,
Woe-worn and pallid is she.

" For skeleton famine is rapidly striding,
Blasting the fruits of the earth,
Many a hovel his victims have died in,
Cursing the hour of their birth.

" Ah ! my First from the heavens has darkly descended,
Wrapping the earth in its gloom ;
The dying lie helpless by corpses extended,
Sullenly waiting their doom.

" And the living watch hopeless the dead and the dying,
All gentler feelings have fled ;
They know not—an hour and they may be lying
Outstretched, and cold with the dead.

" To see their blank features so set and despairing,
To gaze on those dark, tearless eyes
Which look into vacancy listlessly staring,
Might humble the great and the wise.

" Ah ! the great and the wise ! can no way be suggested
By the mighty in power and in soul,
To banish the curse that too long has rested
A shade and a fear on my Whole ? "

III.

" There stood by the stake a sable form,
His grimy arms were bare,
A heavy sledge on his shoulder swung
That had fashioned many a share,
And his dark eyes shone like fiery sparks
From the red-hot iron's glare.

" Open the way !   Fall back !   Fall back !
    And let the victim through,
  To the mocking chant of the bigot priest
    And the muffled drums tattoo ;
  They have tortured him long, but his spirit strong,
    Ne'er cowed 'neath rack or screw.

" My First stepped forth and grasped his arm
    (He felt no muscle shake),
  And led him within the fatal ring ;
    Nor then did his victim quake,
  When a chain was riveted to his waist,
    And round the fatal stake.

" He had seen my Second red with blood
    Of friend and foe and steed,
  He had looked on death in every form,
    He had seen a father bleed ;
  The flames of my Whole were a terrible goal,
    But he could not renounce his creed."

                        IV.

                 (August, 1849.)

" She stood upon the scaffold
    With a firm, undaunted mien,
  Condemned to die a shameful death,
    But yesterday a Queen !
  Ill-fated Jane, how brief thy reign !
    How dark thy closing scene l

" She fearless gazes on my First
    With sable trappings hung,
  And to the bright and glittering axe
    She speaks with jesting tongue :
  ' Fear not to fall, my neck is small,
    Thy work is quickly done.'

" Where are the eyes that fearless gazed ?
    Their lustre now is fled.
  Where is the tongue where hung the jest ?
    Inanimate and dead.
  The snowy neck she used to deck,
    The axe has left it red.

" A ghastly sight it is to see
  My Second bleeding there,
Distorted now those features, erst
  So perfect and so fair ;
No art can dress that gory tress
  Of dark, luxuriant hair.

    *   *   *   *   *

" This is a scene from history's page,
  The triumph of might and wrong ;
That barbarous age has passed away
  With the power of the proud and strong ;
But still in our day by law we slay
  To teach the erring throng.

" To show our abhorrence of shedding blood
  We send the murderer's soul,
Unfit, I ween, to meet his judge,
  To a last and awful goal.
He who can draw good from such law
  Must be my senseless Whole."

SOLUTIONS OF THE ENIGMAS.

1. Donkey. 2. Ireland. 3. Smithfield. 4. Blockhead.

CHAPTER XX

IN LONDON, AND VOYAGE TO SINGAPORE

AMONG the letters preserved and kindly returned to me by
Dr. Spruce is one partly written on board ship on my way
home, giving an account of my somewhat adventurous voyage
while it was fresh in my memory, and containing some details
not given in the narrative in my "Travels on the Amazon."
I will therefore print it here, as no part of it has yet been
made public.

> "Brig *Jordeson*, N. Lat. 49° 30′, W. Long. 20°.
> "Sunday, September 19, 1852.

"MY DEAR FRIEND,

"Having now some prospect of being home in a
week or ten days, I will commence giving you an account of
the peculiar circumstances which have already kept me at sea
seventy days on a voyage which took us only twenty-nine
days on our passage out. I hope you have received the letter
sent you from Para, dated July 9 or 10, in which I informed
you that I had taken my passage in a vessel bound for
London, which was to sail in a few days. On Monday, July
12, I went on board with all my cargo, and some articles purchased or collected on my way down, with the remnant
(about twenty) of my live stock.[1] After being at sea about
a week I had a slight attack of fever, and at first thought I
had got the yellow fever after all. However, a little calomel

---

[1] These consisted of numerous parrots and parrakeets, and several uncommon
monkeys, a forest wild-dog, etc.

set me right in a few days, but I remained rather weak, and
spent most of my time reading in the cabin, which was very
comfortable. On Friday, August 6, we were in N. Lat. 30° 30′,
W. Long. 52°, when, about nine in the morning, just after
breakfast, Captain Turner, who was half-owner of the vessel,
came into the cabin, and said, ' I'm afraid the ship's on fire.
Come and see what you think of it.' Going on deck I found
a thick smoke coming out of the forecastle, which we both
thought more like the steam from heating vegetable matter
than the smoke from a fire. The fore hatchway was immediately opened to try and ascertain the origin of the
smoke, and a quantity of cargo was thrown out, but the
smoke continuing without any perceptible increase, we went
to the after hatchway, and after throwing out a quantity of
piassaba, with which the upper part of the hold was filled,
the smoke became so dense that the men could not stay in
it. Most of them were then set to work throwing in buckets
of water, and the rest proceeded to the cabin and opened the
lazaretto or store-place beneath its floor, and found smoke
issuing from the bulkhead separating it from the hold, which
extended halfway under the fore part of the cabin. Attempts
were then made to break down this bulkhead, but it resisted
all efforts, the smoke being so suffocating as to prevent any
one stopping in it more that a minute at a time. A hole was
then cut in the cabin floor, and while the carpenter was doing
this, the rest of the crew were employed getting out the boats,
the captain looked after his chronometer, sextant, books,
charts and compasses, and I got up a small tin box containing a few shirts, and put in it my drawings of fishes and
palms, which were luckily at hand ; also my watch and a purse
with a few sovereigns. Most of my clothes were scattered
about the cabin, and in the dense suffocating smoke it was impossible to look about after them. There were two boats, the
long-boat and the captain's gig, and it took a good deal of
time to get the merest necessaries collected and put into
them, and to lower them into the water. Two casks of biscuit
and a cask of water were got in, a lot of raw pork and some
ham, a few tins of preserved meats and vegetables, and some

wine. Then there were corks to stop the holes in the boats, oars, masts, sails, and rudders to be looked up, spare spars, cordage, twine, canvas, needles, carpenter's tools, nails, etc. The crew brought up their bags of clothes, and all were bundled indiscriminately into the boats, which, having been so long in the sun, were very leaky and soon became half full of water, so that two men in each of them had to be constantly baling out the water with buckets. Blankets, rugs, pillows, and clothes were all soaked, and the boats seemed overloaded, though there was really very little weight in them. All being now prepared, the crew were again employed pouring water in the cabin and hatchway.

"The cargo of the ship consisted of rubber, cocoa, anatto, balsam-capivi, and piassaba. The balsam was in small casks, twenty stowed in sand, and twenty small kegs in rice-chaff, immediately beneath the cabin floor, where the fire seemed to be. For some time we had heard this bubbling and hissing as if boiling furiously, the heat in the cabin was very great, flame soon broke into the berths and through the cabin floor, and in a few minutes more blazed up through the skylight on deck. All hands were at once ordered into the boats, which were astern of the ship. It was now about twelve o'clock, only three hours from the time the smoke was first discovered. I had to let myself down into the boat by a rope, and being rather weak it slipped through my hands and took the skin off all my fingers, and finding the boat still half full of water I set to baling, which made my hands smart very painfully. We lay near the ship all the afternoon, watching the progress of the flames, which soon covered the hinder part of the vessel and rushed up the shrouds and sails in a most magnificent conflagration. Soon afterwards, by the rolling of the ship, the masts broke off and fell overboard, the decks soon burnt away, the ironwork at the sides became red-hot, and last of all the bowsprit, being burnt at the base, fell also. No one had thought of being hungry till darkness came on, when we had a meal of biscuit and raw ham, and then disposed ourselves as well as we could for the night, which, you may be sure, was by no means a pleasant one. Our boats continued very leaky, and we could not cease an instant from baling; there was a considerable swell, though the day had been remarkably fine, and there were constantly floating around us pieces of the burnt wreck, masts, etc., which might have stove in our boats had we not kept a constant look-out to keep clear of them. We remained near the ship all night in order that we might have the benefit of its flames attracting any vessel that might pass within sight of it.

"It now presented a magnificent and awful sight as it rolled over, looking like a huge caldron of fire, the whole cargo of rubber, etc., forming a liquid burning mass at the bottom. In the morning our little masts and sails were got up, and we bade adieu to the *Helen*, now burnt down to the water's edge, and proceeded with a light east wind towards the Bermudas, the nearest land, but which were more than seven hundred miles from us. As we were nearly in the track of West Indian vessels, we expected to fall in with some ship in a few days.

"I cannot attempt to describe my feelings and thoughts during these events. I was surprised to find myself very cool and collected. I hardly thought it possible we should escape, and I remember thinking it almost foolish to save my watch and the little money I had at hand. However, after being in the boats some days I began to have more hope, and regretted not having saved some new shoes, cloth coat and trousers, hat, etc., which I might have done with a little trouble. My collections, however, were in the hold, and were irretrievably lost. And now I began to think that almost all the reward of my four years of privation and danger was lost. What I had hitherto sent home had little more than paid my expenses, and what I had with me in the *Helen* I estimated would have realized about £500. But even all this might have gone with little regret had not by far the richest part of my own private collection gone also. All my private collection of insects and birds since I left Para was with me, and comprised hundreds of new and beautiful species, which would have rendered (I had fondly hoped) my cabinet, as far as regards American species, one of the finest in Europe.

Fancy your regrets had you lost all your Pyrenean mosses
on your voyage home, or should you now lose all your South
American collection, and you will have some idea of what I
suffer. But besides this, I have lost a number of sketches,
drawings, notes, and observations on natural history, besides
the three most interesting years of my journal, the whole of
which, unlike any pecuniary loss, can never be replaced ; so
you will see that I have some need of philosophic resigna-
tion to bear my fate with patience and equanimity.

"Day after day we continued in the boats. The winds
changed, blowing dead from the point to which we wanted
to go. We were scorched by the sun, my hands, nose, and
ears being completely skinned, and were drenched continually
by the seas or spray. We were therefore almost constantly
wet, and had no comfort and little sleep at night. Our
meals consisted of raw pork and biscuit, with a little
preserved meat or carrots once a day, which was a great
luxury, and a short allowance of water, which left us as
thirsty as before directly after we had drunk it. Ten days
and ten nights we spent in this manner. We were still two
hundred miles from Bermuda, when in the afternoon a vessel
was seen, and by eight in the evening we were on board her,
much rejoiced to have escaped a death on the wide ocean,
whence none would have come to tell the tale. The ship
was the *Jordeson*, bound for London, and proves to be one
of the slowest old ships going. With a favourable wind and
all sail set, she seldom does more than five knots, her average
being two or three, so that we have had a most tedious time
of it, and even now cannot calculate with any certainty as to
when we shall arrive. Besides this, she was rather short of
provisions, and as our arrival exactly doubled her crew, we
were all obliged to be put on strict allowance of bread, meat,
and water. A little ham and butter of the captain's were soon
used up, and we have been now for some time on the poorest
of fare. We have no suet, butter, or raisins with which to
make 'duff,' or even molasses, and barely enough sugar to
sweeten our tea or coffee, which we take with dry, coarse
biscuit, and for dinner, beef or pork of the very worst quality

I have ever eaten or even imagined to exist. This, repeated day after day without any variation, beats even Rio Negro fare, rough though it often was. About a week after we were picked up we spoke and boarded an outward-bound ship, and got from her some biscuits, a few potatoes, and some salt cod, which were a great improvement, but did not last long. We have also occasionally caught some dolphin and a few fish resembling the acarrás of the Rio Negro; but for some time now we have seen none, so that I am looking forward to the 'flesh-pots of Egypt' with as much pleasure as when we were luxuriating daily on farinha and 'fiel amigo.'[1] While we were in the boats we had generally fine weather, though with a few days and nights squally and with a heavy sea, which made me often tremble for our safety, as we heeled over till the water poured in over the boat's side. We had almost despaired of seeing any vessel, our circle of vision being so limited; but we had great hopes of reaching Bermuda, though it is doubtful if we should have done so, the neighbourhood of those islands being noted for sudden squalls and hurricanes, and it was the time of year when the hurricanes most frequently occur. Having never seen a great gale or storm at sea, I had some desire to witness the phenomenon, and have now been completely gratified. The first we had about a fortnight ago. In the morning there was a strong breeze and the barometer had fallen nearly half an inch during the night and continued sinking, so the captain commenced taking in sail, and while getting in the royals and studding-sails, the wind increased so as to split the mainsail, fore-top-sail, fore-trysail, and jib, and it was some hours before they could be got off her, and the main-topsail and fore-sail double reefed. We then went flying along, the whole ocean a mass of boiling foam, the crests of the waves being carried in spray over our decks. The sea did not get up immediately, but by night it was very rough, the ship plunging and rolling most fearfully, the sea pouring in a deluge over the top of her bulwarks, and sometimes up over the cabin skylight. The next

---

[1] This was the name given by our kind host, Señor Henrique, at Barra to dried pirarucú, meaning "faithful friend," always at hand when other food failed.

morning the wind abated, but the ship, which is a very old one, took in a deal of water, and the pumps were kept going nearly the whole day to keep her dry. During this gale the wind went completely round the compass, and then settled nearly due east, where it pertinaciously continued for twelve days, keeping us tacking about, and making less than forty miles a day against it. Three days ago we had another gale, more severe than the former one—a regular equinoctial, which lasted two entire days and nights, and split one of the newest and strongest sails on the ship. The rolling and plunging were fearful, the bowsprit going completely under water, and the ship being very heavily laden with mahogany, fustic, and other heavy woods from Cuba, strained and creaked tremendously, and leaked to that extent that the pumps were obliged to be kept constantly going, and their continued click-clack, click-clack all through the night was a most disagreeable and nervous sound. One day no fire could be made owing to the sea breaking continually into the galley, so we had to eat a biscuit for our dinner ; and not a moment's rest was to be had, as we were obliged to be constantly holding on, whether standing, sitting, or lying, to prevent being pitched about by the violent plunges and lurches of the vessel. The gale, however, has now happily passed, and we have a fine breeze from the north-west, which is taking us along six or seven knots—quicker than we have ever gone yet. Among our other disagreeables here we have no fresh water to spare for washing, and as I only saved a couple of shirts, they are in a state of most uncomfortable dirtiness, but I console myself with the thoughts of a glorious warm bath when I get on shore.

* * * * * *

"*October* 1. Oh, glorious day! Here we are on shore at Deal, where the ship is at anchor. Such a dinner, with our two captains! Oh, beef-steaks and damson tart, a paradise for hungry sinners.

* * * * * *

"*October* 5, London. Here I am laid up with swelled ankles, my legs not being able to stand work after such a

long rest in the ship. I cannot write now at any length—I
have too much to think about. We had a narrow escape in
the Channel. Many vessels were lost in a storm on the night
of September 29, but we escaped. The old 'Iron Duke' is
dead. The Crystal Palace is being pulled down, and is
being rebuilt on a larger and improved plan by a company.
Loddige's collection of plants has been bought entire to
stock it, and they think by heating it in the centre to get a
gradation of climates, so as to be able to have the plants of
different countries, tropical or temperate, in one undivided
building. This is Paxton's plan.

"How I begin to envy you in that glorious country where
'the sun shines for ever unchangeably bright,' where farinha
abounds, and of bananas and plantains there is no lack!
Fifty times since I left Para have I vowed, if I once reached
England, never to trust myself more on the ocean. But
good resolutions soon fade, and I am already only doubtful
whether the Andes or the Philippines are to be the scene of
my next wanderings. However, for six months I am a
fixture here in London, as I am determined to make up for
lost time by enjoying myself as much as possible for awhile.
I am fortunate in having about £200 insured by Mr. Stevens'
foresight, so I must be contented, though it is very hard to
have nothing to show of what I took so much pains to
procure.

"I trust you are well and successful. Kind remembrances
to everybody, everywhere, and particularly to the respectable
Senhor João de Lima of São Joachim.

"Your very sincere friend,
"ALFRED R. WALLACE."

Some of the most alarming incidents, to a landsman, are
not mentioned either in this letter or in my published
"Narrative." The captain had given the only berths in the
cabin to Captain Turner and myself, he sleeping on a sofa in
fine weather, and on a mattress on the floor of the cabin when
rough. On the worst night of the storm I saw him, to my
surprise, bring down an axe and lay it beside him, and on

asking what it was for, he replied, "To cut away the masts in case we capsize in the night." In the middle of the night a great sea smashed our skylight and poured in a deluge of water, soaking the poor captain, and then slushing from side to side with every roll of the ship. Now, I thought, our time is come; and I expected to see the captain rush up on deck with his axe. But he only swore a good deal, sought out a dry coat and blanket, and then lay down on the sofa as if nothing had happened. So I was a little reassured.

Not less alarming was the circumstance of the crew coming aft in a body to say that the forecastle was uninhabitable as it was constantly wet, and several of them brought handfuls of wet rotten wood which they could pull out in many places. This happened soon after the first gale began; so the two captains and I went to look, and we saw sprays and squirts of water coming in at the joints in numerous places, soaking almost all the men's berths, while here and there we could see the places where they had pulled out rotten wood with their fingers. The captain then had the sail-room amid-ships cleared out for the men to sleep in for the rest of the voyage.

One day in the height of the storm, when we were being flooded with spray and enormous waves were coming up behind us, Captain Turner and I were sitting on the poop in the driest place we could find, and, as a bigger wave than usual rolled under us and dashed over our sides, he said quietly to me, "If we are pooped by one of those waves we shall go to the bottom;" then added, "We were not very safe in our two small boats, but I had rather be back in them where we were picked up than in this rotten old tub." It is, therefore, I think, quite evident that we *did* have a very narrow escape. Yet this unseaworthy old ship, which ought to have been condemned years before, had actually taken Government stores out to Halifax, had there been patched up, and sent to Cuba for a cargo of heavy timber, which we were bringing home.

I may here make a few remarks on the cause of the fire, which at the time was quite a mystery to us. We learnt

afterwards that balsam-capivi is liable to spontaneous com-
bustion by the constant motion on a voyage, and it is for that
reason that it is always carried in small kegs and imbedded
in damp sand in the lowest part of the hold. Captain Turner
had never carried any before, and knew nothing of its
properties, and when at the last moment another boat-load
of small kegs of balsam came with no sand to pack them in,
he used rice-chaff which was at hand, and which he thought
would do as well ; and this lot was stored under the cabin
floor, where the flames first burst through and where the fire,
no doubt, originated.

Captain Turner had evidently had no experience of fire
in a ship's cargo, and took quite the wrong way in the
attempt to deal with it. By opening the hatchways to pour
in water he admitted an abundance of air, and this was what
changed a smouldering heat into actual fire. If he had at
once set all hands at work caulking up every crack through
which smoke came out, making the hatchways also air-tight
by nailing tarpaulines over them, no flame could have been
produced, or could have spread far, and the heat due to the
decomposition of the balsam would have been gradually
diffused through the cargo, and in all probability have done
no harm. A few years later a relative of mine returning
home from Australia had a somewhat similar experience, in
which the captain adopted this plan and saved the ship.
When in the Indian Ocean some portion of the cargo was
found to be on fire, by smoke coming out as in our case. But
the captain immediately made all hatches and bulkheads air-
tight ; then had the boats got out and prepared for the worst,
towing them astern ; but he reached Mauritius in safety, and
was there able to extinguish the fire and save the greater
part of the cargo.

On the receipt of my letter Dr. Spruce, who was then,
I think, somewhere on the Rio Negro or Uaupés, wrote to
the "Joao de Lima," referred to by me (and usually mentioned in my "Travels" as Senhor L.), giving him a short
account of my voyage home ; and a few months later he
received a reply from him. He was a Portuguese trader who

had been many years resident in the upper Rio Negro, on whose boat I took a passage for my first voyage up the river, and with whom I lived a long time at Guia. I also went with him on my first voyage up the river Uaupés. He was a fairly educated man, and had an inexhaustible fund of anecdotes of his early life in Portugal, and would also relate many "old-time" stories, usually of the grossest kind, somewhat in the style of Rabelais, or of Chaucer's coarsest Canterbury tales. Old Jeronymo was a quiet old man, a half-bred Indian, or Mameluco as they were called, who lived with Senhor Lima as a humble dependent, assisting him in his business and making himself generally useful. It was these two who were with me during my terrible fever, and who one night gave me up as certain not to live till morning. Dr. Spruce gave me this letter, and as it mainly refers to me, I will here give a nearly literal translation of it.

"San Joaquim, June 7, 1853.

"ILLUSTRISSIMO SENHOR RICARDO SPRUCE,

"I received your greatly esteemed favour dated the 26th April last, and was rejoiced to hear of your honour's health and all the news that you give me, and I was much grieved at the misfortunes which befell our good friend Alfredo! My dear Senhor Spruce, what labours he performed for mankind, and what trouble to lose all his work of four years; but yet his life is saved, and that is the most precious for a man! Do me the favour, when you write to Senhor Alfredo, to give my kind remembrances. The mother of my children also begs you to give her remembrances to Senhor Alfredo, also tell him from me that if he ever comes to these parts again he will find that I shall be to him the same Lima as before, and give him more remembrances from the bottom of my heart, and also to yourself, from

"Yours, with much affection and respect,
"JOAÕ ANTONIO DE LIMA.

"N.B.—Old Jeronymo also asks you to remember him to Senhor Alfredo, and to tell him that he still has the shirt that Senhor Alfredo gave him, and that he is still living a poor wanderer with his friend Lima."

On reaching London in the condition described in my letter to Dr. Spruce, and my only clothing a suit of the thinnest calico, I was met by my kind friend and agent, Mr. Samuel Stevens, who took me first to the nearest ready-made clothes shop, where I got a warm suit, then to his own tailor, where I was measured for what clothes I required, and afterwards to a haberdasher's to get a small stock of other necessaries. Having at that time no relatives in London, his mother, with whom he lived in the south of London—I think in Kennington—had invited me to stay with her. Here I lived most comfortably for a week, enjoying the excellent food and delicacies Mrs. Stevens provided for me, which quickly restored me to my usual health and vigour.

Since I left home, and after my brother John had gone to California in 1849, my sister had married Mr. Thomas Sims, the elder son of my former host at Neath. Mr. Sims had taught himself the then rapidly advancing art of photography, and as my sister could draw very nicely in water-colours, they had gone to live at Weston-super-Mare, and established a small photographic business. As I wished to be with my sister and mother during my stay in England, I took a house then vacant in Upper Albany Street (No. 44), where there was then no photographer, so that we might all live together. While it was getting ready I took lodgings next door, as the situation was convenient, being close to the Regent's Park and Zoological Gardens, and also near the Society's offices in Hanover Square, and within easy access to Mr. Stevens's office close to the old British Museum. At Christmas we were all comfortably settled, and I was able to begin the work which I had determined to do before again leaving England.

In the small tin box which I had saved from the wreck I fortunately had a set of careful pencil drawings of all the different species of palms I had met with, together with notes as to their distribution and uses. I had also a large number of drawings of fish, as already stated, carefully made to scale,

with notes of their colours, their dentition, and their fin-rays
scales, etc. I had also a folio Portuguese note-book containing my diary while on the Rio Negro, and some notes and
observations made for a map of that river and the Uaupés.
With these scanty materials, helped by the letters I had sent
home, I now set to work to write an account of my travels, as
well as a few scientific papers for which I had materials in the
portion of my collections made in Para, Santarem, and the
Lower Rio Negro. These I had sent off before leaving Barra
on my first voyage up the Rio Negro, and they had arrived
home safely; but I had reserved all my private collections
for comparison with future discoveries, and though I left
these to be sent home before starting on my second voyage
up the Rio Negro, they were never despatched, owing to the
Custom House authorities at Barra insisting on seeing the
contents before allowing them to go away. I therefore found
them at Barra on my way home, and they were all lost with
the ship.

I had sent home in 1850 a short paper on the Umbrella
Bird, then almost unknown to British ornithologists, and it
was printed in the Zoological Society's Proceedings for that
year. The bird is in size and general appearance like a
short-legged crow, being black with metallic blue tints on
the outer margins of the feathers. Its special peculiarity is
its wonderful crest. This is formed of a quantity of slender
straight feathers, which grow on the contractile skin of the
top of the head. The shafts of these feathers are white, with
a tufted plume at the end, which is glossy blue and almost
hair-like. When the bird is flying or feeding the crest is laid
back, forming a compact white mass sloping a little upward,
with the terminal plumes forming a tuft behind ; but when
at rest the bird expands the crest, which then forms an
elongated dome of a fine, glossy, deep blue colour, extending
beyond the beak, and thus completely masking the head.
This dome is about five inches long by four or four and a half
inches wide. Another almost equally remarkable feature is a
long cylindrical plume of feathers depending from the lower
part of the neck. These feathers grow on a fleshy tube as

thick as a goose-quill, and about an inch and a half long. They are large and overlap each other, with margins of a fine metallic blue. The whole skin of the neck is very loose and extensible, and when the crest is expanded the neck is inflated, and the cylindrical neck-ornament hangs down in front of it. The effect of these two strange appendages when the bird is at rest and the head turned backwards must be to form an irregular ovate black mass with neither legs, beak, nor eyes visible, so as to be quite unlike any living thing. It may thus be a protection against arboreal carnivora, owls, etc. It is, undoubtedly, one of the most extraordinary of birds, and is an extreme form of the great family of Chatterers, which are peculiar to tropical America. Strange to say, it is rather nearly allied to the curious white bell-bird, so different in colour, but also possessing a fleshy erectile appendage from the base of the upper mandible. The umbrella bird inhabits the lofty forests of the islands of the lower Rio Negro, and some portions of the flooded forests of the Upper Amazon.

About the time when I was collecting these birds (January, 1850) a new species (*Cephalopterus glabricollis*) was brought home by M. Warzewickz from Central America, where a single specimen was obtained on the mountains of Chiriqué at an elevation of eight thousand feet. This is a similar bird, and has a crest of the same form but somewhat less developed ; but the main distinction is that a large patch on the neck is of bare red skin, from the lower part of which hangs the fleshy tube, also red and bare, with only a few feathers, forming a small tuft at its extremity. This species is figured in the " Proceedings of the Zoological Society for 1850 " (p. 92), and will serve to explain my description of the larger species in the same volume (p. 206). Nine years later a third species was discovered in the eastern Andes of Ecuador, which more resembles the original species, but has the feathered dewlap so greatly developed as to be nearly as long as the whole bird. This is figured in *The Ibis* (1859, Pl. III.). The white species which I was told inhabited the Uaupés river has not been found, and may probably have been confounded by my informants with the white bell-bird.

During the two ascents and descents of the Rio Negro and Uaupés in 1850–1852 I took observations with a prismatic compass, not only of the course of the canoe, but also of every visible point, hill, house, or channel between the islands, so as to be able to map this little known river. For the distances I timed our journey by a good watch, and estimated the rate of travel up or down the river, and whether paddling or sailing. With my sextant I determined several latitudes by altitudes of the sun, or of some of the fixed stars. The longitudes of Barra and of San Carlos, near the mouth of the Cassiquiare, had been determined by previous travellers, and my aim was to give a tolerable idea of the course and width of the river between these points, and to map the almost unknown river Uaupés for the first four hundred miles of its course. From these observations I made a large map to illustrate a paper which I read before the Royal Geographical Society. This map was reduced and lithographed to accompany the paper, and as it contains a good deal of information as to the nature of the country along the banks of the rivers, the isolated granite mountains and peaks, with an enlarged map of the river Uaupés, showing the position of the various cataracts I ascended, the Indian tribes that inhabit it, with some of the more important vegetable products of the surrounding forests, it is here given to illustrate this and the two preceding chapters (see p. 320). It will also be of interest to readers who possess my " Travels on the Amazon and Rio Negro," which was published before the map was available.

The great feature of this river is its enormous width, often fifteen or twenty miles, and its being so crowded with islands, all densely forest-clad and often of great extent, that for a distance of nearly five hundred miles it is only at rare intervals that the northern bank is visible from the southern, or *vice versâ*. For the first four hundred and fifty miles of its course the country is a great forest plain, the banks mostly of alluvial clays and sands, though there are occasional patches of sandstone. Then commences the great granitic plateau of the upper river, with isolated mountains and rock-pillars, extending over the watershed to the cataracts of the Orinoko, to the mountains of Guiana, and, perhaps, in some parts up to the foot of the Andes. The other great peculiarity of the river is its dark brown, or nearly black, waters, which are yet perfectly clear and pleasant to drink. This is due, no doubt, to the greater part of the river's basin being an enormous forest-covered plain, and its chief tributaries flowing over granite rocks. It is, in fact, of the same nature as the coffee-coloured waters of our Welsh and Highland streams, which have their sources among peat-bogs. A delightful peculiarity of all these black or clear water rivers is that their shores are entirely free from mosquitoes, as is amusingly referred to in my brother's letter, already quoted in Chapter XVIII.

After my journey the river Uaupés remained unknown to the world for thirty years, when, in 1881 and 1882, Count Ermanno Stradelli, after spending two years in various parts of the Amazon valley, ascending the Purus and Jurua rivers, visited this river to beyond the first cataracts. Having fever he returned to Manaos (Barra), and joined an expedition to determine the boundary between Brazil and Venezuela through an unknown region, and descended the Rio Branco to Manaos. He then went a voyage up the Madeira river, returning home in 1884. In 1887 he again visited South America, ascending the Orinoko, passed through the Cassiquiare to the Rio Negro, and having become much interested in the rock-pictures he had met with in various parts of these rivers, he again made a voyage up the Uaupés, this time penetrating to the Jurupari cataract, which I had failed to reach, and going about a hundred miles beyond it. This last voyage was made in 1890–1891. His only objects seem to have been geographical and anthropological explorations, and he has probably explored a larger number of the great tributaries of the Amazon and Orinoko than any other European.

For a knowledge of this great traveller I am indebted to Mr. Heawood, the librarian of the Royal Geographical Society, who, in reply to my inquiry as to any ascents of the Uaupés since my journey, sent me two volumes of the *Bolletino della Societa Geographica Italiana* (1887 and 1900), which give, so far as he can ascertain, all that is known of Count Stradelli's work. This is most scanty. In the 1887 volume there is a very short abstract of his earlier explorations, with a portion of his journey up the Orinoko in that year. In the volume for 1900 is an article by the Count, almost entirely devoted to a description, with drawings, of all the rock inscriptions which he found in the Uaupés. These drawings are very carefully made, and are twelve in number, each representing a whole rock surface, often containing several groups of forty or fifty distinct figures. It is rather curious that several of the groups in my two plates do not appear in any of the twelve plates of Count Stradelli. Besides these drawings there are several large scale sketch-plans of the portions of the river where they were found, mostly at cataracts or rapids where there are large exposed rock surfaces. The map showing the first three cataracts well illustrates the description of them given at p. 197 of my "Travels." But besides these sketch-plans there is a large folding map of the Uaupés, drawn by Count Stradelli from "compass" bearings during this last journey. There is no reference whatever to this map by the Count himself, except the statement on the title that it is by "compass" observations, as was mine. And as there is no reference to any determinations of longitude the distances could only have been ascertained by estimated rates of canoe-travel, such as I used myself. I therefore compared the two maps with much interest, and found some discrepancies of considerable amount. His map is on a scale rather more than four times that of mine; but my original map, now in the possession of the Geographical Society, is on a larger scale than his. His longitude of the river's mouth is 67° 5′, mine being 68°, more accurate determinations having now been made than were available at the time I prepared my map, more than fifty years ago. On comparing the two maps we see at once a very close agreement in the various curves, sharp bends, loops, and other irregularities of the river's course, so that, omitting the minuter details, the two correspond very satisfactorily. But when we compare the total length of the river to my furthest point, close to the mouth of the Codiary, there is a large difference. The difference of the longitudes

of these two points on the count's map is 2° 22', whereas on
mine it is 3° 45'; my estimate being about 60 per cent. more
than his. By measuring carefully with compasses in lengths
of five miles, with a little allowance for the minuter bends,
his distance is 315 miles, mine 494, mine being thus 55 per
cent. more.

It is unfortunate that Count Stradelli has given us no
information as to how he estimated his distances. In a river
flowing through a densely wooded country, with nowhere
more than a few hundred yards of clear ground on its banks,
with a very crooked and twisted course, and with a current
varying from being scarcely perceptible to such rapidity that
a whole crew of paddlers can hardly make way against it, it
is exceedingly difficult to ascertain the rate of motion in
miles per hour.

Canoes of different sizes do not travel at very different
rates, when each has its complement of men, and I had
taken many opportunities to ascertain this rate in still water.
Then, by noting the time occupied for a particular distance,
say between two of the cataracts, both during the ascent and
descent of the river, the mean of the two would be the time
if there were no current. Making a little allowance for the
load in the canoe, the number or the quality of the rowers,
etc., this time multiplied by the rate of travel in still water
would give the distance. This was the plan I adopted in
making my map of the Uaupés. It is, of course, a mere
approximation, and liable to considerable errors, but I did
not think they would lead to such a large difference of
distance as that between the Count's map and my own. We
have no doubt erred in opposite directions, and the truth lies
somewhere between us; but until some traveller takes a good
chronometer up the river with a sextant for determining local
time, or a telescope of sufficient size to observe eclipses of
Jupiter's satellites, the true length of the river will not be
settled.

In one of the latest atlases, "The Twentieth Century
Citizens' Atlas," by Bartholomew, the position of the Jurupari
fall is 62 per cent. further from the mouth of the river than

on Stradelli's map, which seems to show either that some other traveller has determined the longitude, or that they consider my distances more correct than his.

Another traveller, Dr. T. Koch, only last year (1904) ascended the Uaupés to beyond the Jurupari fall, and also went up the Codiary branch, where he reached an elevated plateau. But it is not stated whether he made any observations to determine the true positions of his farthest point (*The Geographical Journal*, July, 1905, p. 89).

It seems probable, therefore, that the upper course of this great river for a distance of two or three hundred miles is quite unknown. But this is only one indication of the enormous area of country in the central plains of South America, which, except the banks of a few of the larger rivers, is occupied only by widely scattered tribes of Indians, and is as absolutely unknown to civilized man as any portion of the globe. From the Meta river on the north, to the Juambari and Beni rivers on the south, a distance of about twelve hundred miles, and to an equal average distance from the lower slopes of the Andes eastward, is one vast, nearly level, tropical forest, only known or utilized for a few miles from the banks of comparatively few of the rivers that everywhere permeate it. It is to be hoped that in the not remote future this grand and luxuriant country will be utilized, not for the creation of wealth for speculators, but to provide happy homes for millions of families.

As my collections had now made my name well known to the authorities of the Zoological and Entomological Societies, I received a ticket from the former, giving me admission to their gardens while I remained in England, and I was a welcome visitor at the scientific meetings of both societies, which I attended very regularly, and thus made the acquaintance of most of the London zoologists and entomologists. I also went frequently to examine the insect and bird collections in the British Museum (then in Great Russell Street), and also to the Linnean Society, and to the Kew Herbarium to consult works on botany, in order to name my palms.

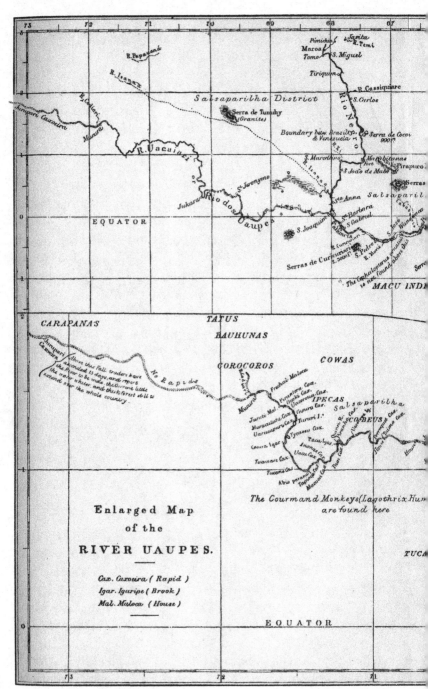

Enlarged Map
of the
RIVER UAUPES.

—

Cax. Caxoeira (Rapid)
Igar. Iguripe (Brook)
Mal. Maloca (House)

—

EQUATOR

The Gourmand Monkeys (Lagothrix Hum
are found here

Published for the Journal of the Royal Geog

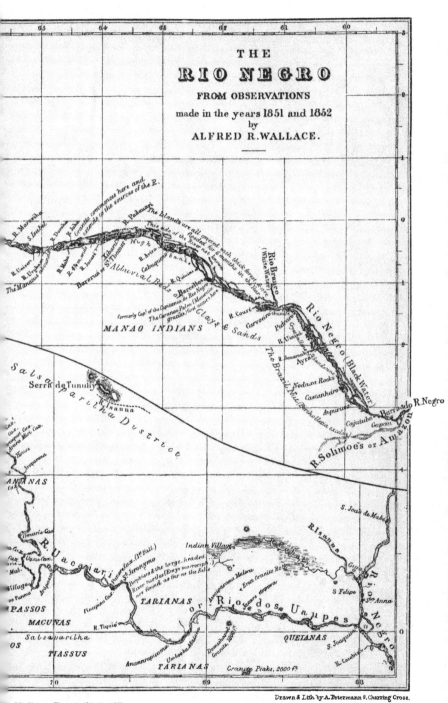

John Murray Albemarle St London 1853.

Drawn & Lith. by A. Petermann 9, Charing Cross.

[To face p. 320, VOL. I.

After discussing the matter with some of my friends, I determined to publish, at my own expense, a small, popular volume on the " Palms of the Amazon and Rio Negro," with an account of their uses and distribution, and figures of all the species from my sketches and specimens of fruits. I arranged with Mr. Walter Fitch of Kew, the first botanical artist of the day, to draw them on stone, adding a few artistic touches to give them life and variety, and in a few cases some botanical details from species living in the gardens. In one of the drawings a large native house on the Uaupés is introduced, with some figures which, I am sorry to say, are as unlike the natives as are the inhabitants of a London slum. I arranged with Mr. Van Voorst to publish this small volume, and it was not thought advisable to print more than 250 copies, the sale of which just covered all expenses.

At the same time I was preparing my " Travels on the Amazon and Rio Negro" from the scanty materials I had saved, supplemented by the letters I had written home. I arranged with Mr. Lovel Reeve for its publication on an agreement for " half profits." Only 750 copies were printed, and when I returned home from the East in 1862, about 250 copies were still unsold, and there were consequently no profits to divide. We agreed, however, to share the remaining copies, and my portion was disposed of by my new publisher, Messrs. Macmillan & Co., and brought me in a few pounds.

I had brought with me vocabularies of about a hundred common words in ten different Indian languages, and as the greatest philologist at that time was the late Dr. R. G. Latham, I obtained an introduction to him, and he kindly offered to write some " Remarks " upon the vocabularies, and these are published in the first edition of my " Travels."

Dr. Latham was at this time engaged in fitting up groups of figures to illustrate the family life and habits of the various races of mankind at the new Crystal Palace at Sydenham, then just completed, and he asked me to meet him there and

see whether any alterations were required in a group of
natives, I think, of Guiana.

I found Dr. Latham among a number of workmen in
white aprons, several life-size clay models of Indians, and a
number of their ornaments, weapons, and utensils. The head
modellers were Italians, and Dr. Latham told me he could
get no Englishmen to do the work, and that these Italians,
although clever modellers of the human figure in any
required attitude, had all been trained in the schools of
classical sculpture, and were unable to get away from this
training. The result was very curious, and often even
ludicrous, a brown Indian man or girl being given the
attitudes and expressions of an Apollo or a Hercules, a
Venus or a Minerva. In those days there were no photographs, and the ethnologist had to trust to paintings or
drawings, usually exaggerated or taken from individuals of
exceptional beauty or ugliness. Under my suggestion
alterations were made both in the features and pose of one
or two of the figures just completed, so as to give them a
little more of the Indian character, and serve as a guide in
modelling others, in which the same type of physiognomy
was to be preserved. I went several times during the work
on the groups of South American origin, but though when
completed, with the real ornaments, clothing, weapons, and
domestic implements, the groups were fairly characteristic
and life-like, yet there remained occasionally details of attitude or expression which suggested classic Greek or Italy
rather than the South American savage.

These ethnological figures, although instructive to the
student, were never very popular, and soon became the
subject of contempt and ridicule. One reason of this was
their arrangement in the open, quite close to the passing
visitor, with nothing to isolate them from altogether incon-
gruous surroundings. Another was, that they were not carefully attended to, and when I saw them after my return
from the East, they had a shabby and dilapidated appearance, and the figures themselves were more or less dusty,
which had a most ludicrous effect in what were intended to

represent living men and women, being so utterly unlike the clear, glossy, living skins of all savage peoples. To be successful and life-like, such groups should be each completely isolated in a deep recess, with three sides representing houses or huts, or the forest, or river-bank, while the open front should be enclosed by a single sheet of plate-glass, and the group should be seen at a distance of at least ten or fifteen feet. In this way, with a carefully arranged illumination from above and an artistic colouring of the figures and accessories, each group might be made to appear as life-like as some of the best figures at Madame Tussaud's, or as the grand interiors of cathedrals, which were then exhibited at the Diorama. In the museum of the future, such groups will find their place in due succession to the groups illustrating the life histories of the other mammalia; but ample space and a very careful attention to details must be given in order to ensure a successful and attractive representation.

It was at this time that I first saw Huxley. At one of the evening meetings of the Zoological Society (in December, 1852) he gave an account of some Echinococci found in the liver of a zebra which died in the gardens. He did not read the paper, but, with the help of diagrams and sketches on the blackboard, showed us clearly its main points of structure, its mode of development, and the strange transformations it underwent when the parent worm migrated from the intestine to other parts of the body of the animal. I was particularly struck with his wonderful power of making a difficult and rather complex subject perfectly intelligible and extremely interesting to persons who, like myself, were absolutely ignorant of the whole group. Although he was two years younger than myself, Huxley had already made a considerable reputation as a comparative anatomist, was a Fellow of the Royal Society, and a few months later was appointed Professor of Natural History and Palæontology at the Royal School of Mines. I was amazed, too, at his complete mastery of the subject, and his great amount of technical knowledge of a kind to which I have never given any attention, the structure and development of the lower forms of animal life. From that time I always looked up to Huxley as being immeasurably superior to myself in scientific knowledge, and supposed him to be much older than I was. Many years afterwards I was surprised to find that he was really younger.

About this time I read before the same Society a few notes on the species of monkeys I had observed on the Amazon, either wild or in a state of captivity, with the particular object of pointing out their peculiarities of distribution. As with butterflies and many birds, I found that both the Amazon and the Rio Negro formed the limit to the range of several species. The rare monkey, *Lagothrix Humboldti*, inhabits the district between the Rio Negro and the Andes, but is quite unknown to the east of that river. A spider-monkey (*Ateles paniscus*) is found in the Guiana district up to the Rio Negro, but not beyond it. The short-tailed *Brachiurus Couxiu* has the same range, while distinct species are found in the Upper Amazon and the Upper Rio Negro. The two species of sloth-monkeys (*Pithecia*) are found one to the north, the other to the south of the Upper Amazon. In several other cases also, as well as with the beautiful trumpeters among birds, the great rivers are found to form the dividing lines between quite distinct species. Four great divisions of eastern equatorial America, which may be termed those of Guiana, Ecuador, Peru, and Brazil, are thus distinctly marked out by the Amazon and its great northern and southern tributaries—the Rio Negro and the Madeira river ; and it seems easy to account for this if we look upon the vast central plains of South America, so little elevated above the sea-level, as having been formerly a gulf or great inland sea which has been gradually filled up by alluvial deposits from the surrounding highlands, and to have been all stocked with forms of life from the three great land-masses of the continent. These would be diversely modified by the different conditions of each of these areas, and as the intervening seas became formed into alluvial plains drained by a great river, that river would naturally

ALFRED R. WALLACE. 1853.

[*To face p.* 324, VOL. I.

form the dividing line between distinct but closely allied species.

It was in the autumn of 1853 that I made my first visit to Switzerland with my friend Mr. George Silk. On our way from London to Dover we had for companion in our compartment a stout, good-humoured American, a New-England manufacturer, going to Paris on business for the first time. He asked us if we could recommend him a good kafe. On telling him we didn't know what a kafe was, he said, "Why, a hotel or eating-house, to be sure; the French call it 'kafe.'" So we told him where we were going for the night, and he went with us. The next day we went on by diligence to Geneva, where we stayed a day, and then walked with our knapsacks to Chamouni; but the heat was so intense that we stayed at a small inn on the way for the night. We walked up to the Flegere to see the grand view of the Aiguilles and Mont Blanc, and the next day joined a party to Montanvert, the Mer de Glace, and the Jardin, having a guide to take care of us. The day was magnificent; we saw the sights of the glacier, its crevasses and ice-tables, and when passing round the precipice of the Couvercle above the ice-fall of the Talefre glacier, there were masses of cloud below us which partially rolled away, revealing the wonderful ice-pinnacles brilliantly illuminated by the afternoon sun, and affording a spectacle the grandeur and sublimity of which I have never since seen equalled. Only a portion of our party reached the Jardin, where I made a hasty collection of the flowers, and by the time we got back to the hotel, having made the steep descent from Montanvert in the dark, we were all pretty well exhausted.

The next day I and my friend walked over the Tete Noir to Martigny. From here we took a chaise to Leuk, and then walked up to Leukerbad and hired a porter to carry our knapsacks up the Gemmi Pass, in order that we might enjoy the ascent of that wonderful mountain road. Before reaching the top snow began to fall, and we reached the little inn on the summit in a snow-storm. It was crowded,

and we had to sleep on the floor. Next day we walked down
to Thun, whence we returned home *viâ* Strasburg and Paris.
Although I enjoyed this my first visit to snowy mountains
and glaciers, I had not at that time sufficient knowledge to
fully appreciate them. The three visits I have since made have
filled me with a deeper sense of the grandeur and the exquisite
scenery of the Alps. My increased general knowledge of
geology, and especially of the glacial theory, have added
greatly to my enjoyment of the great physical features of the
country ; while my continually growing interest in botany
and in the cultivation of plants has invested every detail of
meadow and forest, rock and alp, with beauties and delights
which were almost absent from my early visit. The appreciation of nature grows with years, and I feel to-day more
deeply than ever its mystery and its charms.

During my constant attendance at the meetings of the
Zoological and Entomological Societies, and visits to the
insect and bird departments of the British Museum, I had
obtained sufficient information to satisfy me that the very
finest field for an exploring and collecting naturalist was to
be found in the great Malayan Archipelago, of which just
sufficient was known to prove its wonderful richness, while
no part of it, with the one exception of the island of Java,
had been well explored as regards its natural history. Sir
James Brooke had recently become Rajah of Sarawak, while
the numerous Dutch settlements in Celebes and the Moluccas
offered great facilities for a traveller. So far as known also,
the country was generally healthy, and I determined that it
would be much better for me to go to such a new country than
to return to the Amazon, where Bates had already been successfully collecting for five years, and where I knew there was
a good bird-collector who had been long at work in the upper
part of the river towards the Andes.

As the journey to the East was an expensive one, I was
advised to try and get a free passage in some Government
ship. Through my paper on the Rio Negro, I had made the
acquaintance of Sir Roderick Murchison, then President of

the Royal Geographical Society, and one of the most accessible and kindly of men of science. On calling upon him and stating my wishes, he at once agreed to make an application on my behalf for a passage to some Malayan port, and as he was personally known to many members of the Government and had great influence with them, a passage was promised me on the first ship going to those seas. This was, I think, near the end of the year 1853, when I had published my two books, and had spent much of my spare time at the British Museum, examining the collections, and making notes and sketches, of the rarer and more valuable species of birds, butterflies, and beetles of the various Malay islands.

Among the greatest wants of a collector who wishes to know what he is doing, and how many of his captures are new or rare, are books containing a compact summary with brief descriptions of all the more important known species ; and, speaking broadly, such books did not then nor do now exist. Having found by my experience when beginning botany how useful are even the shortest characters in determining a great number of species, I endeavoured to do the same thing in this case. I purchased the " Conspectus Generum Avium " of Prince Lucien Bonaparte, a large octavo volume of 800 pages, containing a well-arranged catalogue of all the known species of birds up to 1850, with references to descriptions and figures, and the native country and distribution of each species. Besides this, in a very large number—I should think nearly half—a short but excellent Latin description was given, by which the species could be easily determined. In many families (the cuckoos and woodpeckers, for example) every species was thus described, in others a large proportion. As the book had very wide margins I consulted all the books referred to for the Malayan species, and copied out in abbreviated form such of the characters as I thought would enable me to determine each, the result being that during my whole eight years' collecting in the East, I could almost always identify every bird already described, and if I could not do so, was pretty sure that it was a new or undescribed species.

No one who is not a naturalist and collector can imagine the value of this book to me. It was my constant companion on all my journeys, and as I had also noted in it the species not in the British Museum, I was able every evening to satisfy myself whether among my day's captures there was anything either new or rare. Now, such a book is equally valuable to the amateur collector at home in naming and arranging his collections, but to answer the purpose thoroughly it must, of course, be complete—that is, *every* species must be shortly characterized. During the last fifty years it is probable that the described species of birds have doubled in number, yet with slight alteration the whole of these might be included in a volume no larger than that I am referring to. This could be effected by giving only *one name* to each species (that in most general use), whereas Prince Bonaparte has usually given several synonyms and references to figures, so that these occupy fully as much space as the descriptions. These are quite unnecessary for the collector abroad or at home. What he requires is to have a compact and cheap volume by which he can name, if not all, at least all well-marked species. A series of volumes of this character should be issued by the various national museums of the world (each one taking certain groups) and be kept up to date by annual or quinquennial supplements, as in the case of the admirable " List of Plants introduced to Cultivation during the twenty-one years, 1876–1896, issued by the Director of Kew Gardens." In this very compact volume of 420 pages, 7600 species of plants are sufficiently described for identification, while by the use of double columns and thin paper, the volume is only about half the weight of Bonaparte's " Conspectus," in which about the same number of birds are catalogued, but only half of them described. By a division of labour such as is here suggested, the mammals, reptiles, and freshwater fishes might be issued in this form without difficulty. The land and freshwater shells might have separate volumes dealing with the eastern and western hemispheres, or with the separate continents, as might the Diurnal Lepidoptera. The other orders of insects are too extensive to be treated in this

XX]   LONDON: VOYAGE TO SINGAPORE   329

way, but the more attractive families—as the Geodephaga, the Lamellicornes, the Longicornes, and the Buprestidæ among beetles, the bees and wasps among Hysuoptera, might have volumes devoted to them. As these volumes would, if compact and cheap, have a very large sale in every civilized country, they might be issued at a very low price, and would be an immense boon to all amateur collectors, travellers, and residents abroad; and if the chief genera were illustrated by a careful selection of photographic prints, now so easily and economically produced, they would constitute one of the greatest incentives to the study of nature.

The only other book of much use to me was the volume by Boisduval, describing all the known species of the two families of butterflies, the Papilionidæ and Pieridæ. The descriptions by this French author are so clear and precise that every species can be easily determined, and the volume, though dealing with so limited a group, was of immense interest to me. For other families of butterflies and for some of the beetles I made notes and sketches at the British Museum, which enabled me to recognize some of the larger and best known species; but I soon found that so many of the species I collected were new or very rare, that in the less known groups I could safely collect all as of equal importance.

It was, I think, in the latter part of January, 1854, that I received a notification from the Government that a passage had been granted me to Singapore in the brig *Frolic*, shortly sailing for that port, and that I was to communicate with the captain—Commander Nolloth—as to when I should go on board. I think it was about the middle of February that I went to Portsmouth with all necessaries for the voyage, my heavy baggage having been sent off by a merchant ship some time previously. The *Frolic* was anchored at Spithead with a number of other warships. She was about seven hundred tons, and carried, I think, twelve guns. The accommodation was very scanty. I messed with the gun-room officers, and as there was no vacant cabin or berth, the captain very kindly accommodated me in a cot slung in his

cabin, which was a large one, and also provided me with a
small table in one corner where I could write or read quietly.

The captain was a rather small, nervous man, but very
kind and of rather scientific and literary tastes. He wished
to take some deep-sea soundings during the voyage, and to
bring up good samples of the bottom ; and we discussed an
apparatus he was having made for the purpose, in which I
suggested some improvements, which he adopted. Sailing
orders were expected every day, as the ship was quite ready,
with the stores she was taking out to the East all on board ;
but day after day and week after week passed, signals were
exchanged with the admiral, but we seemed no nearer
sailing than when I came on board. It was rather dull work,
but I consoled myself with getting acquainted with the ship
and its ways, the regular routine of which went on, and
everybody seemed as fully occupied as if we were at sea.
The captain had a nice little library in his cabin, among which
the only book I specially remember was a fine Spanish
edition of "Don Quixote." This I intended to read through
during the voyage, as my familiarity with Portuguese and the
small experience of Spanish conversation while in Venezuela
enabled me to understand a good deal of it. But this was
not to be.

Having read almost all Marryat's novels, I was especially
interested in the characters and manners of the various
officers, in whom I found several of Marryat's types reproduced. The captain, as I have said, was nervous, and especially
on everything connected with official etiquette. One day
signals were being made from the admiral's ship, and there
seemed to be some doubt as to what ships it was intended
for. The first-lieutenant asked what they were to do about
it, and the captain was quite excited for fear of a reprimand,
and at last said, "We can only do what the others do. Watch
them and repeat the signals they make." Whether it was
right or not I don't remember. One officer, I think it was
the purser, was the great authority on naval history. His
small cabin had a complete set of the Navy List for fifty years
or more, and every matter in dispute as to what ship was at

a certain station in a given year, or where any particular
officer was stationed, was always referred to him, and if he
could not say off hand, he retired to his cabin for a few
minutes, and then produced the authority, which settled the
question.   The others were nothing remarkable, except the
doctor, who was of the jolly, talkative sort, and seemed
especially to pride himself on his knowledge of seamanship.
One day I remember the captain was summoned by signal to
go on shore to the admiral's office.   It was a cold day with
a strong wind, and there was a very choppy sea on, as there
often is at Spithead.   When the captain's gig came alongside
it was difficult to keep it clear of the ship, it was so tossed
about in sudden and unexpected ways ; and when the captain
had got in, there was a difficulty in getting away, and for a
few moments the boat seemed quite out of command and in
danger of upsetting.   The officers were all looking on with
anxiety, and as soon as the boat had got clear away, it was
the doctor that spoke, and declared that he never saw such
bad seamanship.   They were very near losing the captain !
They were a set of lubbers ! etc. etc.

Finding that I was a bad sailor, I was assured that before
we got to Singapore I should be thoroughly seasoned, for the
brig was what they called a Simonite, a class of ships named
after the designer, which, though stable, were very uncomfortable in bad weather, having a quick jumping motion, which
often made old sailors seasick.   I hoped this was exaggerated,
but looked forward to the ordeal with some dread.   But one
day the captain informed me that he had received fresh
orders to carry stores to the Crimea, where the great war
with Russia was about to commence.   He said that he
regretted the change, because he much preferred the voyage
to Singapore and China, and that he also regretted the loss
of my company ; but as it was, I had better leave the next
morning, and that no doubt the Government would provide
me a passage in some other vessel.   So I bade farewell to
him and his officers, none of whom I ever met again.

On returning to London, I at once call on Sir Roderick
Murchison, and through his representations I received in a

few days a first-class ticket overland to Singapore by the
next Peninsular and Oriental steamer, which sailed in about
a week, so that I did not lose much time. The voyage was
a very interesting one, stopping a few hours at Gibraltar,
passing within sight of the grand Sierra Nevada of Spain,
staying a day at Malta, where the town and the tombs of
the knights were inspected, and then on to Alexandria.
But having by me a long letter I wrote to my schoolfellow, Mr. George Silk, I will here quote from it a few
of the impressions of my journey as they appeared to me
at the time they occurred; and first as to my fellow-
passengers :—

"Our company consists of a few officers and about twenty
cadets for India, three or four Scotch clerks for Calcutta, the
same number of business men for Australia, a Government
interpreter and two or three others for China; a Frenchman;
a Portuguese officer for Goa, with whom I converse; three
Spaniards for the Philippines, very grave; a gentleman and
two ladies, Dutch, going to Batavia; and some English
officers for Alexandria. At Gibraltar we were quarantined
for fear of cholera, then rather prevalent in England, and all
communication with the ship was by means of tongs and a
basin of water, the latter to drop the money in. We had
a morning at Malta, and went on shore from 6 a.m. to 9 a.m.,
walked through the narrow streets, visited the market to
hear the Maltese language, admired the beggar boys and
girls, strolled through the Cathedral of St. John, gorgeous
with marbles and gold and the tombs of the knights. A
clergyman came on board here going to Jerusalem, and a
namesake of my own to Bombay. The latter has a neat
figure, sharp face, and looks highly respectable, not at all like
me! I have found no acquaintance on board who exactly
suits me. One of my cabin mates is going to Australia, and
reads 'How to make Money'—seems to be always thinking
of it, and is very dull and unsociable. The other is one of
the Indian cadets, very aristocratic, great in dressing-case
and jewellery, takes an hour to dress, and persistently studies
the Hindostanee grammar. The Frenchman, the Portuguese,

and the Scotchman I find the most amusing; there is also a little fat Navy lieutenant, who is fond of practical jokes, and has started a Monté Table."

"Steamer *Bengal*, Red Sea, March 26.

"Of all the eventful days in my life (so far), my first in Alexandria was (in some respects) the most exciting. Imagine my feelings when, coming out of the hotel (to which we had been conveyed in an omnibus) with the intention of taking a quiet stroll through the city, I suddenly found myself in the midst of a vast crowd of donkeys and their drivers, all thoroughly determined to appropriate my person to their own use and interest, without in the least consulting my inclinations. In vain with rapid strides and waving arms I endeavoured to clear a way and move forward, arms and legs were seized upon, and even the Christian coat-tails were not sacred from the profane hands of the Mahometan crowd. One would hold together two donkeys by their tails whilst I was struggling between them, and another, forcing their heads together, hoped to compel me to mount one or both of them. One fellow, more impudent than the rest, I laid flat upon the ground, and, sending the little donkey staggering after him, I escaped for a moment midst hideous yells and most unearthly cries. I now beckoned to a fellow more sensible-looking than the rest, and told him that I wished to walk, and would take him as a guide, and now hoped that I might be left at peace. But vain thought! I was in the hands of the Philistines, who, getting me up against a wall, formed around me an impenetrable phalanx of men and brutes, thoroughly determined that I should only escape from the spot upon the four legs of a donkey. So, bethinking myself that donkey-riding was a national institution of venerable antiquity, and seeing a fat Yankee (very like our Paris friend) already mounted, being like myself, hopeless of any other means of escape, I seized upon a bridle in hopes that I should then be left by the remainder of the crowd. But seeing that I was at last going to ride, each one was determined that he alone should profit by the transaction, and a dozen animals were forced suddenly upon me, and a dozen pair of hands tried to lift me upon their respective beasts. But now my patience was exhausted, so, keeping firm hold of the bridle I had first taken with one hand, I hit right and left with the other, and calling upon my guide to do the same, we succeeded in clearing a little space around us. Now, then, behold your long-legged friend mounted upon a jackass in the streets of Alexandria; a boy behind, holding by his tail and whipping him up; Charles, who had been lost sight of in the crowd, upon another; and my guide upon a third; and off we go among a crowd of Jews and Greeks, Turks and Arabs, and veiled women and yelling donkey-boys, to see the city. We saw the bazaars, and the slave market (where I was again nearly pulled to pieces for 'backsheesh'), the mosques with their graceful minarets, and then the pasha's new palace, the interior of which is most gorgeous. We passed lots of Turkish soldiers, walking in comfortable irregularity; and after the consciousness of being dreadful guys for two crowded hours, returned to the hotel, whence we are to start for the canal boats. You may think this little narrative is exaggerated, but it is not so. The pertinacity, vigour, and screams of the Alexandrian donkey-drivers *cannot* be exaggerated. On our way to the boats we passed Pompey's Pillar; for a day we were rowed in small boats on a canal, then on the Nile in barges, with a panorama of mud villages, palm-trees, camels, and irrigating wheels turned by buffaloes,—a perfectly flat country, beautifully green with crops of corn and lentils; endless boats with immense triangular sails. Then the Pyramids came in sight, looking huge and solemn; then a handsome castellated bridge for the Alexandria and Cairo railway; and then Cairo—Grand Cairo! the city of romance, which we reached just before sunset. We took a guide and walked in the city, very picturesque and very dirty. Then to a quiet English hotel, where a Mussulman waiter, rejoicing in the name of Ali-baba, gave us a splendid tea, brown bread and fresh butter. One or two French and English travellers were the only guests, and I could hardly realize my situation. I longed for you to enjoy it with me.

Thackeray's 'First Day in the East' is admirable. Read it again, and you will understand just how I think and feel.

"Next morning at seven we started for Suez in small four-horsed two-wheeled omnibuses, carrying six passengers each. Horses were changed every five miles, and we had a meal every three hours at very comfortable stations. The desert is undulating, mostly covered with a coarse, volcanic-looking gravel. The road is excellent. The skeletons of camels—hundreds of them—lay all along the road ; vultures, sand-grouse, and sand-larks were occasionally seen. We frequently saw the mirage, like distant trees and water. Near the middle station the pasha has a hunting-lodge—a perfect palace. The Indian and Australian mails, about six hundred boxes, as well as all the parcels, goods, and passengers' luggage, were brought by endless trains of camels, which we passed on the way. At the eating-places I took a little stroll, gathering some of the curious highly odoriferous plants that grew here and there in the hollow, which I dried in my pocket-books, and I also found a few land-shells. We enjoyed the ride exceedingly, and reached Suez about midnight. It is a miserable little town, and the bazaar is small, dark, and dirty. There is said to be no water within ten miles. The next afternoon we went on board our ship, a splendid vessel with large and comfortable cabins, and everything very superior to the *Euxine.* Adieu."

I have given this description of my journey from Alexandria to Suez, over the route established by Lieutenant Waghorn, and which was superseded a few years later by the railway, and afterwards by the canal, because few persons now living will remember it, or know that it ever existed. Of the rest of our journey I have no record. We stayed a day at desolate, volcanic Aden, and thence across to Galle, with its groves of cocoa-nut palms, and crowds of natives offering for sale the precious stones of the country ; thence across to Pulo Penang, with its picturesque mountain, its spice-trees, and its waterfall, and on down the Straits of Malacca, with its richly-wooded shores, to our destination, Singapore, where I was to begin the eight years of wandering throughout the Malay Archipelago, which constituted the central and controlling incident of my life.

CHAPTER XXI

THE MALAY ARCHIPELAGO—SINGAPORE, MALACCA, BORNEO

In order not to omit so important a portion of my life as my eight years in the far East, I propose to give a general sketch of my various journeys and their results, told as far as possible in quotations from the few of my letters home that have been preserved, with such connecting facts as may serve to render them intelligible.

Ten days after my arrival at Singapore I wrote home as follows :—" After being a week in a hotel here, I at last got permission to stay with a French Roman Catholic missionary, who lives about eight miles out of town, in the centre of the island, and close to the jungle. The greater part of the inhabitants of Singapore are Chinese, many of whom are very rich, and almost all the villages around are wholly Chinese, who cultivate pepper and gambier, or cut timber. Some of the English merchants have fine country houses. I dined with one, to whom I brought an introduction. His house was spacious, and full of magnificent China and Japan furniture. We are now staying at the mission of Bukit Tima. The missionary (a French Jesuit) speaks English, Malay, and Chinese, and is a very pleasant man. He has built a pretty church here, and has about three hundred Chinese converts."

A month later (May 28th) I wrote—"I am very comfortable here with the missionary. I and Charles go into the jungle every day for insects. The forest here is very similar

to that of South America. Palms are very numerous, but
they are generally small, and very spiny. There are none of
the large majestic species so common on the Amazon. I am
so busy with insects now that I have no time for anything
else. I send now about a thousand beetles to Mr. Stevens,
and I have as many other insects still on hand, which will
form part of my next and principal consignment. Singapore
is rich in beetles, and before I leave I think I shall have a
beautiful collection of them. I will tell you how my day is
now occupied. Get up at half-past five, bath, and coffee.
Sit down to arrange and put away my insects of the day
before, and set them in a safe place to dry. Charles mends
our insect-nets, fills our pin-cushions, and gets ready for the
day. Breakfast at eight; out to the jungle at nine. We
have to walk about a quarter mile up a steep hill to reach
it, and arrive dripping with perspiration. Then we wander
about in the delightful shade along paths made by the
Chinese wood-cutters till two or three in the afternoon,
generally returning with fifty or sixty beetles, some very rare
or beautiful, and perhaps a few butterflies. Change clothes
and sit down to kill and pin insects, Charles doing the flies,
wasps, and bugs; I do not trust him yet with beetles.
Dinner at four, then at work again till six: coffee. Then
read or talk, or, if insects very numerous, work again till eight
or nine. Then to bed."

In July I wrote from "The Jungle, near Malacca:" "We
have been here a week, living in a Chinese house or shed,
which reminds me of some of my old Rio Negro habitations.
We came from Singapore in a small trading schooner, with
about fifty Chinese, Hindoos, and Portuguese passengers, and
were two days on the voyage with nothing but rice and
curry to eat, not having made any special provision, it being
our first experience of the country vessels. Malacca is a
very old Dutch city, but the Portuguese have left the clearest
marks of their possession of it in the common language of
the place being still theirs. I have now two Portuguese
servants, a cook and a hunter, and find myself almost back
in Brazil, owing to the similarity of the language, the people,

and the general aspect of the forest. In Malacca we stayed only two days, being anxious to get into the country as soon as possible. I stayed with a Roman Catholic missionary; there are several here, each devoted to a particular portion of the population—Portuguese, Chinese, and wild Malays of the jungle. The gentleman we were with is building a large church, of which he is architect himself, and superintends the laying of every brick and the cutting of every piece of timber. Money enough could not be raised here, so he took a voyage round the world, and in the United States, California, and India got enough subscribed to finish it. It is a curious and not very creditable thing, that in the English possessions of Singapore and Malacca, there is not a single Protestant missionary; while the conversion, education, and physical and moral improvement of the non-European inhabitants is left entirely to these French missionaries, who, without the slightest assistance from our Government, devote their lives to christianizing and civilizing the varied population under our rule.

"Here the birds are abundant and most beautiful, more so than on the lower Amazon, and I think I shall soon form a fine collection. They are, however, almost all common species, and are of little value, except that I hope they will be better specimens than usually reach England. My guns are both very good, but I find powder and shot actually cheaper in Singapore than in London, so I need not have troubled myself to bring any. So far both I and Charles have had excellent health. He can now shoot pretty well, and is so fond of it that I can hardly get him to do anything else.

"The Chinese here are most industrious. They clear and cultivate the ground with a neatness which I have never seen equalled in the tropics, and they save every particle of manure, both from animals and men, to enrich the ground.

"The country around Malacca is much more beautiful than near Singapore, it being an old settlement with abundance of old fruit and forest trees scattered about. Monkeys of many sorts are abundant; in fact, all animal life *seems* more

abundant than in Brazil. Among the fruits I miss the
delicious oranges of Para and the Amazon. Here they are
scarce and not good, and there is nothing that can replace
them."

I may as well state here that the " Charles " referred to in
the preceding letter was a London boy, the son of a carpenter
who had done a little work for my sister, and whose parents
were willing for him to go with me to learn to be a collector.
He was sixteen years old, but quite undersized for his age,
so that no one would have taken him for more than thirteen
or fourteen. He remained with me about a year and a half,
and learned to shoot and to catch insects pretty well, but not
to prepare them properly. He was rather of a religious turn,
and when I left Borneo he decided to stay with the bishop
and become a teacher. After a year or two, however, he
returned to Singapore, and got employment on some
plantations. About five years later he joined me in the
Moluccas as a collector. He had grown to be a fine young
man, over six feet. When I returned home he remained in
Singapore, married, and had a family. He died some fifteen
years since.

At the end of September I returned to Singapore, whence
I wrote home as follows :—

" I have now just returned to Singapore after two months'
hard work. At Malacca I had a strong touch of fever, with
the old ' Rio Negro ' symptoms, but the Government doctor
made me take large doses of quinine every day for a week,
and so killed it, and in less than a fortnight I was quite well,
and off to the jungle again. I never took half enough quinine
in America to cure me.

"Malacca is a pretty place. Insects are not very abundant
there, still, by perseverance, I got a good number, and many
rare ones. Of birds, too, I made a good collection. I went
to the celebrated Mount Ophir, and ascended to the top,
sleeping under a rock. The walk there was hard work,
thirty miles through jungle in a succession of mud-holes, and
swarming with leeches, which crawled all over us, and sucked

when and where they pleased. We lived a week at the foot
of the mountain, in a little hut built by our men, near a
beautiful rocky stream. I got some fine new butterflies there,
and hundreds of other new or rare insects. Huge centipedes
and scorpions, some nearly a foot long, were common, but we
none of us got bitten or stung. We only had rice, and a little
fish and tea, but came home quite well. The mountain is
over four thousand feet high. Near the top are beautiful
ferns and pitcher-plants, of which I made a small collection.
Elephants and rhinoceroses, as well as tigers, are abundant
there, but we had our usual bad luck in seeing only their
tracks. On returning to Malacca I found the accumulation
of two or three posts—a dozen letters, and about fifty
newspapers. . . . I am glad to be safe in Singapore with
my collections, as from here they can be insured. I have
now a fortnight's work to arrange, examine, and pack
them, and four months hence there will be work for Mr.
Stevens.[1]

"Sir James Brooke is here. I have called on him. He
received me most cordially, and offered me every assistance
at Sarawak. I shall go there next, as the missionary does
not go to Cambodia for some months. Besides, I shall have
some pleasant society at Sarawak, and shall get on in Malay,
which is very easy; but I have had no practice yet, though I
can ask for most common things."

I reached Sarawak early in November, and remained in
Borneo fourteen months, seeing a good deal of the country.
The first four months was the wet season, during which I
made journeys up and down the Sarawak river, but obtained
very scanty collections. In March I went to the Sadong
river, where coal mines were being opened by an English
mining engineer, Mr. Coulson, a Yorkshireman, and I stayed
there nearly nine months, it being the best locality for
beetles I found during my twelve years' tropical collecting,
and very good for other groups. It was also in this place

---

[1] They were sent by sailing ship round the Cape of Good Hope, the overland
route being too costly for goods.

that I obtained numerous skins and skeletons of the orang-
utan, as fully described in my " Malay Archipelago."

In my first letter, dated May, 1855, I gave a sketch of the
country and people :—

" As far inland as I have yet seen this country may be
described as a dead level, and a lofty and swampy forest. It
would, therefore, be very uninviting were it not for a few
small hills which here and there rise abruptly—oases in the
swampy wilderness. It is at one of these that we are located,
a hill covering an area of, perhaps, three or four square miles,
and less than a thousand feet high. In this hill there are
several coal seams ; one of these three feet and a half thick,
of very good coal for steamers, crops out round three-fourths
of the hill, dipping down at a moderate angle. We have here
near a hundred men, mostly Chinese ; ground has been cleared,
and houses built, and a road is being made through the jungle,
a distance of two miles, to the Sadong river, where the coal
will be shipped.

" The jungle here is exceedingly gloomy and monotonous ;
palms are scarce, and flowers almost wanting, except some
species of dwarf gingerworts. It is only high overhead that
flowers can be seen. There are many fine orchids of the
genus cælogyne, with great drooping spikes of white or
yellow flowers, and occasionally bunches of the scarlet flowers
of a magnificent creeper, a species of æschynanthus. Oak
trees are rather common, and I have already noticed three
species having large acorns of a red, brown, and black colour
respectively.

" Our mode of life here is very simple, and we have a continual struggle to get enough to eat, as all fowls and vegetables
grown by the Dyaks go to Sarawak, and I have been obliged
to send there to buy some.

" The old men here relate with pride how many ' heads '
they took in their youth ; and though they all acknowledge
the goodness of the present rajah, yet they think that if they
were allowed to take a few heads, as of old, they would have
better crops. The more I see of uncivilized people, the better
I think of human nature on the whole, and the essential

differences between civilized and savage man seem to disappear. Here we are, two Europeans, surrounded by a population of Chinese, Malays, and Dyaks. The Chinese are generally considered, and with some amount of truth, to be thieves, liars, and reckless of human life, and these Chinese are coolies of the lowest and least educated class, though they can all read and write. The Malays are invariably described as being barbarous and bloodthirsty ; and the Dyaks have only recently ceased to think head-taking a necessity of their existence. We are two days' journey from Sarawak, where, though the government is nominally European, it only exists with the consent and by the support of the native population. Yet I can safely say that in any part of Europe where the same opportunities for crime and disturbance existed, things would not go so smoothly as they do here. We sleep with open doors, and go about constantly unarmed ; one or two petty robberies and a little fighting have occurred among the Chinese, but the great majority of them are quiet, honest, decent sort of people. They did not at first like the strictness and punctuality with which the English manager kept them to their work, and two or three ringleaders tried to get up a strike for shorter hours and higher wages, but Mr. Coulson's energy and decision soon stopped this by discharging the ringleaders at once, and calling all the Malays and Dyaks in the neighbourhood to come up to the mines in case any violence was attempted. It was very gratifying to see how rapidly they obeyed the summons, knowing that Mr. Coulson represented the rajah, and this display of power did much good, for since then everything has gone on smoothly. Preparations are now making for building a 'joss-house,' a sure sign that the Chinese have settled down contentedly."

In my next letter, a month later, I gave the following account of an interesting episode :—

"I must now tell you of the addition to my household of an orphan baby, a curious little half-nigger baby, which I have nursed now more than a month. I will tell you presently how I came to get it, but must first relate my inventive

skill as a nurse. The little innocent was not weaned, and I
had nothing proper to feed it with, so was obliged to give it
rice-water. I got a large-mouthed bottle, making two holes
in the cork, through one of which I inserted a large quill so
that the baby could suck. I fitted up a box for a cradle
with a mat for it to lie upon, which I had washed and
changed every day. I feed it four times a day, and wash it
and brush its hair every day, which it likes very much, only
crying when it is hungry or dirty. In about a week I gave
it the rice-water a little thicker, and always sweetened it to
make it nice. I am afraid you would call it an ugly baby,
for it has a dark brown skin and red hair, a very large mouth,
but very pretty little hands and feet. It has now cut its two
lower front teeth, and the uppers are coming. At first it
would not sleep alone at night, but cried very much ; so I
made it a pillow of an old stocking, which it likes to hug, and
now sleeps very soundly. It has powerful lungs, and sometimes screams tremendously, so I hope it will live.

"But I must now tell you how I came to take charge of
it. Don't be alarmed ; I was the cause of its mother's death.
It happened as follows :—I was out shooting in the jungle
and saw something up a tree which I thought was a large
monkey or orang-utan, so I fired at it, and down fell this little
baby—in its mother's arms. What she did up in the tree of
course I can't imagine, but as she ran about the branches
quite easily, I presume she was a wild 'woman of the woods ;'
so I have preserved her skin and skeleton, and am trying to
bring up her only daughter, and hope some day to introduce
her to fashionable society at the Zoological Gardens. When
its poor mother fell mortally wounded, the baby was plunged
head over ears in a swamp about the consistence of pea-
soup, and when I got it out looked very pitiful. It clung to
me very hard when I carried it home, and having got its little
hands unawares into my beard, it clutched so tight that I
had great difficulty in extricating myself. Its mother, poor
creature, had very long hair, and while she was running
about the trees like a mad woman, the little baby had to
hold fast to prevent itself from falling, which accounts for the

remarkable strength of its little fingers and toes, which catch hold of anything with the firmness of a vice. About a week ago I bought a little monkey with a long tail, and as the baby was very lonely while we were out in the daytime, I put the little monkey into the cradle to keep it warm. Perhaps you will say that this was not proper. 'How could you do such a thing?' But, I assure you, the baby likes it exceedingly, and they are excellent friends. When the monkey wants to run away, as he often does, the baby clutches him by the tail or ears and drags him back; and if the monkey does succeed in escaping, screams violently till he is brought back again. Of course, baby cannot walk yet, but I let it crawl about on the floor to exercise its limbs; but it is the most wonderful baby I ever saw, and has such strength in its arms that it will catch hold of my trousers as I sit at work, and hang under my legs for a quarter of an hour at a time without being the least tired, all the time trying to suck, thinking, no doubt, it has got hold of its poor dear mother. When it finds no milk is to be had, there comes another scream, and I have to put it back in its cradle and give it 'Toby'—the little monkey—to hug, which quiets it immediately. From this short account you will see that my baby is no common baby, and I can safely say, what so many have said before with much less truth, 'There never was such a baby as my baby,' and I am sure nobody ever had such a dear little duck of a darling of a little brown hairy baby before."

In a letter dated Christmas Day, 1855, I gave my impressions of the Dyaks, and of Sir James Brooke, as follows:—

"I have now lived a month in a Dyak's house, and spent a day or two in several others, and I have been very much pleased with them. They are a very kind, simple, hospitable people, and I do not wonder at the great interest Sir James Brooke takes in them. They are more communicative and more cheerful than the American Indians, and it is therefore more agreeable to live with them. In moral character they are far superior to either the Malays or the Chinese, for though head-taking was long a custom among them, it was only as a trophy of war. In their own villages crimes are very rare. Ever since Sir James Brooke has been rajah, more than twelve years, there has only been one case of murder in a Dyak tribe, and that was committed by a stranger who had been adopted into the tribe. One wet day I produced a piece of string to show them how to play 'cat's cradle,' and was quite astonished to find that they knew it much better than I did, and could make all sorts of new figures I had never seen. They were also very clever at tricks with string on their fingers, which seemed to be a favourite amusement. Many of the remoter tribes think the rajah cannot be a man. They ask all sorts of curious questions about him—Whether he is not as old as the mountains; whether he cannot bring the dead to life; and I have no doubt, for many years after his death, he will be held to be a deity and expected to come back again.

"I have now seen a good deal of Sir James, and the more I see of him the more I admire him. With the highest talents for government he combines in a high degree goodness of heart and gentleness of manner. At the same time, he has so much self-confidence and determination that he has put down with the greatest ease the conspiracies of one or two of the Malay chiefs against him. It is a unique case in the history of the world for a private English gentleman to rule over two conflicting races—a superior and an inferior —with their own consent, without any means of coercion, but depending solely upon them both for protection and support, while at the same time he introduces some of the best customs of civilization, and checks all crimes and barbarous practices that before prevailed. Under his government 'running-a-muck,' so frequent in other Malay countries, has never taken place, and in a population of about 30,000 Malays, almost all of whom carry their *kris*, and were accustomed to revenge an insult with a stab, murders only occur once in several years. The people are never taxed except with their own consent, and in the manner most congenial to them, while almost the whole of the rajah's private fortune has been spent in the improvement of the country or for its

benefit. Yet this is the man who has been accused in England of wholesale murder and butchery of unoffending tribes
to secure his own power!"

In my next letter (from Singapore in February, 1856) I
say—"I have now left Sarawak, where I began to feel quite
at home, and may perhaps never return to it again, but I
shall always look back with pleasure to my residence there
and to my acquaintance with Sir James Brooke, who is a
gentleman and a nobleman in the truest and best sense of
those words."

At the end of this letter I make some remarks on the
Crimean War, then almost concluded, and though I afterwards saw reason to change my opinion as regards this
particular war, my views then as to the menace of Russian
power to civilization are not altogether inapplicable at the
present day. I say—"The warlike stores found in Sebasto-
pol are alone a sufficient justification of the war. For what
purpose were four thousand cannon and other stores in proportion accumulated there for if not to take Constantinople,
get a footing in the Mediterranean, and ultimately to subju-
gate Europe? And why do such tremendous fortresses exist
in every part of the frontiers of Russia, if not to render herself invulnerable from the attacks which she has determined
by her ambitious designs to bring upon her? Russia is per-
petually increasing her means both of defence and of aggres-
sion ; if she had continued unmolested for a few years longer,
it would have cost still greater sacrifices to subdue her. The
war, therefore, is absolutely necessary as the only means of
teaching Russia that Europe will not submit to the indefinite
increase of her territory and power, and the constant menace
of her thousands of cannons and millions of men. It is the
only means of saving Europe from a despotism as much worse
than that of Napoleon as the Russian people are behind the
French in civilization."

There is a certain amount of truth in this, but to avoid
misconception I wish to state that I think the danger does
not arise from the Russian Government being any worse
than our own, or than the Governments of Germany or

France. All have the same insatiable craving for extending their territories and ruling subject peoples for the benefit of their own upper classes. Russia is only the most dangerous because she is already so vast, and each fresh extension of her territory adds to her already too large population, from which to create enormous armies, which she can and will use for further aggrandizement. It is a disgrace to Europe that they have allowed Russia to begin the dismemberment of China, and to leave to Japan the tremendous task of putting a check to her progress.

A later letter from Singapore touches on two matters of some interest. " I quite enjoy being a short time in Singapore again. The scene is at once so familiar and yet so strange. The half-naked Chinese coolies, the very neat shopkeepers, the clean, fat, old, long-tailed merchants, all as pushing and full of business as any Londoners. Then the handsome, dark-skinned *klings* from southern India, who always ask double what they will take, and with whom it is most amusing to bargain. The crowd of boatmen at the ferry, a dozen begging and disputing for a farthing fare ; the tall, well-dressed Armenians ; the short, brown Malays in their native dress ; and the numerous Portuguese clerks in black, make up a scene doubly interesting to me now that I know something about them, and can talk to them all in the common language of the place—Malay. The streets of Singapore on a fine day are as crowded and busy as Tottenham Court Road, and from the variety of nationalities and occupations far more interesting. I am more convined than ever that no one can appreciate a new country by a short visit. After two years in the East I only now begin to understand Singapore, and to thoroughly appreciate the life and bustle, and the varied occupations of so many distinct nationalities on a spot which a short time ago was an uninhabited jungle. A volume might be written upon it without exhausting its humours and its singularities. . . .

" I have been spending three weeks with my old friend the French Jesuit missionary at Bukit Tima, going daily into the jungle, and every Friday fasting on omelet and vegetables, a most wholesome custom, which the Protestants erred in leaving off. I have been reading Huc's ' Travels ' in French, and talking a good deal with one of the missionaries just arrived from Tonquin, who can speak no English. I have thus obtained a good deal of information about these countries, and about the extent of the Catholic missions in them, which is really astonishing. How is it that they do their work so much more thoroughly than most Protestant missions ? In Cochin China, Tonquin, and China, where Christian missionaries are obliged to live in secret, and are subject to persecution, expulsion, or death, every province, even those farthest in the interior of China, has its regular establishment of missionaries constantly kept up by fresh supplies, who are all taught the languages of the countries they are going to at Penang or Singapore. In China there are near a million of Catholics, in Tonquin and Cochin China more than half a million. One secret of their success is their mode of living. Each missionary is allowed about £30 a year, on which he lives in whatever country he may be. This has two good results. A large number of missionaries can be kept on limited funds, and the people of the country in which they reside, seeing that they live in poverty and with none of the luxuries of life, are convinced that they are sincere. Most of them are Frenchmen, and those I have seen or heard of are well-educated men, who give up their lives to the good of the people they live among. No wonder they make converts, among the lower orders principally ; for it must be a great blessing to these poor people to have a man among them to whom they can go in any trouble or distress, whose sole object is to advise and help them, who visits them in sickness and relieves them in want, and whom they see living in continual danger of persecution and death only for *their* benefit."

Before leaving Singapore I wrote a long letter to my old fellow traveller and companion, Henry Walter Bates, then collecting on the Upper Amazon, almost wholly devoted to entomology, and especially giving my impressions of the comparative richness of the two countries. As this comparison is of interest not only to entomologists but to all students of the geographical distribution of animals, I give it here almost entire. The letter is dated April 30, 1856:—

"I must first inform you that I have just received the *Zoologist* containing your letters up to September 14, 1855 (Ega), which have interested me greatly, and have almost made me long to be again on the Amazon, even at the cost of leaving the unknown Spice Islands still unexplored. I have been here since February waiting for a vessel to Macassar (Celebes), a country I look forward to with the greatest anxiety and with expectations of vast treasures in the insect world. Malacca, Sumatra, Java, and Borneo form but one zoological province, the *majority* of the species in all classes of animals being common to two or more of these countries. There is decidedly less difference between them than between Para and Santarem or Barra. I have therefore as yet only visited the best known portion of the Archipelago, and consider that I am now about to commence my real work. I have spent six months in Malacca and Singapore, and fifteen months in Borneo (Sarawak), and have therefore got a good idea of what this part of the Archipelago is like. Compared with the Amazon valley, the great and striking feature here is the excessive poverty of the Diurnal Lepidoptera. The glorious Heliconidæ are represented here by a dozen or twenty species of generally obscure-coloured Euplæas, the Nymphalidæ containing nothing comparable with Epicalias, Callitheas, Catagrammas, etc., either in variety or abundance to make up for their want of brilliancy. A few species of Adolias, Limentis, and Charaxes are the most notable forms. The Satyridæ have nothing to be placed by the side of the lovely Hæteras of the Amazon. Your glorious Erycinidæ are represented by half a dozen rather inconspicuous species, and even the Lycænidæ, though more numerous and comprising some lovely species, do not come up to the Theclas of Para. Even the dull Hesperidæ are almost wanting here, for I do not think I have yet exceeded a dozen species of this family. All this is very miserable and discouraging to one who has wandered in the forest-paths around Para or on the sandy shores of the Amazon or Rio Negro. The only group in which we may consider the two countries to be about equal is that of the true Papilios (including Ornithoptera), though even in these I think you have more species. Including Ornithoptera and Leptocircus, I have found as yet only thirty species, five of which I believe are new. Among these is the magnificent *Ornithoptera Brookeana*, perhaps the most elegant butterfly in the world.

" To counterbalance this dearth of butterflies there should be an abundance of other orders, or you will think I have made a change for the worse, and compared with Para only perhaps there is, though it is doubtful whether at Ega you have not found Coleoptera quite as abundant as they are here. But I will tell you my experience so far and then you can decide the question, and let me know *how* you decide it. You must remember that it is now just two years since I reached Singapore, and out of that time I have lost at least six months by voyages and sickness, besides six months of an unusually wet season at Sarawak. However, during the dry weather at Sarawak I was very fortunate in finding a good locality for beetles, at which I worked hard for five or six months. At Singapore and Malacca I collected about a thousand species of beetles, at Sarawak about two thousand, but as about half my Singapore species occurred also at Sarawak, I reckon that my total number of species may be about 2500. The most numerous group is (as I presume with you) the Rhyncophora (weevils, etc.), of which I have at least 600 species, perhaps many more. The majority of these are very small, and all are remarkably obscure in their colours, being in this respect inferior to some of our British species. There are, however, many beautiful and interesting forms, especially among the Anthribidæ, of one of which—a new genus—I send a rough sketch. The group next in point of numbers and, to me, of the highest interest are the Longicorns. Of these I obtained fifty species in the first ten days at Singapore, and when in a good locality I seldom passed a day without getting a new one. At Malacca and Singapore I collected about 160 species, at Sarawak 290, but as only about fifty from the former places occurred at the latter, my Longicorns must now reach about 400 species. . . . As to size, I have only about thirty species which exceed an inch in length, the majority being from one half to three quarters of an inch, while a considerable number are two or three lines only. I see you say you must have near 500 species of Longicorns ; but I do not know if this refers to Ega only, or to your whole South American collections.

"The Geodephaga, always rare in the tropics, we must expect to be still more so in a level forest country so near the equator, yet I have found more species than I anticipated— as nearly as I can reckon, a hundred—twenty-four being Cicindelidæ (tiger beetles) of various groups.

"Lamellicorns are very scarce, about one hundred and forty species in all, of which twenty-five are Cetoniidæ, all rare, and about the same number of Lucanidæ. Elaters are rather plentiful, but with few exceptions small and obscure. I have one hundred and forty species, one nearly three inches long, and several of one and a half inch. The Buprestidæ are exceedingly beautiful, but the larger and finer species are very rare. I have one hundred and ten species, of which half are under one-third of an inch long, though one, *Catoxantha bicolor*, is two and a half inches. Two genera of Cleridæ are rather abundant, others rare ; but I have obtained about fifty species, which, compared with the very few previously known, is very satisfactory. Of the remaining groups, in which I took less interest, I have not accurately noted the number of species.

"The individual abundance of beetles is not, however, so large as the number of species would indicate. I hardly collect on an average more than fifty beetles a day, in which number there will be from thirty to forty species. Often, in fact, twenty or thirty beetles are as much as I can scrape together, even when giving my whole attention to them, for butterflies are too scarce to distract it. Of the other orders of insects, I have no accurate notes ; the species, however, of all united (excluding Lepidoptera) about equal those of the beetles. I found one place only where I could collect moths, and have obtained altogether about one thousand species, mostly of small or average size. My total number of species of insects, therefore, I reckon at about six thousand, and of specimens collected about thirty thousand. From these data I think you will be able to form a pretty good judgment of the comparative entomological riches of the two countries. The matter, however, will not be definitely settled till I have visited Celebes, the Moluccas, etc., which I hope to find as much superior to the western group of islands as the Upper is to the Lower Amazon.

" In other branches of Natural History I have as yet done little. The birds of Malacca and Borneo, though beautiful, are too well known to be worth collecting largely. With the orang-utans I was successful, obtaining fifteen skins and skeletons, and proving, I think, the existence of two species, hitherto a disputed question. The forests here are scarcely to be distinguished from those of Brazil, except by the frequent presence of the various species of Calamus (Rattan palms) and the Pandani (Screw pines), and by the rarity of those Leguminous trees with finely divided foliage, which are so frequent in the Amazonian forests. The people and their customs I hardly like as well as those of Brazil, but the comparatively new settlements of Singapore and Sarawak are not quite comparable with the older towns of the Amazon. Here provisions and labour are dear, and travelling is both tedious and expensive. Servants' wages are high, and the customs of the country do not permit you to live in the free-and-easy style of Brazil.

*   *   *   *   *   *

" I must tell you that the fruits of the East are a delusion. Never have I seen a place where fruits are more scarce and poor than at Singapore. In Malacca and Sarawak they are more abundant, but there is nothing to make up for the deficiency of oranges, which are so poor and sour that they would hardly be eaten even in England. There are only two good fruits, the mangosteen and the durian. The first is a

very delicate juicy fruit, but hardly worthy of the high place
that has been given it; the latter, however, is a wonderful
fruit, quite unique of its kind, and worth coming to the Malay
Archipelago to enjoy; it is totally unlike every other fruit.
A thick glutinous, almond-flavoured custard is the only thing
it can be compared to, but which it far surpasses. These
two fruits, however, can only be had for about two months in
the year, and everywhere, except far into the interior, they
are dear. The plantains and bananas even are poor, like the
worst sorts in South America.

"*May* 10*th*.—The ship for which I have been waiting
nearly three months is in at last, and in about a week I hope
to be off for Macassar. The monsoon, however, is against
us, and we shall probably have a long passage, perhaps forty
days. Celebes is quite as unknown as was the Upper
Amazon before your visit to it, perhaps even more so. In
the British Museum catalogues of Cetoniidæ, Buprestidæ,
Longicorns, and Papilionidæ, not a single specimen is
recorded from Celebes, and very few from the Moluccas;
but the fine large species described by the old naturalists,
some of which have recently been obtained by Madame
Reiffer, give promise of what systematic collection may
produce."

Before giving a general sketch of my life and work in less
known parts of the Archipelago, I must refer to an article I
wrote while in Sarawak, which formed my first contribution
to the great question of the origin of species. It was written
during the wet season, while I was staying in a little house
at the mouth of the Sarawak river, at the foot of the Santu-
bong mountain. I was quite alone, with one Malay boy as
cook, and during the evenings and wet days I had nothing to
do but to look over my books and ponder over the problem
which was rarely absent from my thoughts. Having always
been interested in the geographical distribution of animals
and plants, having studied Swainson and Humboldt, and
having now myself a vivid impression of the fundamental
differences between the Eastern and Western tropics; and

having also read through such books as Bonaparte's "Conspectus," already referred to, and several catalogues of insects and reptiles in the British Museum (which I almost knew by heart), giving a mass of facts as to the distribution of animals over the whole world, it occurred to me that these facts had never been properly utilized as indications of the way in which species had come into existence. The great work of Lyell had furnished me with the main features of the succession of species in time, and by combining the two I thought that some valuable conclusions might be reached. I accordingly put my facts and ideas on paper, and the result seeming to me to be of some importance, I sent it to *The Annals and Magazine of Natural History,* in which it appeared in the following September (1855). Its title was "On the Law which has regulated the Introduction of New Species," which law was briefly stated (at the end) as follows : "*Every species has come into existence coincident both in space and time with a pre-existing closely-allied species.*" This clearly pointed to some kind of evolution. It suggested the *when* and the *where* of its occurrence, and that it could only be through natural generation, as was also suggested in the "Vestiges"; but the *how* was still a secret only to be penetrated some years later.

Soon after this article appeared, Mr. Stevens wrote me that he had heard several naturalists express regret that I was "theorizing," when what we had to do was to collect more facts. After this, I had in a letter to Darwin expressed surprise that no notice appeared to have been taken of my paper, to which he replied that both Sir Charles Lyell and Mr. Edward Blyth, two very good men, specially called his attention to it. I was, however, rewarded later, when in Huxley's chapter, "On the Reception of the Origin of Species," contributed to the "Life and Letters," he referred to this paper as—"his powerful essay," adding—"On reading it afresh I have been astonished to recollect how small was the impression it made" (vol. ii. p. 185). The article is reprinted in my "Natural Selection and Tropical Nature."

CHAPTER XXII

CELEBES, THE MOLUCCAS, NEW GUINEA, TIMOR, JAVA, AND SUMATRA

HAVING been unable to find a vessel direct to Macassar, I took passage to Lombok, whence I was assured I should easily reach my destination. By this delay, which seemed to me at the time a misfortune, I was enabled to make some very interesting collections in Bali and Lombok, two islands which I should otherwise never have seen. I was thus enabled to determine the exact boundary between two of the primary zoological regions, the Oriental and the Australian, and also to see the only existing remnant of the Hindu race and religion, and of the old civilization which had erected the wonderful ruined temples in Java centuries before the Mohammedan invasion of the archipelago.

After two months and a half in Lombok, I found a passage to Macassar, which I reached the beginning of September, and lived there nearly three months, when I left for the Aru Islands in a native prau. The country around Macassar greatly disappointed me, as it was perfectly flat and all cultivated as rice fields, the only sign of woods being the palms and fruit trees in the suburbs of Macassar and others marking the sites of native villages. I had letters to a Dutch merchant who spoke English as well as Malay and the Bugis language of Celebes, and who was quite friendly with the native rajah of the adjacent territory. Through his good offices I was enabled to stay at a native village about eight miles inland, where there were some patches

NATIVE HOUSE, WOKAN, ARU ISLANDS

of forest, and where I at once obtained some of the rare birds and insects peculiar to Celebes. After about a month I returned to Macassar, and found that I could obtain a passage to the celebrated Aru Islands, where at least two species of birds of paradise are found, and which had never been visited by an English collector. This was a piece of good fortune I had not expected, and it was especially fortunate because the next six months would be wet in Celebes, while it would be the dry season in the Aru Islands. This journey was the most successful of any that I undertook, as is fully described in my book; and as no letters referring to it have been preserved, I shall say no more about it here.

The illustration opposite is from a photograph of a native house in the island of Wokan, which was given me by the late Professor Moseley of the *Challenger* expedition, because it so closely resembles the hut in which I lived for a fortnight, and where I obtained my first King bird of paradise, that I feel sure it must be the same, especially as I saw no other like it. It is described at the beginning of chapter xxxi. of my "Malay Archipelago," and will be of interest to such of my readers as possess that work.

Several months later I arrived again at Macassar, and after arranging and despatching my Aru collections, I went to an estate a few days' journey north, the property of a brother of my kind friend Mr. Mesman. I had a house built for me in a patch of forest where I lived with two Malay servants for three months making very interesting collections both of birds and insects; and I have rarely enjoyed myself so much as I did here. About the end of November I returned to Macassar, and in December embarked on the Dutch mail steamer for Amboyna, calling by the way at Timor and at Banda.

At Amboyna I made the acquaintance of a German and a Hungarian doctor, both entomologists, and in a fortnight's visit to an estate in the interior surrounded by virgin forest I obtained some of the lovely birds and gorgeous insects which have made the island celebrated. The only letter I

possess which indicates something of my opinions and antici-
pations at this period of my travels is one to Bates, dated
Amboyna, January 4, 1858, from which I will make a few
extracts. The larger portion is occupied with remarks on the
comparative riches of our respective regions in the various
families of beetles, founded on a letter I had received from
him a few months before, which, though very interesting to
entomologists, are not suitable for reproduction here. I then
touched on the subject of my paper referred to at the end of
the last chapter.

"To persons who have not thought much on the subject
I fear my paper on the 'Succession of Species' will not
appear so clear as it does to you. That paper is, of course,
merely the announcement of the theory, not its development.
I have prepared the plan and written portions of a work
embracing the whole subject, and have endeavoured to prove
in detail what I have as yet only indicated. It was the promulgation of Forbes's theory of 'polarity' which led me to
write and publish, for I was annoyed to see such an ideal
absurdity put forth, when such a simple hypothesis will
explain all the facts. I have been much gratified by a letter
from Darwin, in which he says that he agrees with 'almost
every word' of my paper. He is now preparing his great
work on 'Species and Varieties,' for which he has been collecting materials twenty years. He may save me the trouble of
writing more on my hypothesis, by proving that there is no
difference in nature between the origin of species and of
varieties; or he may give me trouble by arriving at another
conclusion; but, at all events, his facts will be given for me to
work upon. Your collections and my own will furnish most
valuable material to illustrate and prove the universal applic-
ability of the hypothesis. The connection between the
succession of affinities and the geographical distribution of a
group, worked out species by species, has never yet been
shown as we shall be able to show it.

"In this archipelago there are two distinct faunas rigidly
circumscribed, which differ as much as do those of Africa
and South America, and more than those of Europe and

North America ; yet there is nothing on the map or on the face of the islands to mark their limits. The boundary line passes between islands closer together than others belonging to the same group. I believe the western part to be a separated portion of continental Asia, while the eastern is a fragmentary prolongation of a former west Pacific continent. In mammalia and birds the distinction is marked by genera, families, and even orders confined to one region ; in insects by a number of genera, and little groups of peculiar species, the families of insects having generally a very wide or universal distribution."

This letter proves that at this time I had not the least idea of the nature of Darwin's proposed work nor of the definite conclusions he had arrived at, nor had I myself any expectation of a complete solution of the great problem to which my paper was merely the prelude. Yet less than two months later that solution flashed upon me, and to a large extent marked out a different line of work from that which I had up to this time anticipated.

I finished the letter after my arrival at Ternate (January 25, 1858), and made the following observation : "If you go to the Andes I think you will be disappointed, at least in the *number of species*, especially of Coleoptera. My experience here is that the low grounds are *much* the most productive, though the mountains generally produce a few striking and brilliant species." This rather hasty generalization is, I am inclined still to think, a correct one, at all events as regards the individual collector. I doubt if there is any mountain station in the world where so many species of butterflies can be collected within a walk as at Para, or more beetles than at my station in Borneo and Bates' at Ega. Yet it may be the case that many areas of about a hundred miles square in the Andes and in the Himalayas actually contain a larger number of species than any similar area in the lowlands of the Amazon or of Borneo. In other parts of this letter I refer to the work I hoped to do myself in describing, cataloguing, and working out the distribution of my insects. I had in fact been bitten by the passion for species and their description, and if neither Darwin nor myself had hit upon "Natural Selection," I might have spent the best years of my life in this comparatively profitless work. But the new ideas swept all this away. I have for the most part left others to describe my discoveries, and have devoted myself to the great generalizations which the laborious work of species-describers had rendered possible. In this letter to Bates I enclosed a memorandum of my estimate of the number of distinct species of insects I had collected up to the time of writing—three years and a half, nearly one year of which had been lost in journeys, illnesses, and various delays. The totals were as follows :—

Butterflies	620 species
Moths	2000 ,,
Beetles	3700 ,,
Bees, wasps, etc.	750 ,,
Flies	660 ,,
Bugs, cicadas, etc.	500 ,,
Locusts, etc.	160 ,,
Dragonflies, etc.	110 ,,
Earwigs, etc.	40 ,,
Total	8540 species of Insects.

It was while waiting at Ternate in order to get ready for my next journey, and to decide where I should go, that the idea already referred to occurred to me. It has been shown how, for the preceding eight or nine years, the great problem of the origin of species had been continually pondered over, and how my varied observations and study had been made use of to lay the foundation for its full discussion and elucidation. My paper written at Sarawak rendered it certain to my mind that the change had taken place by natural succession and descent—one species becoming changed either slowly or rapidly into another. But the exact process of the change and the causes which led to it were absolutely unknown and appeared almost inconceivable. The great difficulty was to understand how, if one species was gradually changed into another, there continued to be so many quite distinct species, so many which differed from their nearest allies by slight yet perfectly definite and constant characters. One would expect that if it was a law of nature that species were continually changing so as to become in time new and distinct species, the world would be full of an inextricable mixture of various slightly different forms, so that the well-defined and constant species we see would not exist. Again, not only are species, as a rule, separated from each other by distinct external characters, but they almost always differ also to some degree in their food, in the places they frequent, in their habits and instincts, and all these characters are quite as definite and constant as are the external characters. The problem then was, not only how and why do species change, but how and why do they change into new and well-defined species, distinguished from each other in so many ways; why and how do they become so exactly adapted to distinct modes of life ; and why do all the intermediate grades die out (as geology shows they have died out) and leave only clearly defined and well-marked species, genera, and higher groups of animals.

Now, the new idea or principle which Darwin had arrived at twenty years before, and which occurred to me at this time, answers all these questions and solves all these difficulties, and it is because it does so, and also because it is in itself self-evident and absolutely certain, that it has been accepted by the whole scientific world as affording a true solution of the great problem of the origin of species.

At the time in question I was suffering from a sharp attack of intermittent fever, and every day during the cold and succeeding hot fits had to lie down for several hours, during which time I had nothing to do but to think over any subjects then particularly interesting me. One day something brought to my recollection Malthus's " Principles of Population," which I had read about twelve years before. I thought of his clear exposition of "the positive checks to increase "— disease, accidents, war, and famine—which keep down the population of savage races to so much lower an average than that of more civilized peoples. It then occurred to me that these causes or their equivalents are continually acting in the

case of animals also ; and as animals usually breed much more
rapidly than does mankind, the destruction every year from
these causes must be enormous in order to keep down the
numbers of each species, since they evidently do not increase
regularly from year to year, as otherwise the world would long
ago have been densely crowded with those that breed most
quickly. Vaguely thinking over the enormous and constant
destruction which this implied, it occurred to me to ask the
question, Why do some die and some live ? And the answer
was clearly, that on the whole the best fitted live. From the
effects of disease the most healthy escaped ; from enemies,
the strongest, the swiftest, or the most cunning ; from famine,
the best hunters or those with the best digestion ; and so on.
Then it suddenly flashed upon me that this self-acting process
would necessarily *improve the race,* because in every generation the inferior would inevitably be killed off and the
superior would remain—that is, *the fittest would survive.*
Then at once I seemed to see the whole effect of this, that
when changes of land and sea, or of climate, or of food-
supply, or of enemies occurred—and we know that such
changes have always been taking place—and considering the
amount of individual variation that my experience as a
collector had shown me to exist, then it followed that all
the changes necessary for the adaptation of the species to
the changing conditions would be brought about ; and as
great changes in the environment are always slow, there
would be ample time for the change to be effected by the
survival of the best fitted in every generation. In this way
every part of an animal's organization could be modified
exactly as required, and in the very process of this modification the unmodified would die out, and thus the *definite*
characters and the clear *isolation* of each new species would
be explained. The more I thought over it the more I
became convinced that I had at length found the long-sought-
for law of nature that solved the problem of the origin of
species. For the next hour I thought over the deficiencies in
the theories of Lamarck and of the author of the " Vestiges,"
and I saw that my new theory supplemented these views and

obviated every important difficulty. I waited anxiously for
the termination of my fit so that I might at once make notes
for a paper on the subject. The same evening I did this
pretty fully, and on the two succeeding evenings wrote it
out carefully in order to send it to Darwin by the next post,
which would leave in a day or two.

I wrote a letter to him in which I said that I hoped the
idea would be as new to him as it was to me, and that it
would supply the missing factor to explain the origin of
species. I asked him if he thought it sufficiently important
to show it to Sir Charles Lyell, who had thought so highly of
my former paper.

The subsequent history of this article is fully given in the
"Life and Letters," volume ii., and I was, of course, very
much surprised to find that the same idea had occurred to
Darwin, and that he had already nearly completed a large
work fully developing it. The paper is reprinted in my
"Natural Selection and Tropical Nature," and in reading it
now it must be remembered that it was but a hasty first sketch,
that I had no opportunity of revising it before it was printed
in the journal of the Linnean Society, and, especially, that
at that time nobody had any idea of the constant variability
of *every* common species, in every part and organ, which has
since been proved to exist. Almost all the popular objections to Natural Selections are due to ignorance of this fact,
and to the erroneous assumption that what are called
"favourable variations" occur only rarely, instead of being
abundant, as they certainly are, in every generation, and
quite large enough for the efficient action of "survival of the
fittest" in the improvement of the race.

During the first months of my residence at Ternate I
made two visits to different parts of the large island of Gilolo,
where my hunters obtained a number of very fine birds, but
owing to the absence of good virgin forest and my own ill-
health, I obtained very few insects. At length, on March 25,
I obtained a passage to Dorey Harbour, on the north coast
of New Guinea, in a trading schooner, which left me there,

and called for me three or four months later to bring me back to Ternate. I was the first European who had lived alone on this great island; but partly owing to an accident which confined me to the house for a month, and partly because the locality was not a good one, I did not get the rare species of birds of paradise I had expected. I obtained, however, a number of new and rare birds and a fine collection of insects, though not so many of the larger and finer kinds as I expected. The weather had been unusually wet, and the place was unhealthy. I had four Malay servants with me, three of whom had fever as well as myself, and one of my hunters died, and though I should have liked to have stayed longer, we were all weak or unwell, and were very glad when the schooner arrived and took us back to Ternate. Here wholesome food and a comfortable house soon restored us to good health.

When I unpacked and examined my collections I found that the birds I had obtained were very numerous and beautiful, and as my journey and residence in New Guinea had created much interest among my numerous Dutch friends in Ternate, I determined to make a little exhibition of them. I accordingly let it be known that I would be glad to see visitors on the next Sunday afternoon. I had a long table in the verandah which I had covered with new "trade" calico, and on this I laid out the best specimens of all my most showy or strange birds. There were numbers of gorgeous lories, parrots, and parrakeets, white and black cockatoos, exquisite fruit-pigeons of a great variety of colours, many fine king-fishers from the largest to the most minute, as well as the beautiful raquet-tailed species, beautiful black, green, and blue ground-thrushes, some splendid specimens of the Papuan and King paradise-birds, and many beautiful bee-eaters, rollers, fly-catchers, grakles, sun-birds, and paradise-crows, making altogether such an assemblage of strange forms and brilliant colours as no one of my visitors had ever imagined to exist so near them. Even I myself was surprised at the beauty of the show when thus brought together and displayed on the white table, which so well set off their varied and brilliant colours.

I now received letters informing me of the reception of the paper on "Varieties," which I had sent to Darwin, and in a letter home I thus refer to it: "I have received letters from Mr. Darwin and Dr. Hooker, two of the most eminent naturalists in England, which have highly gratified me. I sent Mr. Darwin an essay on a subject upon which he is now writing a great work. He showed it to Dr. Hooker and Sir Charles Lyell, who thought so highly of it that they had it read before the Linnean Society. This insures me the acquaintance of these eminent men on my return home." I also refer to my next voyage as follows :—" I am now about to start for a place where there are some soldiers, and a doctor, and an engineer who can speak English, so if it is good for collecting I shall stay there some months. It is called 'Batchian,' an island on the south-west side of Gilolo, and three or four days' sail from Ternate. I have now quite recovered from the effects of my New Guinea voyage, and am in good health."

I reached Batchian on October 21, and about a month afterwards, there being a Government boat going to Ternate, I took the opportunity of writing to my school-fellow and oldest friend, Mr. George Silk. As he knew nothing whatever of natural history, I wrote to him on subjects more personal to myself, and which may therefore be more suitable to quote here :—

"I have just received yours of August 3 with reminiscences of Switzerland. To you it seems a short time since we were there together, to me an immeasurable series of ages ! In fact, Switzerland and the Amazon now seem to me quite unreal—a sort of former existence or long-ago dream. Malays and Papuans, beetles and birds, are what now occupy my thoughts, mixed with financial calculations and hopes for a happy future in old England, where I may live in solitude and seclusion, except from a few choice friends. You cannot, perhaps, imagine how I have come to love solitude. I seldom have a visitor but I wish him away in an hour. I find it very favourable to reflection ; and if you have any acquaintance who is a fellow of the Linnean

Society, borrow the *Journal of Proceedings* for August last, and in the last article you will find some of my latest lucubrations, and also some complimentary remarks thereon by Sir Charles Lyell and Dr. Hooker, which (as I know neither of them) I am *a little* proud of. As to politics, I hate and abominate them. The news from India I now never read, as it is all an inextricable confusion without good maps and regular papers. Mine come in lumps—two or three months at a time, often with alternate issues stolen or lost. I therefore beg you to write no more politics—nothing public or newspaperish. Tell me about yourself, your own private doings, your health, your visits, your new and old acquaintances (for I know you pick up half a dozen every week *à la Barragan*). But, above all, tell me of what you read. Have you read the ' Currency ' book I returned you, ' Horne Tooke,' ' Bentham,' *Family Herald* leading articles? Give me your opinions on any or all of these. Follow the advice in *Family Herald* Article on ' Happiness,' *Ride a Hobby*, and you will assuredly find happiness in it, as I do. Let ethnology be your hobby, as you seem already to have put your foot in the stirrup, but *ride it hard*. If I live to return I shall come out strong on Malay and Papuan races, and shall astonish Latham, Davis, & Co. ! By-the-by, I have a letter from Davis ; [1] he says he sent my last letter to you, and it is lost mysteriously. Instead, therefore, of sending me a reply to my ' poser,' he repeats what he has said in *every* letter I have had from him, that ' myriads of miracles are required to people the earth from one source.' I am sick of him. You must read ' Pritchard ' through, and Lawrence's ' Lectures on Man ' carefully ; but I am convinced no man can be a good ethnologist who does not travel, and not *travel* merely, but reside, as I do, months and years with each race, becoming well acquainted with their average physiognomy and their character, so as to be able to detect cross-breeds, which totally mislead the hasty traveller, who thinks they are transitions ! Latham, I am sure, is quite wrong on many points.

" When I went to New Guinea, I took an old copy of

---

[1] J. Barnard Davis, the well-known craniologist.

'Tristram Shandy,' which I read through about three times. It is an annoying and, you will perhaps say, a very gross book; but there are passages in it that have never been surpassed, while the character of Uncle Toby has, I think, never been equalled, except perhaps by that of Don Quixote. I have lately read a good many of Dumas's wonderful novels, and they *are* wonderful, but often very careless and some quite unfinished. 'The Memoirs of a Physician' is a wonderful wild mixture of history, science, and romance; the second part, the Queen's Necklace, being the most wonderful and, perhaps, the most true. You should read it, if you have not yet done so, when you are horribly bored!

"In reference to your private communication, it seems to me that marriage has a wonderful effect in brightening the intellect. For example, John used not to be considered witty; yet in his last letter he begs me to write to him 'semi-occasionally,' or 'oftener if I have time,' and I send a not bad extract from his letter. By this mail I send more than a dozen letters, for my correspondence is increasing."

On my return to Ternate in April, 1859, after spending nearly six months in Batchian, where I had made fairly good though not very large collections, including a new and very peculiar bird of paradise and a grand new butterfly of the largest size and most gorgeous colouring, I determined to go next to Timor for a short time, and afterward to Menado, at the north-eastern extremity of Celebes, from which place some of the most interesting birds and mammalia had been obtained. I had, of course, my usual large batch of letters to reply to. One of these from my brother-in-law, Mr. Thomas Sims, urged me very strongly to return home before my health was seriously affected, and for many other reasons. In my reply I gave full expressions to my ideas and feelings compelling me to remain a few years longer, and as these are a part of the history of my life and character, I will give them here.

"Your ingenious arguments to persuade me to come home are quite unconvincing. I have much to do yet before

I can return with satisfaction of mind; were I to leave now
I should be ever regretful and unhappy. That alone is an
all-sufficient reason. I feel that my work is *here* as well as
my pleasure; and why should I not follow out my vocation?
As to materials for work at home, you are in error. I have,
indeed, materials for a life's study of entomology, as far as
the forms and structure and affinities of insects are concerned; but I am engaged in a wider and more general
study—that of the relations of animals to space and time, or,
in other words, their geographical and geological distribution
and its causes. I have set myself to work out this problem
in the Indo-Australian Archipelago, and I must visit and
explore the largest number of islands possible, and collect
materials from the greatest number of localities, in order to
arrive at any definite results. As to health and life, what
are they compared with peace and happiness? and happiness
is admirably defined in the *Family Herald* as to be best
obtained by 'work with a purpose, and the nobler the purpose the greater the happiness.' But besides these weighty
reasons there are others quite as powerful—pecuniary ones.
I have not yet made enough to live upon, and I am likely to
make it quicker here than I could in England. In England
there is only one way in which I could live, by returning to
my old profession of land-surveying. Now, though I always
liked surveying, I like collecting better, and I could never
now give my whole mind to any work apart from the study
to which I have devoted my life. So far from being angry
at being called an enthusiast (as you seem to suppose), it is
my pride and glory to be worthy to be so called. Who ever
did anything good or great who was not an enthusiast? The
majority of mankind are enthusiasts only in one thing—in
money-getting; and these call others enthusiasts as a term
of reproach because they think there is something in the
world better than money-getting. It strikes me that the
power or capability of a man in getting rich is in an *inverse*
proportion to his reflective powers and in *direct* proportion
to his impudence. It is perhaps good to *be* rich, but not to
get rich, or to be always trying to get rich, and few men

are less fitted to get rich, if they did try, than myself."
The rest of the letter is devoted to new discoveries in
photography and allied subjects.

I left Ternate by the Dutch mail steamer on May 1,
1859, calling at Amboyna and spending two days at Banda,
where I visited the celebrated nutmeg plantations, reaching
Coupang, at the west end of Timor, on the 13th. The country
round proving almost a desert for a collector, I went to the
small island of Semau, where I obtained a few birds, but
little else. I therefore returned to Coupang after a week
and determined to go back the way I came by Amboyna
and Ternate to Menado, in order to lose no time, and arrived
there on June 10. Here I remained for four months in one
of the most interesting districts in the whole archipelago. I
visited several localities in the interior, and obtained a
number of the rare and peculiar species of birds and a considerable collection of beetles and butterflies, mostly rare or
new, but by no means so numerous as I had obtained in
other good localities.

In October I returned to Amboyna in order to visit the
almost unknown island of Ceram, which, however, I found
very unproductive and unhealthy. While there I wrote a
short letter to Bates, congratulating him on his safe return
to England, discussing great schemes for the writing and
publication of works on our respective collections, adding,
"I have sent a paper lately to the Linnean Society which
gives my views of the principles of geographical distribution in the archipelago, of which I hope some day to work
out the details." [1]

In December, being almost starved, I returned to Amboyna
to recruit, and in February started on another journey to
Ceram, with the intention, if possible, of again reaching the
Ké Islands, which I had found so rich during the few days I
stayed there on my voyage to the Aru Islands. I visited
several places on the coast of Ceram, and spent three days
very near its centre, where a very rough mountain path

[1] The title of this paper was "On the Zoological Geography of Malay
Archipelago," and it was published in 1860.

crosses from the south to the north coast. But never in the whole of my tropical wanderings have I found a luxuriant forest so utterly barren of almost every form of animal life. Though I had three guns out daily, I did not get a single bird worth having; beetles, too, were totally wanting; and the very few butterflies seen were most difficult to capture. Those who imagine that a tropical forest in the very midst of so rich a region as the Moluccas *must* produce abundance of birds and insects, would have been woefully disillusioned if they could have been with me here. After immense difficulties I reached Goram, about fifty miles beyond the east end of Ceram, where I purchased a boat and started for Ké; but after getting half-way, the weather was so bad and the winds so adverse that I was obliged to return to the Matabello Islands, and thence by way of Goram and the north coast of Ceram to the great island of Waigiou. This was a long and most unfortunate voyage, as fully described elsewhere. I found there, however, what I chiefly went for—the rare red bird of paradise (*Paradisea rubra*); but during the three months I lived there, often with very little food, I obtained only about seventy species of birds, mostly the same as those from New Guinea, though a few species of parrots, pigeons, kingfishers, and other birds were new. Insects were never abundant, but by continued perseverance I obtained rather more species of both butterflies and beetles than at New Guinea, though fewer, I think, of the more showy kinds.

The voyage from Waigiou back to Ternate was again most tedious and unfortunate, occupying thirty-eight days, whereas with reasonably favourable weather it should not have required more than ten or twelve. Taking my whole voyage in this canoe from Goram to Waigiou and Ternate, I thus summarize my account of it in my "Malay Archipelago": " My first crew ran away in a body; two men were lost on a desert island, and only recovered a month later after twice sending in search of them; we were ten times run aground on coral reefs; we lost four anchors; our sails were devoured by rats; our small boat was lost astern; we were thirty-eight days on a voyage which should not have taken twelve; we

were many times short of food and water ; we had no com-
pass-lamp owing to there being not a drop of oil in Waigiou
when we left ; and, to crown all, during our whole voyage from
Goram by Ceram to Waigiou, and from Waigiou to Ternate,
occupying in all seventy-eight days (or only twelve days
short of three months), all in what was supposed to be the
favourable season, we had *not one single day of fair wind.*
We were always close braced up, always struggling against
wind, currents, and leeway, and in a vessel that would scarcely
sail nearer than eight points from the wind !   Every seaman
will admit that my first (and last) voyage in a boat of my
own was a very unfortunate one."

While living at Bessir, the little village where I went to
get the red paradise birds, I wrote a letter to my friend
George Silk, which I finished and posted after my arrival at
Ternate.   As such letters as this, absolutely familiar and confidential, exhibit my actual feelings, opinions, and ideas at
the time, I reproduce it here :—

"Bessir, September 1, 1860.

"MY DEAR GEORGE,

"It is now ten months since the date of my last
letter from England.   You may fancy therefore that, in the
expressive language of the trappers, I am 'half froze' for
news.   No such thing !   Except for my own family and
personal affairs I care not a straw and scarcely give a thought
as to what may be uppermost in the political world.   In my
situation old newspapers are just as good as new ones, and
I enjoy the odd scraps, in which I do up my birds (advertisements and all), as much as you do your *Times* at breakfast.
If I live to return to Ternate in another month, I expect to
get such a deluge of communications that I shall probably
have no time to answer any of them.   I therefore bestow one
of my solitary evenings on answering yours beforehand.   By-
the-by, you do not yet know where I am, for I defy all the
members of the Royal Geographical Society in full conclave
to tell you where is the place from which I date this letter.
I must inform you, therefore, that it is a village on the

south-west coast of the island of Waigiou, at the north-west
extremity of New Guinea. How I came here would be too
long to tell, the details I send to my mother and refer you
to her. While hon. members are shooting partridges I am
shooting, or trying to shoot, birds of paradise—red at that,
as our friend Morris Haggar would say. But enough of this
nonsense. I meant to write you of matters more worthy of a
naturalist's pen. I have been reading of late two books of
the highest interest, but of most diverse characters, and I wish
to recommend their perusal to you if you have time for anything but work or politics. They are Dr. Leon Dufour's
' Histoire de la Prostitution ' and Darwin's ' Origin of Species.'
If there is an English translation of the first, pray get it.
Every student of men and morals should read it, and if many
who talk glibly of putting down the ' social evil ' were first
to devote a few days to its study, they would be both much
better qualified to give an opinion and much more diffident
of their capacity to deal with it. The work is truly a history,
and a great one, and reveals pictures of human nature more
wild and incredible than the pen of the romancist ever dared
to delineate. I doubt if many classical scholars have an idea
of what were really the habits and daily life of the Romans
as here delineated. Again I say, read it.

    " The other book you may have heard of and perhaps read,
but it is not one perusal which will enable any man to
appreciate it. I have read it through five or six times, each
time with increasing admiration. It will live as long as the
' Principia ' of Newton. It shows that nature is, as I before
remarked to you, a study that yields to none in grandeur and
immensity. The cycles of astronomy or even the periods of
geology will alone enable us to appreciate the vast depths of
time we have to contemplate in the endeavour to understand
the slow growth of life upon the earth. The most intricate
effects of the law of gravitation, the mutual disturbances of
all the bodies of the solar system, are simplicity itself compared with the intricate relations and complicated struggle
which have determined what forms of life shall exist and in
what proportions. Mr. Darwin has given the world a *new*

science, and his name should, in my opinion, stand above that of every philosopher of ancient or modern times. The force of admiration can no further go!!!"

"On board steamer from Ternate to Timor, January 2, 1861.

"I have come home safe to Ternate and left it again. For two months I was stupefied with my year's letters, accounts, papers, magazines, and books, in addition to the manipulation, cleaning, arranging, comparing, and packing for safe transmission to the other side of the world of about 16,000 specimens of insects, birds, and shells. This has been intermingled with the troubles of preparing for new voyages, laying in stores, hiring men, paying or refusing to pay their debts, running after them when they try to run away, going to the town with lists of articles *absolutely necessary* for the voyage, and finding that none of them could be had for love or money, conceiving impossible substitutes and not being able to get *them* either,—and all this coming upon me when I am craving repose from the fatigues and privations of an unusually dangerous and miserable voyage, and you may imagine that I have not been in any great humour for letter-writing.

"I think I may promise you that in eighteen months, more or less, we may meet again, if nothing unforeseen occurs.

"Yours,
"A. R. W."

Just before leaving Ternate I also wrote to Bates, chiefly about the "Origin of Species" and some of my results on geographical distribution.

"Ternate, December 24, 1860.

"DEAR BATES,

"Many thanks for your long and interesting letter. I have myself suffered much in the same way as you describe, and I think more severely. The kind of *tædium vitæ* you mention I also occasionally experience here. I impute it to a too monotonous existence.

"I know not how, or to whom, to express fully my admiration of Darwin's book. To *him* it would seem flattery, to others self-praise; but I do honestly believe that with however much patience I had worked and experimented on the subject, I could *never have approached* the completeness of his book, its vast accumulation of evidence, its overwhelming argument, and its admirable tone and spirit. I really feel thankful that it has *not* been left to me to give the theory to the world. Mr. Darwin has created a new science and a new philosophy; and I believe that never has such a complete illustration of a new branch of human knowledge been due to the labours and researches of a single man. Never have such vast masses of widely scattered and hitherto quite unconnected facts been combined into a system and brought to bear upon the establishment of such a grand and new and simple philosophy.

"I am surprised at your joining the north and south banks of the lower Amazon into one region. Did you not find a sufficiency of distinct species at Obydos and Barra to separate them from Villa Nova and Santarem? I am now convinced that insects, on the whole, do not give such true indications of zoological geography as birds and mammals, because, first, they have such immensely greater means of dispersal across rivers and seas; second, because they are so much more influenced by surrounding circumstances; and third, because the species seem to change more quickly, and therefore disguise a comparatively recent identity. Thus the insects of adjacent regions, though originally distinct, may become rapidly amalgamated, or portions of the same region may come to be inhabited by very distinct insect-faunas owing to differences of soil, climate, etc. This is strikingly shown here, where the insect-fauna from Malacca to New Guinea has a very large amount of characteristic uniformity, while Australia, from its distinct climate and vegetation, shows a wide difference. I am inclined to think, therefore, that a preliminary study of, first, the mammals, and then the birds, is indispensable to a correct understanding of the geographical and physical changes on which the present insect-distribution depends.[1] . . .

"In a day or two I leave for Timor, where, if I am lucky in finding a good locality, I expect some fine and interesting insects."

I reached Delli, the chief place in the Portuguese part of the island, on January 12, 1861, and stayed there about three months and a half. I lived with an Englishman, Captain Hart, who had a coffee plantation about a mile out of the town; and there was also another Englishman, Mr. Geach, a mining engineer, who had come out to open copper mines for the Portuguese Government, but as no copper ore could be found, he was waiting for an opportunity to return to Singapore. They were both very pleasant people, and I enjoyed myself while there, though the collecting was but poor, owing to the excessive aridity of the climate and the absence of forests. I obtained, however, some rare birds and a few very rare and beautiful butterflies by the side of a stream in a little rocky valley shaded by a few fine trees and bushes. Of beetles, however, there were absolutely none worth collecting.

Leaving Timor at the end of April, I went by the Dutch mail steamer to Cajeli in Bouru, the last of the Molucca Islands which I visited. Here I stayed two months, but was again disappointed, since the country was almost as unproductive as Ceram. For miles round the town there were only low hills covered with coarse grass and scattered trees, less productive of insects than a bare moor in England. Some patches of wood here and there and the fruit trees around the town produced a few birds of peculiar species. I went to a place about twenty miles off, where there was some forest, and remained there most of my time; but insects were still very scarce, and birds almost equally so. I obtained, however, about a dozen quite new species of birds and others which were very rare, together with a small collection of beetles; and then, about the end of June, took the mail

[1] These ideas were thoroughly worked out in my book on "The Geographical Distribution of Animals," published in 1876.

steamer by Ternate and Menado to Sourabaya, the chief town in eastern Java.

I stayed here about a month, spending most of the time at the foot of the celebrated mount Arjuna; but the season was too dry, and both birds and insects very scarce. I therefore went on to Batavia and thence to Buitenzorg and to the Pangerango mountain, over ten thousand feet high. At a station about four thousand feet above sea-level, where the main road passes through some virgin forest, I stayed some weeks, and made a tolerable collection of birds and butterflies, though the season was here as much too wet as East Java was too dry. I next went to Palembang in Sumatra, which I reached by way of Banka on November 8. Here the country was mostly flooded, and I had to go up the river some distance to where a military road starts for the interior and across the mountains to Bencoolen. On this road, about seventy miles from Palembang, I came to a place called Lobo Raman, surrounded with some fine virgin forest and near the centre of East Sumatra. Here, and at another station on the road, I stayed about a month, and obtained a few very interesting birds and butterflies; but it was the height of the wet season, and all insects were scarce. I therefore returned to Palembang and Banka, and thence to Singapore, on my way home. While waiting here for the mail steamer, two living specimens of the smaller paradise bird (*Paradisea papuana*) were brought to Singapore by a trader, and I went to see them. They were in a large cage about five or six feet square, and seemed in good health, but the price asked for them was enormous, as they are so seldom brought, and the rich Chinese merchants or rich natives in Calcutta are always ready to purchase them. As they had never been seen alive in Europe I determined to take the risk and at once secured them, and with some difficulty succeeded in bringing them home in safety, where they lived in the Zoological Gardens for one and two years respectively.

While living in the wilds of Sumatra I wrote two letters, to my friends Bates and Silk, which, being the last I wrote before reaching home, may be of interest as showing what subjects were then uppermost in my mind. The first from which I will quote is that to Mr. Bates, and referring to a paper on the Papilios of the Amazon which he had sent me I make some remarks on the distribution of animals in South America, which I do not think I have published anywhere.

"Your paper is in every respect an admirable one, and proves the necessity of minute and exact observation over a wide extent of country to enable a man to grapple with the more difficult groups, unravel their synonymy, and mark out the limits of the several species and varieties. All this you have done, and have, besides, established a very interesting fact in zoological geography, that of the southern bank of the lower river having received its fauna from Guayana, and not from Brazil. There is, however, another fact, I think, of equal interest and importance which you have barely touched upon, and yet I think your own materials in this very paper establish it, viz. that the river, in a great many cases, limits the range of species or of well-marked varieties. This fact I considered was proved by the imperfect materials I brought home, both as regards the Amazon and Rio Negro. In a paper I read on 'The Monkeys of the Lower Amazon and Rio Negro' I showed that the species were often different on the opposite sides of the river. Guayana species came up to the east bank, Columbian species to the west bank, and I stated that it was therefore important that travellers collecting on the banks of large rivers should note from which side every specimen came. Upon this Dr. Gray came down upon me with a regular floorer. 'Why,' said he, 'we have specimens collected by Mr. Wallace himself marked "Rio Negro" only.' I do not think I answered him properly at the time, that those specimens were sent from Barra before I had the slightest idea myself that the species were different on the two banks. In mammals the fact was not so much to be wondered at, but few persons would credit that it would apply also to birds and winged insects. Yet I am convinced it does, and I only regret that I had not collected and studied birds there with the same assiduity as I

have here, as I am sure they would furnish some most interesting results. Now, it seems to me that a person having no
special knowledge of the district would have no idea from
your paper that the species did not in almost every instance
occur on both banks of the river. In only one case do you
specially mention a species being found *only* on the north
bank. In other cases, except where the insect is local and
confined to one small district, no one can tell whether they
occur on one or both banks. Obydos you only mention once,
Barra and the Tunantins not at all. I think a list of the
species or varieties occurring on the south bank or north
bank only should have been given, and would be of much
interest as establishing the fact that large rivers do act
as *limits* in determining the range of species. From the
localities you give, it appears that of the sixteen species of
papilio peculiar to the Amazon, fourteen occur only on the
south bank ; also, that the Guayana species all pass to the
south bank. These facts I have picked out. They are not
stated by you. It would seem, therefore, that Guayana
forms, having once crossed the river, have a great tendency
to become modified, and then never recross. Why the
Brazilian species should not first have taken possession of
their own side of the river is a mystery. I should be
inclined to think that the present river bed is comparatively
new, and that the southern lowlands were once continuous
with Guayana ; in fact, that Guayana is older than north
Brazil, and that after it had pushed out its alluvial plains
into what is now north Brazil, an elevation on the Brazilian
side made the river cut a new channel to the northward,
leaving the Guayana species isolated, exposed to competition
with a new set of species from further south, and so becoming modified, as we now find them. . . . The whole district
is, I fear, too little known geologically to test this supposition.
The mountains of north Brazil are, however, said to be of the
cretaceous period, and if so their elevation must have occurred
in tertiary times, and may have continued to a comparatively
recent period. Now if there are no proofs of such recent
upheaval in the southern mountains of Guayana, the theory

would thus far receive support. I regret that your time was not more equally divided between the north and south banks, but I suppose you found the south so much more productive in new and fine things. . . .

"I am here making what I intend to be my last collections, but am doing very little in insects, as it is the wet season and all seems dead. I find in those districts where the seasons are strongly contrasted the good collecting time is very limited—only about a month or two at the beginning of the dry, and a few weeks at the commencement of the rains. It is now two years since I have been able to get any beetles, owing to bad localities and bad weather, so I am becoming disgusted. When I do find a good place it is generally very good, but such are dreadfully scarce. In Java I had to go forty miles in the eastern part and sixty miles in the western to reach a bit of forest, and then I got scarcely anything. Here I had to come a hundred miles inland, by Palembang, and though in the very centre of Eastern Sumatra, the forest is only in patches, and it is the height of the rains, so I get nothing. A longicorn is a rarity, and I suppose I shall not have as many species in two months as I have obtained in three or four days in a really good locality. I am getting, however, some sweet little blue butterflies (*Lycænidæ*), which is the only thing that keeps up my spirits."

The letter to my friend Silk will be, perhaps, a little more amusing, and perhaps not less instructive.

"Lobo Roman, Sumatra, December 22, 1861.

"MY DEAR GEORGE,

"Between eight and nine years ago, when we were concocting that absurd book, 'Travels on the Amazon and Rio Negro,' you gave me this identical piece of waste paper with sundry others, and now having scribbled away my last sheet of 'hot-pressed writing,' and being just sixty miles from another, I send you back your gift, with interest; so you see that a good action, sooner or later, find its sure reward.

"I now write you a *letter*, I hope for the last time, for I trust our future letters may be *vivâ voce*, as an Irishman would say, while our epistolary correspondence will be confined to *notes*. I really do now think and believe that I am coming home, and as I am quite uncertain when I may be able to send you this letter, I may possibly arrive not very long after it. Some fine morning I expect to walk into 79, Pall Mall, and shall, I suppose, find things just the same as if I had walked out yesterday and come in to-morrow! There will you be seated on the same chair, at the same table, surrounded by the same account books, and writing upon paper of the same size and colour as when I last beheld you. I shall find your inkstand, pens, and pencils in the same places, and in the same beautiful order, which my idiosyncrasy compels me to admire, but forbids me to imitate. (Could you see the table at which I am now writing, your hair would stand on end at the reckless confusion it exhibits!) I suppose you have now added a few more secretaryships to your former multifarious duties. I suppose that you still walk every morning from Kensington and back in the evening, and that things at the archdeacon's go on precisely and identically as they did eight years ago.[1] I feel almost inclined to parody the words of Cicero, and to ask indignantly, 'How long, O Georgius, will you thus abuse our patience? How long will this sublime indifference last?' But I fear the stern despot, habit, has too strongly riveted your chains, and as, after many years of torture the Indian fanatic can at last sleep only on his bed of spikes, so perhaps now you would hardly care to change that daily routine, even if the opportunity were thrust upon you. Excuse me, my dear George, if I express myself too strongly on this subject, which is truly no business of mine, but I cannot see, without regret, my earliest friend devote himself so entirely, mind and body, to the service of others.

"I am here in one of the places unknown to the Royal Geographical Society, situated in the very centre of East

---

[1] Mr. Silk was private secretary and reader to the then Archdeacon Sinclair, Vicar of Kensington.

Sumatra, about one hundred miles from the sea in three directions. It is the height of the wet season, and the rain pours down strong and steady, generally all night and half the day. Bad times for me, but I walk out regularly three or four hours every day, picking up what I can, and generally getting some little new or rare or beautiful thing to reward me. This is the land of the two-horned rhinoceros, the elephant, the tiger, and the tapir; but they all make themselves very scarce, and beyond their tracks and their dung, and once hearing a rhinoceros *bark* not far off, I am not aware of their existence. This, too, is the very land of monkeys; they swarm about the villages and plantations, long-tailed and short-tailed, and with no tail at all, white, black, and grey; they are eternally racing about the tree-tops, and gambolling in the most amusing manner. The way they jump is amazing. They throw themselves recklessly through the air, apparently sure, with one or other of their four hands, to catch hold of something. I estimated one jump by a long-tailed white monkey at thirty feet horizontal, and sixty feet vertical, from a high tree on to a lower one; he fell through, however, so great was his impetus, on to a lower branch, and then, without a moment's stop, scampered away from tree to tree, evidently quite pleased with his own pluck. When I startle a band, and one leader takes a leap like this, it is amusing to watch the others—some afraid and hesitating on the brink till at last they pluck up courage, take a run at it, and often roll over in the air with their desperate efforts. Then there are the long-armed apes, who never walk or run upon the trees, but travel altogether by their long arms, swinging themselves from bough to bough in the easiest and most graceful manner possible.

"But I must leave the monkeys and turn to the men, who will interest you more, though there is nothing very remarkable in them. They are Malays, speaking a curious, half-unintelligible Malay dialect—Mohammedans, but retaining many pagan customs and superstitions. They are very ignorant, very lazy, and live almost absolutely on rice alone, thriving upon it, however, just as the Irish do, or did, upon

potatoes. They were a bad lot a few years ago, but the
Dutch have brought them into order by their admirable
system of supervision and government. By-the-by, I hope
you have read Mr. Money's book on Java. It is well worth
while, and you will see that I had come to the same conclusions as to Dutch colonial government from what I saw in
Menado. Nothing is worse and more absurd than the sneer-
ing prejudiced tone in which almost all English writers speak
of the Dutch government in the East. It never has been
worse than ours has been, and it is now very much better;
and what is greatly to their credit and not generally known,
they take nearly the same pains to establish order and good
government in those islands and possessions which are an
annual loss to them, as in those which yield them a revenue.
I am convinced that their system is *right* in principle, and
ours *wrong*, though, of course, in the practical working there
may and must be defects; and among the Dutch themselves,
both in Europe and the Indies, there is a strong party
against the present system, but that party consists mostly of
merchants and planters, who want to get the trade and
commerce of the country made free, which in my opinion
would be an act of suicidal madness, and would, moreover,
seriously injure instead of benefiting the natives.

" Personally, I do not much like the Dutch out here, or the
Dutch officials; but I cannot help bearing witness to the
excellence of their government of native races, gentle yet
firm, respecting their manners, customs, and prejudices, yet
introducing everywhere European law, order, and industry."

" Singapore, January 20, 1862.

" I cannot write more now. I do not know how long I
shall be here; perhaps a month. Then, ho! for England!"

When I was at Sarawak in 1855 I engaged a Malay boy
named Ali as a personal servant, and also to help me to
learn the Malay language by the necessity of constant communication with him. He was attentive and clean, and
could cook very well. He soon learnt to shoot birds, to skin

MY FAITHFUL MALAY BOY—ALI. 1855–1862.

[*To face p.* 383, VOL. I.

them properly, and latterly even to put up the skins very neatly. Of course he was a good boatman, as are all Malays, and in all the difficulties or dangers of our journeys he was quite undisturbed and ready to do anything required of him. He accompanied me through all my travels, sometimes alone, but more frequently with several others, and was then very useful in teaching them their duties, as he soon became well acquainted with my wants and habits. During our residence at Ternate he married, but his wife lived with her family, and it made no difference in his accompanying me wherever I went till we reached Singapore on my way home. On parting, besides a present in money, I gave him my two double-barrelled guns and whatever ammunition I had, with a lot of surplus stores, tools, and sundries, which made him quite rich. He here, for the first time, adopted European clothes, which did not suit him nearly so well as his native dress, and thus clad a friend took a very good photograph of him. I therefore now present his likeness to my readers as that of the best native servant I ever had, and the faithful companion of almost all my journeyings among the islands of the far East.

The two birds of paradise which I had purchased gave me a good deal of trouble and anxiety on my way home. I had first to make an arrangement for a place to stand the large cage on deck. A stock of food was required, which consisted chiefly of bananas; but to my surprise I found that they would eat cockroaches greedily, and as these abound on every ship in the tropics, I hoped to be able to obtain a good supply. Every evening I went to the storeroom in the fore part of the ship, where I was allowed to brush the cockroaches into a biscuit tin. The ship stayed three or four days at Bombay to discharge and take in cargo, coal, etc., and all the passengers went to a hotel, so I brought the birds on shore and stood them in the hotel verandah, where they were a great attraction to visitors. While staying at Bombay a small party of us had the good fortune to visit the celebrated cave-temple of Elephanta on a grand festival day, when it was crowded with thousands of natives—men, women, and children, in ever-changing crowds, kneeling or praying before the images or the altars, making gifts to the gods or the priests, and outside cooking and eating—a most characteristic and striking scene.

The journey to Suez offered no particular incident, and the birds continued in good health ; as did two or three lories I had brought. But with the railway journey to Alexandria difficulties began. It was in February, and the night was clear and almost frosty. The railway officials made difficulties, and it was only by representing the rarity and value of the birds that I could have the cage placed in a box-truck. When we got into the Mediterranean the weather became suddenly cold, and worse still, I found that the ship was free from cockroaches. As I thought that animal food was perhaps necessary to counteract the cold, I felt afraid for the safety of my charge, and determined to stay a fortnight at Malta in order to reach England a little later, and also to lay in a store of the necessary food. I accordingly arranged to break my voyage there, went to a hotel, and found that I could get unlimited cockroaches at a baker's close by.

At Marseilles I again had trouble, but at last succeeded in getting them placed in a guard's van, with permission to enter and feed them *en route*. Passing through France it was a sharp frost, but they did not seem to suffer ; and when we reached London I was glad to transfer them into the care of Mr. Bartlett, who conveyed them to the Zoological Gardens.

Thus ended my Malayan travels.

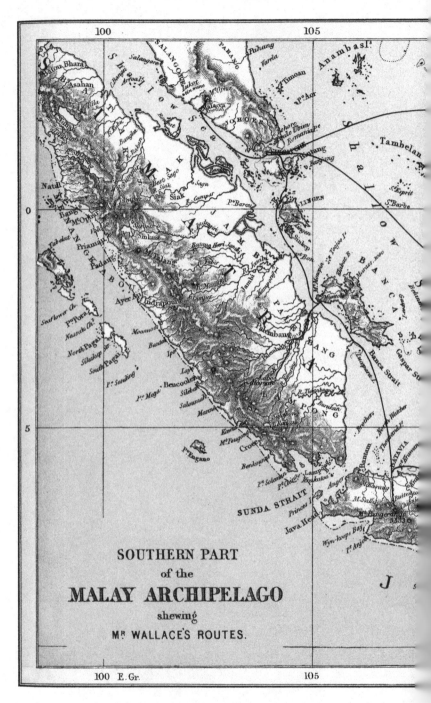

SOUTHERN PART
of the
MALAY ARCHIPELAGO
shewing
Mᴿ WALLACE'S ROUTES.

Mᴿ Wallace's routes shewn thus —————— Mᴿ Allen's routes shewn

Active Volcanoes shewn thus   ●

London : Chapma

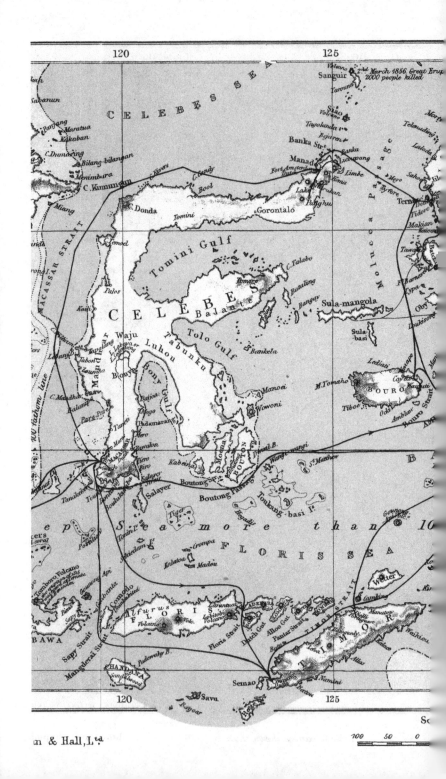

120

125

CELEBES SEA

Sanguir

2nd March 1856. Great Erup.
2000 people killed

Sabanun

Tarouna

Panjang
Maratua
Kakaban

Siao
Volcano

Telaudji

Tayohanda I.

C.Dumaring

Bilang-bilangan

Banka Str.

Banka

Laloda

Menimbara

Manado

Ranowang

Morh

Miang

C.Kimunam

Fort Amsterdam
Dutch

Limbe

Meyo

Sahoe

Jil

Kema

C.Bondy

Laki
Kuban

Panghu

Tawgu

Ternate
Tidore
Malcian
Kaiowa

C.Donda

Bool

Tomini

Gorontalo

Batchian

Tomini Gulf

P.Barnadt
Lypa

Cemoel

C.Talabo

Oby

Tulos

Tomaa

Butaling

Sula-mangola

Doukisore

Kaili

C E L E B E S

Bangay

Balantz

Bankela

Sula-
basi

Waju

Tolo Gulf

Labuan

Tabunku

Luhou

Sengeri

Boni Gulf

Lediati

Passage

M.Tomaho

Casdtc
Waduti

BOURO

C.Maudha

Kalamil

Baioa

Manoei

Tiboe

Oddi

Amblaw

Bouro Strait

Par & Parei

Tango

Togo
Indamarang

Tnao

Wowoni

MACASSAR STRAIT

Pinrang
Zeumba

Blonji

MACASSAR

Pero
Bonthain

Buton

Moena

Tiero
Biro

Kabaena

Buangi

S.Matthew

B

Tanakeke

Saleyer

Boutong Sp.

Wetter

Kamboee

Boutong Passage

Toukang-basi I.

Pombian

Salayer

Bagadi

Gounong
Api Volc.

Tiga I.

10

Deep Sea more than 100

Tombora Volcano

Ninedam

Crompa

FLORIS SEA

Gounong Api

Kalatoa Madou

Kanica

Manggarai Strait

ADE

TIMOR STRAIT

Sumbawa

Condon
Volcano

FLORIS

Volcano

Ischitobi
Volcano

Lohitobi

Alloo Gut

Dutch Gut

Tambing

Kanutov

Waikiou

Sapy Strait

Toro

Pantar Strait

Lora I.

Allas

CHANDANA
Sandal Wood

Padomaly B.

Floris Strait

Semao

Savu

Saggar

Namini

120

125

m & Hall, Ltd.

So

100   50   0

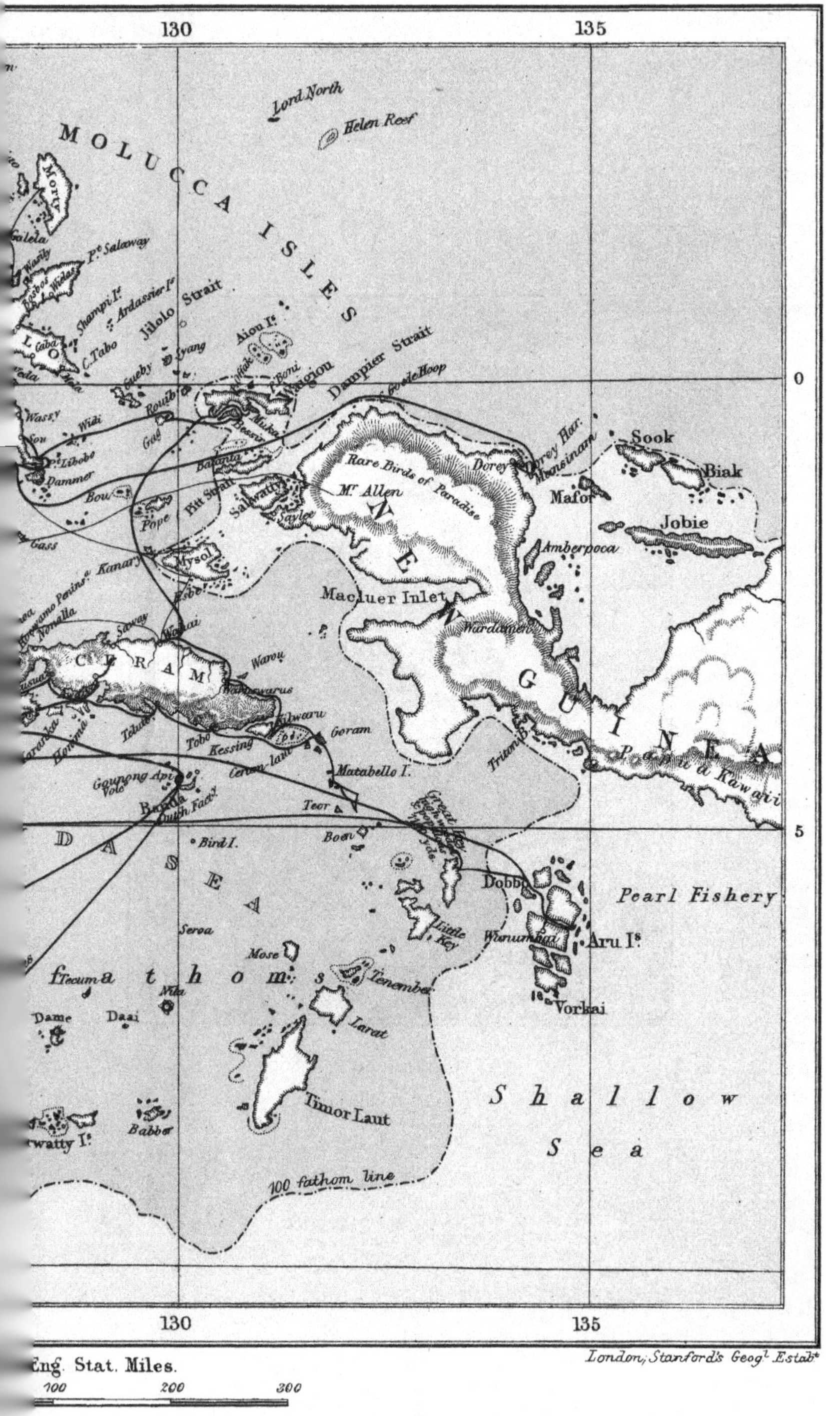
130
135
0
5
MOLUCCA ISLES
Lord North
Helen Reef
Morty
Galela
Pt. Salaway
Jilolo Strait
Aiou Is.
Dampier Strait
Waigiou
Batanta
Salwatty
Pitt Strait
Pope
Mysol
Dammer
Sook
Biak
Mafor
Jobie
Amberpoca
Dorey
Rare Birds of Paradise
Mt. Allen
NEW GUINEA
Macluer Inlet
Papua Kawaii
CERAM
Warou
Goram
Matabello I.
Teor
Banda
Bird I.
Seroa
Mose
Nila
Dame
Daai
Babber
Little Key
Tenember
Larat
Timor Laut
Dobbo
Aru Is.
Vorkai
Pearl Fishery
Shallow Sea
100 fathom line
Eng. Stat. Miles.
100
200
300
London; Stanford's Geog.l Estab.t

A. R. WALLACE. 1869.

[*To face p.* 385, VOL. I.

CHAPTER XXIII

LIFE IN LONDON, 1862–1871—SCIENTIFIC AND LITERARY WORK

On reaching London in the spring of 1862 I went to live with my brother-in-law, Mr. Thomas Sims, and my sister Mrs. Sims, who had a photographic business in Westbourne Grove. Here, in a large empty room at the top of the house, I brought together all the collections which I had reserved for myself and which my agent, Mr. Stevens, had taken care of for me. I found myself surrounded by a quantity of packing-cases and store-boxes, the contents of many of which I had not seen for five or six years, and to the examination and study of which I looked forward with intense interest.

From my first arrival in the East I had determined to keep a complete set of certain groups from every island or distinct locality which I visited for my own study on my return home, as I felt sure they would afford me very valuable materials for working out the geographical distribution of animals in the archipelago, and also throw light on various other problems. These various sets of specimens were sent home regularly with the duplicates for sale, but either packed separately or so distinctly marked " Private " that they could be easily put aside till my return home. The groups thus reserved were the birds, butterflies, beetles, and land-shells, and they amounted roughly to about three thousand bird skins of about a thousand species, and, perhaps, twenty thousand beetles and butterflies of about seven thousand species.

As I reached home in a very weak state of health, and could not work long at a time without rest, my first step was to purchase the largest and most comfortable easy-chair I could find in the neighbourhood, and then engage a carpenter to fit up one side of the room with movable deal shelves, and to make a long deal table, supported on trestles, on which I could unpack and assort my specimens. In order to classify and preserve my bird skins I obtained from a manufacturer about a gross of cardboard boxes of three sizes, which, when duly labelled with the name of the genus or family, and arranged in proper order upon the shelves, enabled me to find any species without difficulty. For the next month I was fully occupied in the unpacking and arranging of my collections, while I usually attended the evening meetings of the Zoological, Entomological, and Linnean Societies, where I met many old friends and made several new ones, and greatly enjoyed the society of people interested in the subjects that now had almost become the business of my life.

As soon as I began to study my birds I had to pay frequent visits to the bird-room of the British Museum, then in charge of Mr. George Robert Gray, who had described many of my discoveries as I sent them home, and also to the library of the Zoological Society to consult the works of the older ornithologists. In this way the time passed rapidly, and I became so interested in my various occupations, and saw so many opportunities for useful and instructive papers on various groups of my birds and insects, that I came to the conclusion to devote myself for some years to this work, and to put off the writing of a book on my travels till I could embody in it all the more generally interesting results derived from the detailed study of certain portions of my collections. This delay turned out very well, as I was thereby enabled to make my book not merely the journal of a traveller, but also a fairly complete sketch of the whole of the great Malayan Archipelago from the point of view of the philosophic naturalist. The result has been that it long continued to be the most popular of my books, and that even now, thirty-six years after its publication, its sale is equal to that of any of the others.

Having, as already described, brought home two living birds of paradise, which were attracting much notice at the Zoological Gardens, I thought it would be of interest to the Fellows of the Society to give an outline of my various journeys in search of these wonderful birds, and of the reasons why I was, comparatively speaking, so unsuccessful. This was the first paper I wrote after my return, and I read it to the society on May 11. As it gives an account of how I pursued this special object, and summarizes a number of voyages, the description of which occupies six or seven chapters of my " Malay Archipelago," and as it is not accessible to general readers, I give the larger portion of it here.

NARRATIVE OF SEARCH AFTER BIRDS OF PARADISE.

Having visited most of the islands inhabited by the paradise birds, in the hope of obtaining good specimens of many of the species, and some knowledge of their habits and distribution, I have thought that an outline of my several voyages, with the causes that have led to their only partial success, might not prove uninteresting.

At the close of the year 1856, being then at Macassar, in the island of Celebes, I was introduced to the master of a prau trading to the Aru Islands, who assured me that two sorts of birds of paradise were abundant there—the large yellow and the small red kinds—the Paradisea apoda and P. regia of naturalists.

He seemed to think there was no doubt but I could obtain them either by purchase from the natives or by shooting them myself. Thus encouraged, I agreed with him for a passage there and back (his stay being six months), and made all my preparations to start by the middle of December.

Our vessel was a Malay prau of about 100 tons burthen, but differing widely from anything to be seen in European waters. The deck sloped downwards towards the bows, the two rudders were hung by rattans and ropes on the quarters, the masts were triangles standing on the decks, and the huge mat sail, considerably longer than the vessel, with its yard of bamboos, rose upwards at a great angle, so as to make up for the lowness of the mast. In this strange vessel, which, under very favourable circumstances, plunged along at nearly five miles an hour, and with a Buginese crew, all of whom seemed to have a voice in cases of difficulty or danger, we made the voyage of about a thousand miles in perfect safety, and very agreeably ; in fact, of all the sea voyages I have made, this was one of the pleasantest.

On reaching the Bugis trading settlement of Dobbo, I found that the small island on which it is situated does not contain any paradise birds. Just as I was trying to arrange a trip to the larger island, a fleet of Magindano pirates made their appearance, committing great devastations, and putting the whole place in an uproar ; and it was only after they had been some time gone that confidence began to be restored, and the natives could be persuaded to take the smallest voyage. This delayed me two months in Dobbo without seeing a paradise bird.

When, however, I at length reached the main island and ascended a small stream to a native village, I soon obtained a specimen of the lovely king bird of paradise, which, when first brought me, excited greater admiration and delight than I have experienced on any similar occasion. The larger species was still not to be seen, and the natives assured me that it would be some months before their plumage arrived at perfection, when they were accustomed to congregate together and could be more easily obtained. This proved to be correct, for it was about four months after my arrival at Dobbo that I obtained my first full-plumaged specimen of the great paradise bird. This was near the centre of the large island of Aru ; and there, with the assistance of the natives, I procured the fine series which first arrived in England.

While at Dobbo I had frequent conversations with the Bugis traders and with the Rajah of Goram, who all assured me that in the northern parts of New Guinea I could travel with safety, and that at Mysol, Waigiou, Salwatty, and Dorey I could get all the different sorts of Paradiseæ. Their accounts excited me so much that I could think of nothing else ; and after another excursion in Celebes I made my way to Ternate, as the best headquarters for the Moluccas and New Guinea. Finding a schooner about to sail on its annual trading voyage to the north coast of New Guinea, I agreed for a passage to Dorey, and to be called for on the return of the vessel after an interval of three or four months. We arrived there, after a tedious voyage, in April, 1858, and I began my second search after the birds of paradise.

I went to Dorey in full confidence of success, and thought myself extremely fortunate in being able to visit that particular locality ; for it was there that Lesson, in the French discovery ship *Coquille*, purchased from the natives the skins of at least eight spec es, viz. Paradisea papuana, with regia, magnifica, superba, and sexsetacea, Astrapia nigra, Epimachus magnus, and Sericulus aureus. Here was a prospect for me ! The very anticipation of it made me thrill with expectation.

My disappointment, therefore, may be imagined when, shortly after my arrival, I found all these bright hopes fade away. In vain I inquired for the native bird-hunters ; none were to be found there ; and the inhabitants assured me that not a single bird of paradise of any kind was ever prepared by the Dorey people, and that only the common yellow one (P.

papuana) was found in the district. This turned out to be the case ; for I could get nothing but this species sparingly, a few females of the king-bird and one young male of the twelve-wired bird of paradise, a species Lesson does not mention. Nevertheless, Lesson did undoubtedly obtain all the birds he names at Dorey ; but the natives are great traders in a petty way, and are constantly making voyages along the coast and to the neighbouring islands, where they purchase birds of paradise and sell them again to the Bugis praus, Molucca traders, and whale-ships which annually visit Dorey harbour. Lesson must have been there at a good time, when there happened to be an accumulation of bird-skins ; I, at a bad one, for I could not buy a single rare bird all the time I was there. I also suffered much by the visit of a Dutch surveying steamer, which, for want of coals, lay in Dorey harbour for a month ; and during that time I got nothing from the natives, every specimen being taken on board the steamer, where the commonest birds and insects were bought at high prices. During this time two skins of the black paradise bird (Astrapia nigra) were brought by a Bugis trader and sold to an amateur ornitho-logist on board, and I never had another chance of getting a skin of this rare and beautiful bird.

The Dorey people all agreed that Amerbaki, about one hundred miles west, was the place for birds of paradise, and that almost all the different sorts were to be found there. Determined to make an effort to secure them, I sent my two best men with ten natives and a large stock of goods to stay there a fortnight, with instructions to shoot and buy all they could. They returned, however, with absolutely nothing. They could not buy any skins but those of the common P. papuana, and could not find any birds but a single specimen of P. regia. They were assured that the birds all came from two or three days' journey in the interior, over several ridges of mountains, and were never seen near the coast. The coast people never go there themselves, nor do the mountaineers, who kill and preserve them, ever come to the coast, but sell them to the inhabitants of intermediate villages, where the coast people go to buy them. These sell them to the Dorey people, or any other native traders ; so that the specimens Lesson purchased had already passed through three or four hands.

These disappointments, with a scarcity of food sometimes approaching starvation, and almost constant sickness both of myself and men, one of whom died of dysentery, made me heartily glad when the schooner re-turned and took me away from Dorey. I had gone there with the most brilliant hopes, which, I think, were fully justified by the facts known before my visit ; and yet, as far as my special object (the birds of paradise) was concerned, I had accomplished next to nothing.

My ardour for New Guinea voyages being now somewhat abated, for the next year and a half I occupied myself in the Moluccas ; but in January, 1860, being joined (when at Amboyna) by my assistant, Mr. Charles Allen, I arranged a plan for the further exploration of the country of the Paradiseas, by sending Mr. Allen to Mysol, while I myself, after making the circuit of the island of Ceram, was to visit him with stores and provisions and proceed to Waigiou, both returning independently to meet at Ternate in the autumn.

I had been assured by the Goram and Bugis traders that Mysol was the very best country for the birds of paradise, and that they were finer and more abundant there than anywhere else. For Waigiou I had, besides the authority of the native traders, that of Lesson also, who visited the north coast for a few days, and mentions seven species of paradise birds purchased there by him.

These two promising expeditions turned out unfortunately in every respect. On reaching Goram, after much difficulty and delay, I found it impossible to make the voyage I had projected without a vessel of my own. I therefore purchased a small native prau of about eight tons, and after spending a month in strengthening and fitting it up, and having with great difficulty secured a native crew, paid them half their wages in advance, and overcome all the difficulties and objections which every one of them made to starting when all was ready, we at length got away, and I congratulated myself on my favourable prospects. Touching at Ceramlaut, the rendezvous of the New Guinea traders, I invested all my spare cash in goods for barter with the natives, and then proceeded towards Mysol.

The very next day, however, being obliged to anchor on the east coast of Ceram on account of bad weather, my crew all ran away during the night, leaving myself and my two Amboyna hunters to get on as we could. With great difficulty I procured other men to take us as far as Wahai, on the north coast of Ceram, opposite to Mysol, and there by a great chance succeeded in picking up a make-shift crew of four men willing to go with me to Mysol, Waigiou, and Ternate. I here found a letter from Mr. Allen, telling me he was much in want of rice and other necessaries, and was waiting my arrival to go to the north coast of Mysol, where alone the Paradiseæ could be obtained.

On attempting to cross the strait, seventy miles wide, between Ceram and Mysol, a strong east wind blew us out of our course, so that we passed to the westward of that island without any possibility of getting back to it. Mr. Allen, finding it impossible to live without rice, had to return to Wahai, much against his will, and there was kept two months waiting a supply from Amboyna. When at length he was able to return to Mysol, he had only a fortnight at the best place on the north coast, when the last boat of the season left, and he was obliged to take his only chance of getting back to Ternate.

Through this unfortunate series of accidents he was only able to get a single specimen of P. papuana, which is there finer than in most other places, a few of the Cicinnurus regius, and of P. magnifica only a native skin, though this beautiful little species is not rare in the island, and during a longer stay might easily have been obtained.

My own voyage was beset with misfortunes. After passing Mysol, I lost two of my scanty crew on a little desert island, our anchor breaking while they were on shore, and a powerful current carrying us rapidly away. One of them was our pilot; and, without a chart or any knowledge of the coasts, we had to blunder our way short-handed among the rocks and reefs and innumerable islands which surround the rocky coasts of Waigiou. Our little vessel was five times on the rocks in the space of twenty-four hours, and a little more wind or sea would in several cases have caused our destruction. On at length reaching our resting-place on the south coast of Waigiou, I immediately sent a native boat after my lost sailors, which, however, returned in a week without them, owing to bad weather. Again they were induced to make the attempt, and this time returned with them in a very weak and emaciated condition, as they had lived a month on a mere sand-bank, about a mile in diameter, subsisting on shell-fish and the succulent shoots of a wild plant.

I now devoted myself to an investigation of the natural history of Waigiou, having great expectations raised by Lesson's account, who says that he purchased the three true Paradiseas, as well as P. magnifica and P. sexsetacea, with Epimachus magnus and Sericulus aureus, in the island, and also mentions several rare Psittaci as probably found there. I soon ascertained, however, from the universal testimony of the inhabitants, afterwards confirmed by my own observation, that none of these species exist on the island, except P. rubra, which is the sole representative of the two families, Paradiseidæ and Epimachidæ, and is strictly limited to this one spot.

With more than the usual amount of difficulties, privations, and hunger, I succeeded in obtaining a good series of this beautiful and extraordinary bird; and three months' assiduous collecting produced no other species at all worthy of attention. The parrots and pigeons were all of known species; and there was really nothing in the island to render it worth visiting by a naturalist, except the P. rubra, which can be obtained nowhere else.

Our two expeditions to two almost unknown Papuan islands have thus added but one species to the Paradiseas which I had before obtained from Aru and Dorey. These voyages occupied us nearly a year; for we parted company in Amboyna in February, and met again at Ternate in November, and it was not till the following January that we were either of us able to start again on a fresh voyage.

At Waigiou I learned that the birds of paradise all came from three places on the north coast, between Salwatty and Dorey—Sorong, Maas, and Amberbaki. The latter I had tried unsuccessfully from Dorey; at Maas, the natives who procured the birds were said to live three days' journey in the interior, and to be cannibals; but at Sorong, which was near Salwatty, they were only about a day from the coast, and were less dangerous to visit. At Mysol, Mr. Allen had received somewhat similar information; and we therefore resolved that he should make another attempt at Sorong, where we were assured all the sorts could be obtained. The whole of that country being under the jurisdiction of the Sultan of Tidore, I obtained, through the Dutch resident at Ternate, a Tidore

lieutenant and two soldiers to accompany Mr. Allen as a protection, and
to facilitate his operations in getting men and visiting the interior.

Notwithstanding these precautions, Mr. Allen met with difficulties in
this voyage which we had not encountered before. To understand these,
it is necessary to consider that the birds of paradise are an article of
commerce, and are the monopoly of the chiefs of the coast villages, who
obtain them at a low rate from the mountaineers, and sell them to the
Bugis traders. A portion of the skins is also paid every year as tribute
to the Sultan of Tidore. The natives are therefore very jealous of a
stranger, especially a European, interfering in their trade, and above
all of his going into the interior to deal with the mountaineers themselves. They, of course, think he will raise the prices in the interior,
and lessen the demand on the coast, greatly to their disadvantage ; they
also think their tribute will be raised if a European takes back a
quantity of the rare sorts ; and they have, besides, a vague and very
natural dread of some ulterior object in a white man's coming at so
much trouble and expense to their country only to get birds of paradise, of which they know he can buy plenty at Ternate, Macassar, or
Singapore.

It thus happened that when Mr. Allen arrived at Sorong and explained his intentions of going to seek birds of paradise in the interior,
innumerable objections were raised. He was told it was three or four
days' journey over swamps and mountains ; that the mountaineers were
savages and cannibals, who would certainly kill him ; and, lastly, that
not a man in the village could be found who dare go with him. After
some days spent in these discussions, as he still persisted in making the
attempt, and showed them his authority from the Sultan of Tidore to go
where he pleased and receive every assistance, they at length provided
him with a boat to go the first part of the journey up a river ; at the same
time, however, they sent private orders to the interior villages to refuse
to sell any provisions, so as to compel him to return. On arriving at the
village where they were to leave the river and strike inland, the coast
people returned, leaving Mr. Allen to get on as he could. Here he called
on the Tidore lieutenant to assist him, and procure men as guides and to
carry his baggage to the villages of the mountaineers. This, however,
was not so easily done ; a quarrel took place, and the natives, refusing to
obey the somewhat harsh orders of the lieutenant, got out their knives
and spears to attack him and his soldiers, and Mr. Allen himself was
obliged to interfere to protect those who had come to guard him. The
respect due to a white man and the timely distribution of a few presents
prevailed ; and on showing the knives, hatchets, and beads he was
willing to give to those who accompanied him, peace was restored, and
the next day, travelling over a frightfully rugged country, they reached
the villages of the mountaineers. Here Mr. Allen remained a month,
without any interpreter through whom he could understand a word or
communicate a want. However, by signs and presents and a pretty
liberal barter, he got on very well, some of them accompanying him every

day in the forest to shoot and receiving a small present when he was successful.

In the grand matter of the paradise birds, however, little was done. Only one additional species was found, the Seleucides alba (or twelve-wired bird of paradise), of which he had already obtained a specimen on the island of Salwatty on his way to Sorong; so that at this much-vaunted place in the mountains, and among the bird-catching natives, nothing fresh was obtained. The P. magnifica, they said, was found there, but was rare; the Sericulus aureus also rare; Epimachus magnus, Astrapia nigra, Parotia sexsetacea, and Lophorina superba not found there, but only much further in the interior, as well as the lovely little lory, Charmosyna papuana. Moreover, neither at Sorong nor at Salwatty could he obtain a single native skin of the rarer species.

Thus ended my search after these beautiful birds. Five voyages to different parts of the district they inhabit, each occupying in its preparation and execution the larger part of a year, have produced me only five species out of the thirteen known to exist in New Guinea. The kinds obtained are those that inhabit the districts near the coasts of New Guinea and its islands, the remainder seeming to be strictly confined to the central mountain ranges of the northern peninsula; and our researches at Dorey and Amberbaki, near one end of this peninsula, and at Salwatty and Sorong, near the other, enable me to decide with some certainty on the native country of these rare and lovely birds, good specimens of which have never yet been seen in Europe. It must be considered as somewhat extraordinary that during five years' residence and travel in Celebes, the Moluccas, and New Guinea I should never have been able to purchase skins of half the species which Lesson, forty years ago, obtained during a few weeks in the same countries. I believe that all, except the common species of commerce, are now much more difficult to obtain than they were even twenty years ago; and I impute it principally to their having been sought after by the Dutch officials through the Sultan of Tidore. The chiefs of the annual expeditions to collect tribute have had orders to get all the rare sorts of paradise birds; and as they pay little or nothing for them (it being sufficient to say they are for the Sultan), the head men of the coast villages would for the future refuse to purchase them from the mountaineers, and confine themselves instead to the commoner species, which are less sought after by amateurs, but are to them a profitable merchandise. The same causes frequently lead the inhabitants of uncivilized countries to conceal any minerals or other natural products with which they may become acquainted, from the fear of being obliged to pay increased tribute, or of bringing upon themselves a new and oppressive labour.

I have given this short sketch of my search after the birds of paradise, barely touching on the many difficulties and dangers I experienced, because I fear that the somewhat scanty results of my exertions may have led to the opinion that they failed for want of judgment or perseverance. I trust, however, that the mere enumeration of my voyages

will show that patience and perseverance were not altogether wanting ;
but I must plead guilty to having been misled, first by Lesson and then
by all the native traders, it never having occurred to me (and I think it
could not have occurred to any one), that in scarcely a single instance
would the birds be found to inhabit the districts in which they are most
frequently to be purchased.   Yet such is the case ; for neither at Dorey,
nor at Salwatty, nor Waigiou, nor Mysol are any of the rarer species to
be found alive.   Not only this, but even at Sorong, where the Waigiou
chiefs go every year and purchase all kinds of birds of paradise, it has
turned out that most of the specimens are brought from the central
mountain ranges by the natives, and reach the shore in places where it is
not safe for trading praus to go, owing to the want of anchorage on an
exposed rocky coast.

Nature seems to have taken every precaution that these, her choicest
treasures, may not lose value by being too easily obtained.   First, we find
an open, harbourless, inhospitable coast, exposed to the full swell of the
Pacific Ocean ; next, a rugged and mountainous country, covered with
dense forests, offering in its swamps and precipices and serrated ridges
an almost impassable barrier to the central regions ; and lastly, a race of
the most savage and ruthless character, in the very lowest stage of
civilization.   In such a country and among such a people are found
these wonderful productions of nature.   In those trackless wilds do they
display that exquisite beauty and that marvellous development of
plumage, calculated to excite admiration and astonishment among the
most civilized and most intellectual races of men.   A feather is itself a
wonderful and a beautiful thing.   A bird clothed with feathers is almost
necessarily a beautiful creature.   How much, then, must we wonder at and
admire the modification of simple feathers into the rigid, polished, wavy
ribbons which adorn Paradisea rubra, the mass of airy plumes on P. apoda,
the tufts and wires of Seleucides alba, or the golden buds borne upon
airy stems that spring from the tail of Cicinnurus regius ; while gems and
polished metals can alone compare with the tints that adorn the breast
of Parotia sexsetacea and Astrapia nigra, and the immensely developed
shoulder-plumes of Epimachus magnus.

My next work was to describe five new birds from New
Guinea obtained by my assistant, Mr. Allen, during his
last visit there, and also seven new species obtained during
his visit to the north of Gilolo and Morty Island.   I also
described three new species of the beautiful genus Pitta,
commonly called ground-thrushes, but more nearly allied to
the South American ant-thrushes (Formicariidæ), or perhaps
to the Australian lyre-birds.   I also began a series of papers
dealing with the birds of certain islands or groups of islands
for the purpose of elucidating the geographical distribution

of animals in the archipelago. The first of these was a list of the birds from the Sula or Xulla Islands, situated between Celebes and the Moluccas, but by their position seeming to belong more to the latter. I believe that not a single species of bird was known from these small islands, and I should probably not have thought them worth visiting had I not been assured by native traders that a very pretty little parrot was found there and nowhere else. I therefore sent Mr. Allen there for two months, and he obtained a small but very interesting collection, consisting of forty-eight species of birds, of which seven were entirely new, including the little parrakeet which I named *Loriculus sclateri*, and which is one of the most beautiful of the genus. But the most interesting feature of the collection was that it proved indisputably that these islands, though nearer to Bouru and the Batchian group than to Celebes, really formed outlying portions of the latter island, since no less than twenty of the species were found also in Celebes and only ten in the Moluccas, while of the new species five were closely allied to Celebesian types, while only two were nearest to Molluccan species. This very curious and interesting result has led other naturalists to visit these islands as well as all the other small islands which cluster around the strangely formed large island. The result has been that considerable numbers of new species have been discovered, while the intimate connection of these islands with Celebes, so clearly shown by this first small collection, has been powerfully enforced.

During the succeeding five years I continued the study of my collections, writing many papers, of which more than a dozen related to birds, some being of considerable length and involving months of continuous study. But I also wrote several on physical and zoological geography, six on various questions of anthropology, and five or six on special applications of the theory of natural selection. I also began working at my insect collections, on which I wrote four rather elaborate papers. As several of these papers discussed matters of considerable interest and novelty, I will here give a brief

summary of the more important of them in the order in
which they were written.

The first of these, read in January, 1863, at a meeting of
the Zoological Society, was on my birds from Bouru, and was
chiefly important as showing that this island was undoubtedly
one of the Moluccan group, every bird found there which was
not widely distributed being either identical with or closely
allied to Moluccan species, while none had special affinities
with Celebes. It was clear, then, that this island formed the
most westerly outlier of the Moluccan group.

My next paper of importance, read before the same society
in the following November, was on the birds of the chain of
islands extending from Lombok to the great island of Timor.
I gave a list of one hundred and eighty-six species of birds,
of which twenty-nine were altogether new; but the special
importance of the paper was that it enabled me to mark out
precisely the boundary line between the Indian and Australian
zoological regions, and to trace the derivation of the rather
peculiar fauna of these islands, partly from Australia and
partly from the Moluccas, but with a strong recent migration
of Javanese species due to the very narrow straits separating
most of the islands from each other. The following table will
serve to illustrate this :—

	Lombok.	Flores.	Timor.
Species derived from Java ...	... 34	28	17
Species derived from Australia	... 7	14	36

This table shows how two streams of immigration have
entered these islands, the one from Java diminishing in
intensity as it passed on farther and farther to Timor; the
other from Australia entering Timor and diminishing still
more rapidly towards Lombok. This indicates, as its
geological structure shows, that Timor is the older island and
that it received immigrants from Australia at a period when,
probably, Lombok and Flores had not come into existence
or were uninhabitable. This is also indicated by the fact that
the Australian immigrants have undergone greater modification than the Javan. If we compare the birds of the whole

chain of islands according as they are of Javan or Australian
origin, we have the following results :—

Javan species 36	Australian species ... 13
Javan allied species ... 11	Australian allied species 35
47	48

We thus see that while the proportion of the birds derived
from each source is almost exactly equal, about three-fourths
of those from Java have remained unchanged, while three-
fourths of those from Australia have become so modified as
to be very distinct species. This shows us how the distribution of birds can, when carefully studied, give us information
as to the past history of the earth.

We can also feel confident that Timor has not been
actually connected with Australia, because it has none of the
peculiar Australian mammalia, and also because many of the
commonest and most widespread groups of Australian birds
are entirely wanting. And we are equally certain that
Lombok and the islands further east have never been united
to Bali and Java, because four Australian or papuan genera
of parrots and cockatoos are found in them, but not in Java,
as are several species of honeysuckers (Meliphagidæ), a
family of birds confined to the Australian region. On the
other hand, a large number of genera which extend over the
whole of the true Malay islands, from Sumatra to Java, never
pass the narrow straits into Lombok. Among these are the
long-tail parrakeets (Palæornis), the barbets (Megalæmidæ),
the weaver-birds (Ploceus), the ground starlings (Sturno-
pastor), several genera of woodpeckers, and an immense
number of genera of flycatchers, tits, gapers, bulbuls, and
other perching birds which abound everywhere in Borneo
and Java.

Two other papers dealt with the parrots and the pigeons
of the whole archipelago, and are among the most important
of my studies of geographical distribution. That on parrots
was written in 1864, and read at a meeting of the Zoological

Society in June. Although the Malay Archipelago as a whole is one of the richest countries in varied forms of the parrot tribe, that richness is almost wholly confined to its eastern or Australian portion, for while there are about seventy species between Celebes and the Solomon Islands, there are only five in the three large islands, Java, Borneo, and Sumatra, together with the Malay peninsula, while the Philippine Islands have twelve. This extreme richness of the Moluccas and New Guinea is also characteristic of the Pacific Islands and Australia, so that the Australian region, with its comparatively small area of land, contains nearly as many species of this tribe of birds as the rest of the globe, and considerably more than the vast area of tropical America, the next richest of all the regions.

No two groups of birds can well be more unlike in structure, form, and habits than parrots and pigeons, yet we find that the main features of the distribution of the former, as just described, are found also, though in a less marked degree, in the latter. The Australian region by itself contains three-fourths as many pigeons as the whole of the rest of the globe ; tropical America, the next richest, having only about half the number ; while tropical Africa and Asia are as poor, comparatively, in this group as they are in parrots. Turning now to our special subject, the Malay Archipelago, we find that it contains about one hundred and twenty species of pigeons, of which more than two-thirds (about ninety species) belong to the eastern or Austro-Malayan portion of it, which portion thus contains considerably more species, and much more varied forms and colours, than the whole of South America, Mexico, and the West Indies, forming the next richest area on the globe.

But this is not the only feature in which the parrots and the pigeons resemble each other. Both have characteristic forms and colours, which prevail generally over the whole world. In parrots this may be said to be green, varying into yellow, grey, red, and more rarely blue, and, except for a lengthened tail, having rarely any special developments of plumage. In pigeons, soft ashy lilac or brown tints are

characteristic of the whole group, often with metallic reflections; while soft greens, and sometimes metallic greens,
occur in the forest regions of tropical Africa and Asia, but
rarely anything approaching to crests or other developments
of plumage.

But as soon as we reach the Moluccas and New Guinea
we find a new type of coloration appearing in both groups.
Among the lories we find vivid red and crimson, sometimes
with a remnant of green on the wings and tail, but often
covering the whole plumage, varied with bands or patches
of equally vivid blue or yellow, while the red sometimes
deepens into a blackish-purple. Among the cockatoos we
have pure whites and deep black, with highly developed
crests, often of great beauty, so that in these two families
we seem to depart altogether from the usual parrot type of
coloration.

Still more remarkably is this the case with the pigeons.
In the extensive genus of small fruit-pigeons (*Ptilonopus*)
the usual ground colour is a clear soft green, variegated by
blue, purple, or yellow breasts, and crowns of equally brilliant
colours. Besides these, we have larger fruit-pigeons almost
wholly cream white, while the very large ground pigeons
of New Guinea possess flat vertical crests, which are unique
in this order of birds. The wonderfully brilliant golden
green Nicobar pigeon is probably a native of the Austro-
Malayan islands, and may have been carried westward by
Malay traders, and have become naturalized on a few small
islands.

These peculiarities of distribution and coloration in two
such very diverse groups of birds interested me greatly, and
I endeavoured to explain them in accordance with the laws
of natural selection. In the paper on Pigeons (published
in *The Ibis* of October, 1865) I suggest that the excessive
development of both these groups in the Moluccas and the
Papuan islands has been due primarily to the total absence
of arboreal, carnivorous, or egg-destroying mammals,
especially of the whole monkey tribe, which in all other
tropical forest regions are exceedingly abundant, and are

very destructive to eggs and young birds. I also point out
that there are here comparatively few other groups of fruit-
eating birds like the extensive families of chatterers, tanagers,
and toucans of America, or the barbets, bulbuls, finches,
starlings, and many other groups of India and Africa, while
in all those countries monkeys, squirrels, and other arboreal
mammals consume enormous quantities of fruits. It is clear,
therefore, that in the Australian region, especially in the
forest-clad portions of it, both parrots and pigeons have
fewer enemies and fewer competitors for food than in other
tropical regions, the result being that they have had freer
scope for development in various directions leading to the
production of forms and styles of colouring unknown
elsewhere. It is also very suggestive that the only other
country in which *black* pigeons and *black* parrots are found
is Madagascar, an island where also there are neither
monkeys nor squirrels, and where arboreal carnivora or fruit-
eating birds are very scarce. The satisfactory solution of
these curious facts of distribution gave me very great
pleasure, and I am not aware that the conclusions I arrived
at have been seriously objected to.

Before I had written these two papers I had begun the
study of my collection of butterflies, and in March, 1864,
I read before the Linnean Society a rather elaborate paper
on "The Malayan Papilionidæ, as illustrating the Theory of
Natural Selection." This was published in the Society's
Transactions, vol. xxv., and was illustrated by fine coloured
plates drawn by Professor Westwood. I reprinted the intro-
ductory portion of this paper in the first edition of my "Con-
tributions to the Theory of Natural Selection" in 1870, but
in later editions it was omitted, as being rather too technical
for general readers, and not easily followed without the
coloured plates. I will therefore give a short outline of its
purport here.

I may state for the information of non-entomological
readers that the Papilionidæ form one of the most extensive
families of butterflies, and from their large size, elegant forms,

and splendid colours were considered by all the older writers to be the princes of the whole lepidopterous order. They are usually known by the English term "Swallow-tailed butterflies," because the only British species, as well as a great many of the tropical forms, have the hind wings tailed. They are pretty uniformly distributed over all the warmer regions, but are especially abundant in the tropical forests, of which they form one of the greatest ornaments. In coloration they are wonderfully varied. The ground colour is very frequently black, on which appear bands, spots, or large patches of brilliant colours—pale or golden yellow, rich crimsons or gorgeous metallic blues and greens, which colours sometimes spread over nearly the whole wing surface. Some are thickly speckled with golden green dots and adorned with large patches of intense metallic green or azure blue, others are simply black and white in a great variety of patterns many very striking and beautiful, while others again have crimson or golden patches, which when viewed at certain angles change to quite different opalescent hues, unsurpassed by the rarest gems.

But it is not this grand development of size and colour that constitutes the attraction of these insects to the student of evolution, but the fact that they exhibit, in a remarkable degree, almost every kind of variation, as well as some of the most beautiful examples of polymorphism and of mimicry. Besides these features, the family presents us with examples of differences of size, form, and colour, characteristic of certain localities, which are among the most singular and mysterious phenomena known to naturalists. A short statement of the nature of these phenomena will be useful to show the great interest of the subject.

In all parts of the world there are certain insects which, from a disagreeable smell or taste, are rarely attacked or devoured by enemies. Such groups are said to be "protected," and they almost always have distinctive and conspicuous colours. In the Malay Archipelago there are several groups of butterflies which have this kind of protection; and one group is coloured black, with rich blue glosses and

ornamented with white bands or spots. These are excessively
abundant, and, having few enemies, they fly slowly. Now
there are also several different kinds of papilios, which in
colour are so exactly like these, that when on the wing they
cannot be distinguished, although they frequent the same
places and are often found intermingled. Other protected
butterflies are of paler colours with dark stripes, and these
are also closely imitated by other papilios. Altogether there
are about fifteen species which thus closely resemble protected butterflies externally, although in structure and transformations they have no affinity with them. In some cases
both sexes possess this resemblance, or "mimicry," as it is
termed, but most frequently it is the female only that is thus
modified, especially when she lays her eggs on low-growing
plants; while the male, whose flight is stronger and can take
care of himself, does not possess it, and is often so different
from his mate as to have been considered a distinct species.

This leads us to the phenomenon of dimorphism and
polymorphism, in which the females of one species present
two or three different forms. Several such cases occur in the
Malay Archipelago, in which there are two distinct kinds of
females, sometimes even three, to a single male, which differs
from either of them. In one case four females are known to
one male, though only two of them appear to occur in one
locality. These have been almost always described as
distinct species, but observation has now proved them to be
one, and it has further been noticed that each of the females,
which are very unlike the male, resembles more or less
closely some "protected" species. It has also been proved
by experimental breeding that eggs laid by any one of these
females are capable of producing butterflies of all the
different forms, which in the few cases recorded are quite
distinct from each other, without intermediate gradations.

The local diversities of form are illustrated by outline
figures (as regards two species of papilio from Celebes) in
my "Malay Archipelago" (p. 216), and similar local peculiarities of colour, both in papilio and other groups, are
described in my " Natural Selection and Tropical Nature"

(pp. 384, 385), while extraordinary development of size in
Amboyna is referred to at p. 307 of my " Malay Archipelago."

This brief outline of the paper will, perhaps, enable my
readers to understand the intense interest I felt in working
out all these strange phenomena, and showing how they
could almost all be explained by that law of " Natural
Selection " which Darwin had discovered many years before,
and which I had also been so fortunate as to hit upon.

The only other groups of insects upon which I did any
systematic work were the families of Pieridæ among butterflies and Cetoniidæ among beetles. Of the former family,
which contains our common whites, our brimstone and orange
tip butterflies, I gave a list of all known from the Indian and
Australian regions, describing fifty new species, mostly from
my own collection. This paper is in the " Transactions of
the Entomological Society for 1867," and is illustrated by four
coloured plates. The other paper, which is contained in the
same volume, is a catalogue of the Cetoniidæ (or Rose-
chafers, named after our common species) of the Malay
Archipelago, in which I described seventy new species, the
majority of which were collected by myself, and it is illustrated by four coloured plates, beautifully executed by the
late Mr. E. W. Robinson, in which thirty-two of the species
are figured. These two papers, filling about 200 pages of the
society's " Transactions," occupied me for several months,
and if I had not had wider and more varied interests—
evolution, distribution, physical geography, anthropology,
the glacial period, geological time, sociology, and several
others—I might have spent the rest of my life upon similar
work, for which my own collection afforded ample materials,
and thus settled down into a regular " species-monger." For
even in this humble occupation there is a great fascination ;
constant difficulties are encountered in unravelling the
mistakes of previous describers who have had imperfect
materials, while the detection of those minute differences,
which often serve to distinguish allied species, and the many
curious modifications of structure which characterize genera
or their subdivisions, become intensely interesting, especially

when, after weeks of study, a whole series of specimens, which seemed at first hardly distinguishable, are gradually separated into well-defined species, and order arises out of chaos.

The series of papers on birds and insects now described, together with others on the physical geography of the archipelago and its various races of man, furnished me with the necessary materials for that general sketch of the natural history of the islands and of the various interesting problems which arise from its study, which has made my "Malay Archipelago" the most popular of my books. At the same time it opened up so many fields of research as to render me indisposed for further technical work in the mere description of my collections, which I should certainly never have been able to complete. I therefore now began to dispose of various portions of my insects to students of special groups, who undertook to publish lists of them with descriptions of the new species, reserving for myself only a few boxes of duplicates to serve as mementoes of the exquisite or fantastic organisms which I had procured during my eight years' wanderings.

In order that my scientific friends might be able to see the chief treasures which I had brought home, I displayed a series of the rarest and most beautiful of my birds and butterflies in Mr. Sims's large photographic gallery in the same manner as I had found so effective with my New Guinea collections at Ternate. The entire series of my parrots, pigeons, and paradise birds, when laid out on long tables covered with white paper, formed a display of brilliant colours, strange forms, and exquisite texture that could hardly be surpassed; and when to these were added the most curious and beautiful among the warblers, flycatchers, drongos, starlings, gapers, ground thrushes, woodpeckers, barbets, cuckoos, trogons, kingfishers, hornbills, and pheasants, the general effect of the whole, and the impression it gave of the inexhaustible variety and beauty of nature in her richest treasure houses, was far superior to that of any collection of stuffed and mounted birds I have ever seen.

This mode of exhibiting bird skins is especially suitable for artificial light, and I believe that if a portion of the enormous wealth of the national collection in unmounted bird skins were used for evening display in the public galleries, it would be exceedingly attractive. Different regions or subregions might be illustrated by showing specimens of all the most distinct and remarkable species that characterize them, and each month during the winter a fresh series might be shown, and thus all parts of the world in turn represented. And in the case of insects the permanent series shown in the public galleries might be thus arranged, those of each region or of the well-marked subregions being kept quite separate. This would be not only more instructive, but very much more interesting, because such large numbers of persons have now visited or resided in various foreign countries, and a still larger number have friends or relatives living abroad, and all these would be especially interested in seeing the butterflies, beetles, and birds which are found there. In this way it would be possible to supply the great want in all public museums—a *geographical* rather than a purely *systematic* arrangement for the bulk of the collections exhibited to the public. The systematic portion so exhibited might be limited to the most distinctive types of organization, and these might be given in a moderate-sized room.

Having thus prepared the way by these preliminary studies, I devoted the larger portion of my time in the years 1867 and 1868 to writing my " Malay Archipelago." I had previously read what works I could procure on the islands, and had made numerous extracts from the old voyagers on the parts I myself was acquainted with. These added much to the interest of my own accounts of the manners and character of the people, and by means of a tolerably full journal and the various papers I had written, I had no difficulty in going steadily on with my work. As my publishers wished the book to be well illustrated, I had to spend a good deal of time in deciding on the plates and getting them drawn, either from my own sketches, from photographs, or from actual specimens, and having obtained the services of the best artists and wood engravers then in London, the result was, on the whole, satisfactory. I would particularly indicate the frontispiece by Wolf as a most artistic and spirited picture, while the two plates of beetles by Robinson, the "twelve-wired" and "king" birds of paradise by Keulemaus, and the head of the black cockatoo by Wood, are admirable specimens of life-like drawing and fine wood engraving. I was especially indebted to Mr. T. Baines, the well-known African traveller, and the first artist to depict the Victoria Falls and numerous scenes of Kaffir life, for the skill with which he has infused life and movement into an outline sketch of my own, of "Dobbo in the Trading Season."

The book was published in 1869, but during its progress, and while it was slowly passing through the press, I wrote several important papers, among which was one in the *Quarterly Review* for April, 1889, on "Geological Climates and the Origin of Species," which was in large part a review and eulogy of Sir Charles Lyell's great work, "The Principles of Geology," which greatly pleased him as well as Darwin. A considerable part of this article was devoted to a discussion of Mr. Croll's explanation of the glacial epoch, and, by a combination of his views with those of Lyell on the great effect of changed distribution of sea and land, or of differences in altitude, I showed how we might arrive at a better explanation than either view by itself could give us. As the article was too long, a good deal of it had to be cut out, but it served as the foundation for my more detailed examination of the whole question when writing my "Island Life," twelve years later.

As soon as the proofs of the "Malay Archipelago" were out of my hands, I began the preparation of a small volume of my scattered articles dealing with various aspects of the theory of Natural Selection. Many of these had appeared in little known periodicals, and were now carefully revised, or partially rewritten, while two new ones were added.   The longest article, occupying nearly a quarter of the volume, was one which I had written in 1865–6, but which was not published (in the *Westminster Review*) till July, 1867, and was entitled "Mimicry, and other Protective Resemblances among Animals."   In this article I endeavoured to give a general account of the whole subject of protective resemblance, of which theory, what was termed by Bates "mimicry," is a very curious special case.   I called attention to the wide extent of the phenomenon, and showed that it pervades animal life from mammals to fishes and through every grade of the insect tribes.   I pointed out that the whole series of phenomena depend upon the great principle of the utility of every character, upon the need of protection or of concealment by almost all animals, and upon the known fact that no character are so variable as colour, and that therefore concealment has been most easily obtained by colour modification.

Coming to the subject of "mimicry" I gave a popular account of its principle, with numerous illustrations of its existence in all the chief groups of insects, not only in the tropics, but even in our own country.   I also showed, I think for the first time, that it occurs among birds in a few well-marked cases, and also in at least one instance among mammalia, and I explained why we could not expect it to occur more frequently among these higher animals.

Two other articles which may be just mentioned are those entitled "A Theory of Birds' Nests" and "The Limits of Natural Selection applied to Man."   In the first I pointed out the important relation that exists between concealed nests and the bright colours of female birds, leading to conclusions adverse to Mr. Darwin's theory of colours and ornaments in the males being the result of female choice. In the other (the last in the volume) I apply Darwin's principle of natural selection, acting solely by means of "utilities," to show that certain physical modifications and mental faculties in man could not have been acquired through the preservation of useful variations, because there is some direct evidence to show that they *were not* and *are not* useful in the ordinary sense, or, as Professor Lloyd Morgan well puts it, not of " life-preserving value," while there is absolutely no evidence to show that they were so. In reply, Darwin has appealed to the effects of female choice in developing these characteristics, of which, however, not a particle of evidence is to be found among existing savage races.

Besides the literary and scientific work now described, in the last three years of the period now dealt with I contributed about twenty letters or short papers to various periodicals, delivered several lectures, and reviewed a dozen books, including such important works as Darwin's " Descent of Man " and Galton's " Hereditary Genius." I also gave a Presidential Address to the Entomological Society in January, 1871, in which I discussed the interesting problems arising from the peculiarities of insular insects as especially illustrated by the beetles of Madeira.

As it was during the ten years of which I have now sketched my scientific and literary work that I saw most of my various scientific friends and acquaintances, and it was also in this period that the course of my future life and work was mainly determined, I will devote the next five chapters to a short summary of my more personal affairs, together with a few recollections of those friends with whom I became most familiar.

CHAPTER XXIV

HOME LIFE—MY FRIENDS AND ACQUAINTANCES—SIR CHARLES LYELL

SOON after my return home in the spring of 1862, my oldest friend and schoolfellow, Mr. George Silk, introduced me to a small circle of his friends, who had formed a private chess club, and thereafter, while I lived in the vicinity of Kensington, I was invited to attend the meetings of the club. One of these friends was a Mr. L——, a widower with two daughters, and a son who was at Cambridge University. I sometimes went there with Silk on Sunday afternoons, and after a few months was asked to call on them whenever I liked in the evening to play a game with Mr. L——. On these occasions the young ladies were present, and we had tea or supper together, and soon became very friendly. The eldest Miss L—— was, I think, about seven or eight and twenty, very agreeable though quiet, pleasant looking, well educated, and fond of art and literature, and I soon began to feel an affection for her, and to hope that she would become my wife. In about a year after my first visit there, thinking I was then sufficiently known, and being too shy to make a verbal offer, I wrote to her, describing my feelings and asking if she could in any way respond to my affection. Her reply was a negative, but not a very decided one. Evidently my undemonstrative manner had given her no intimation of my intentions. She concluded her letter, which was a very kind one, by begging that I would not allow her refusal to break off my visits to her father.

At first I was inclined not to go again, but on showing

the letter to my sister and mother, they thought the young lady was favourably disposed, and that I had better go on as before, and make another offer later on. Another year passed, and thinking I saw signs of a change in her feelings towards me, but fearing another refusal, I wrote to her father, stating the whole circumstances, and asking him to ascertain his daughter's wishes, and, if she was now favourable, to grant me a private interview. In reply I was asked to call on Mr. L——, who inquired as to my means, etc., told me that his daughter had a small income of her own, and asked that I should settle an equal amount on her. This was satisfactorily arranged, and at a subsequent meeting we were engaged.

Everything went on smoothly for some months. We met two or three times a week, and after delays, owing to Miss L——'s ill-health and other causes, the wedding day was fixed and all details arranged. I had brought her to visit my mother and sister, and I was quite unaware of any cause of doubt or uncertainty when one day, on making my usual call, I was informed by the servant that Miss L—— was not at home, that she had gone away that morning, and would write. I came home completely staggered, and the next morning had a letter from Mr. L——, saying that his daughter wished to break off the engagement and would write to me shortly. The blow was very severe, and I have never in my life experienced such intensely painful emotion.

When the letter came I was hardly more enlightened. The alleged cause was that I was silent as to myself and family, that I seemed to have something to conceal, and that I had told her nothing about a widow lady, a friend of my mother's, that I had almost been engaged to. All this was to me the wildest delusion. The lady was the widow of an Indian officer, very pleasant and good-natured, and very gossipy, but as utterly remote in my mind from all ideas of marriage as would have been an aunt or a grandmother. As to concealment, it was the furthest thing possible from my thoughts ; but it never occurs to me at any time to talk about myself, even my own children say that they know nothing about my early life ; but if any one asks me and wishes to

XXIV]   HOME LIFE   411

know, I am willing to tell all that I know or remember. I was dreadfully hurt. I wrote, I am afraid, too strongly, and perhaps bitterly, trying to explain my real feelings towards her, and assuring her that I had never had a moment's thought of any one but her, and hoping that this explanation would suffice. But I received no reply, and from that day I never saw, or heard of, any of the family.

While these events were in progress, my dear friend, Dr. Richard Spruce, came home from Peru in very weak health, and, after staying a short time in London, went to live at Hurstpierpoint, in Sussex, in order to be near Mr. William Mitten, then the greatest English authority on mosses, and who had undertaken to describe his great collections from South America. This was in the autumn of 1864, and in the spring of 1865 I took a small house for myself and my mother, in St. Mark's Crescent, Regent's Park, quite near the Zoological Gardens, and within a pleasant walk across the park of the society's library in Hanover Square, where I had to go very often to consult books of reference. Here I lived five years, having Dr. W. B. Carpenter for a near neighbour, and it was while living in this house that I saw most of my few scientific friends.

During the summer and autumn I often went to Hurstpierpoint to enjoy the society of my friend, and thus became intimate with Mr. Mitten and his family. Mr. Mitten was an enthusiastic botanist and gardener, and knew every wild plant in the very rich district which surrounds the village, and all his family were lovers of wild flowers. I remember my delight, on the occasion of my first or second visit there, at seeing a vase full of the delicate and fantastic flowers of the large butterfly-orchis and the curious fly-orchis, neither of which I had ever seen before, and which I was surprised to hear were abundant in the woods at the foot of the downs. It was an immense delight to me to be taken to these woods, and to some fields on the downs where the bee-orchis and half a dozen other species grew abundantly, with giant cowslips nearly two feet high, the dyers' broom, and many other interesting plants. The richness of this district may be judged by the fact that within a walk more than twenty species of orchises have been found. This similarity of taste led to a close intimacy, and in the spring of the following year I was married to Mr. Mitten's eldest daughter, then about eighteen years old.

After a week at Windsor we came to live in London, and in early autumn went for a month to North Wales, staying at Llanberris and Dolgelly. I took with me Sir Andrew Ramsay's little book on "The Old Glaciers of Switzerland and North Wales," and thoroughly enjoyed the fine examples of ice-groovings and striations, smoothed rock-surfaces, *roches moutonnées*, moraines, perched blocks, and rock-basins, with which the valleys around Snowdon abound. Every day revealed some fresh object of interest as we climbed among the higher *cwms* of Snowdon ; and from what I saw during that first visit the Ice Age became almost as much a reality to me as any fact of direct observation. Every future tour to Scotland, to the lake district, or to Switzerland became doubled in interest. I read a good deal of the literature of the subject, and have, I believe, in my later writings been able to set forth the evidence in favour of the glacial origin of lake-basins more forcibly than it has ever been done before. As a result of my observations I wrote my first article on the subject, "Ice-marks in North Wales," which appeared in the *Quarterly Journal of Science* of January, 1867. In this paper I gave a sketch of the more important phenomena, which were then by no means so well known as they are now ; and I also gave reasons for doubting the conclusions of Mr. Macintosh in the *Journal of the Geological Society*, that most of the valleys and rocky *cwms* of North Wales had been formed by the action of the sea. I also gave, I think for the first time, a detailed explanation of *how* glaciers can have formed lake-basins, by grinding due to unequal pressure, not by "scooping out," as usually supposed.

In 1867 I spent the month of June in Switzerland with my wife, staying at Champery, opposite the beautiful Dent

du Midi, where at first we were the only visitors in a huge
new hotel, but for the second week had the company of an
English clergyman, his wife, and son. We greatly enjoyed
the beautiful subalpine flowers then in perfection, and one
day I went with the clergyman and his son, a boy of about
thirteen, to see how far we could get on the way to the great
mountain's summit. On the alp above the pine forest we
had our lunch at a cow-herd's hut, with a large jug of cream,
and then got the man to act as guide. He took us over a
ravine filled with snow, and then up a zigzag path among
the rocks along a *mauvais pas*, where an iron bar was fixed
on the face of a precipice, and then up to an ice-smoothed
plateau of limestone rock, still partly snow-clad, all the
crevices of which were full of alpine flowers. I was just
beginning to gather specimens of these and thought to enjoy
an hour's botanizing when our guide warned us that a snow-
storm was coming, and we must return directly, and the
black clouds and a few snowflakes made us only too willing
to follow him. We got back safely, but I have always
regretted that hasty peep of the alpine rock-flora at a time
of year when I never afterwards had an opportunity of
seeing it.

We then went by Martigny over the St. Bernard, reaching
the hospice after dark through deep snow, and next day
walked down to Aosta, a place which had been recommended
to me by Mr. William Mathews, a well-known Alpine
climber. It was a very hot place, and its chief interest
to us was an excursion on mules to the Becca de Nona,
which took us a long day, going up by the easiest and
descending the most precipitous road—the latter a mere
staircase of rock. The last thousand feet I walked up alone,
and was highly delighted with the summit and the wonderful
scene of fractured rocks, ridges, and peaks all around, but
more especially with the summit itself, hardly so large as
that of Snowdon and exhibiting far grander precipices and
rock-masses, all in a state of visible degradation, and showing
how powerfully the atmospheric forces of denudation are in
constant action at this altitude—10,380 feet. Hardly less

interesting were the charming little alpine plants in the
patches of turf and the crevices in the rocks, among which
were two species of the exquisite Androsaces, the true gems
of the primrose tribe. I also one day took a lonely walk up
a wild valley which terminated in the glacier that descends
from Mount Emilius ; and on another day we drove up the
main valley to Villeneuve, and then walked up a little way
into the Val Savaranches. This is one of those large open
valleys which have been the outlet of a great glacier, and in
which the subglacial torrent has cut a deep narrow chasm
through hard rocks at its termination, through which the
river now empties itself into the main stream of the Dora
Baltea. This was the first of the kind I had specially noticed,
though I had seen the Gorge of the Trient on my first visit
to Switzerland at a time when I had barely heard of the
glacial epoch.

Returning over the St. Bernard we went to Interlachen
and Grindelwald, saw the glaciers there, and then went over
the Wengern Alp, staying two days at the hotel to see the
avalanches and botanize among the pastures and moraines.
Then down to Lauterbrunnen to see the Staubbach, and
thence home.

As I had found that amid the distractions and excitement
of London, its scientific meetings, dinner parties and sight-
seeing, I could not settle down to work at the more scientific
chapters of my " Malay Archipelago," I let my house in
London for a year, from Midsummer, 1867, and went to live
with my wife's family at Hurstpierpoint. There, in perfect
quiet, and with beautiful fields and downs around me, I was
able to work steadily, having all my materials already
prepared. Returning to London in the summer of 1868,
I was fully occupied in arranging for the illustrations and
correcting the proofs. The work appeared at the end of the
year, and my volume on " Natural Selection " in the following
March.

I may here state that although the proceeds of my eight
years' collecting in the East brought me in a sufficient income

to live quietly as a single man, I was always on the lookout
for some permanent congenial employment which would yet
leave time for the study of my collections. The possibility
of ever earning anything substantial either by lecturing or
by writing never occurred to me. My deficient organ of
language prevented me from ever becoming a good lecturer
or having any taste for it, while the experience of my first
work on "The Amazon" did not encourage me to think that
I could write anything that would much more than pay
expenses. The first vacancy that occurred was the assistant
secretaryship of the Royal Geographical Society, for which
Bates and myself were candidates. Bates had just published
his "Naturalist on the Amazon," and was, besides, much
better qualified than myself by his business experience and
his knowledge of German, which he had taught himself when
abroad. Besides, the confinement and the London life
would, I am sure, have soon become uncongenial to me,
and would, I feel equally certain, have greatly shortened
my life. I am therefore glad I did not get it, and I do not
think I felt any disappointment at the time; and as it
brought Bates to live in London, I was able to see him
frequently in his private room and occasionally at his home,
and talk over old times or of scientific matters that interested
us both, while we frequently met at the Entomological or
other societies' evening meetings. This was in 1864, and I
was too busy with my descriptive work and writings to
think much more on the subject till 1869, when it was
decided by the Government to establish a branch museum
in Bethnal Green which should combine art and natural
history for the instruction of the people. I thought this
would suit me very well if I could get the directorship.
Lord Ripon, then Lord President of the Council, was a
friend of Sir Charles Lyell, and after an interview with him
he promised to help me with the Government, while Huxley
(I think) introduced me to Sir Henry Cole, then head of the
Science and Art Department at South Kensington. I also
had the kind assistance of several other friends, but though
the museum was built and opened, I think, in 1872, it was

managed from South Kensington and no special director
was required. Partly because (in my inexperience of such
matters) I felt rather confident of getting this appointment,
and also because I was becoming tired of London and wished
for a country life, I took a small house at Barking in 1870,
and in 1871 leased four acres of ground at Grays, including a
very picturesque well-timbered old chalk-pit, above which I
built a house having a very fine view across to the hills of
North Kent and down a reach of the Thames to Gravesend.

Seven years later, in 1878, when Epping Forest had been
acquired by the Corporation of London, a superintendent
was to be appointed to see to its protection and improvement
while preserving its "natural aspect" in accordance with the
Act of Parliament which restored it to the public. This
position would have suited me exactly, and if I had obtained
it and had been allowed to utilize the large extent of open
unwooded land in the way I suggested in my article in the
Fortnightly Review ("Epping Forest, and how best to deal
with it"), an experiment in illustration of the geographical
distribution of plants would have been made which would
have been both unique and educational, as well as generally
interesting. I obtained recommendations and testimonials
from the presidents of all the natural history societies in
London, from numerous residents near the forest and in
London, from many eminent men and members of Parliament—seventy in all; but the City merchants and tradesmen
with whom the appointment lay wanted a "practical man"
to carry out their own ideas, which were to utilize all the
open spaces for games and sports, to build a large hotel
close to Queen Elizabeth's hunting lodge, and to encourage
excursions and school treats, allowing swings, round-abouts,
and other such amusements more suited to a beer-garden or
village fair than to a tract of land secured at enormous cost
and much hardship to individuals in order to preserve an
example of the wild natural woodland wastes of our country
for the enjoyment and instruction of successive generations of
nature-lovers.

I still think it is much to be regretted that no effort is

made to carry out my suggestion in the article above referred to (reprinted in my "Studies," vol. ii., under the title, "Epping Forest and Temperate Forest Regions"). There still remains in the open moors and bare wastes, forming outlying parts of the New Forest, ample space on which to try the experiment, and at all events to extend the forest character of the scenery.

My failure to obtain the post at Epping Forest was certainly a disappointment to me, but I am inclined to think now that even that was really for the best, since it left me free to do literary work which I should certainly not have done if I had had permanent employment so engrossing and interesting as that at Epping. In that case I should not have gone to lecture in America, and should not have written "Darwinism," perhaps none of my later books, and very few of the articles contained in my "Studies." This body of literary and popular scientific work is, perhaps, what I was best fitted to perform, and if so, neither I nor my readers have any reason to regret my failure to obtain the post of superintendent and guardian of Epping Forest.

Among the eminent men of science with whom I became more or less intimate during the period of my residence in London, I give the first place to Sir Charles Lyell, not only on account of his great abilities and his position as one of the brightest ornaments of the nineteenth century, but because I saw more of him than of any other man at all approaching him as a thinker and leader in the world of science, while my correspondence with him was more varied in the subjects touched upon, and in some respects of more general interest, than my more extensive correspondence with Darwin. My friend, Sir Leonard Lyell, has kindly lent me a volume containing the letters from his scientific correspondence which have been preserved, and I am therefore able to see what subjects I wrote about, and to give such portions of the letters as seem to be of general interest.

Early in 1864 Sir Charles was preparing his presidential address for the meeting of the British Association at Bath, and wishing to introduce a paragraph as to the division of the Malay Archipelago into two regions, and the relation of this division to the races of man, and also as to the probable rate of change of insects, he asked me for a short statement of my conclusions on these subjects. On the latter point I wrote :—

"As regards insects changing rapidly, I see nothing improbable in it, because, though in a totally different way, they are as highly specialized as are birds or mammals, and, through the transformations they undergo, have still more complicated relations with the organic and inorganic worlds. For instance, they are subject to different kinds of danger in their *larva, pupa,* and *imago* state ; they have different enemies and special means of protection in each of these states, and changes of climate may probably affect them differently in each state. We may therefore expect very slight changes in the proportions of other animals, in physical geography, or in climate, to produce an immediate change in their numbers, and often in their organization. The fact that they *do* change rapidly is, I think, shown by the large number of peculiar species of insects in Madeira as compared with the birds and plants ; the same thing occurs in Corsica, where there are many peculiar species of insects ; also, we see the very limited of range of many insects as found by Bates and myself. Again, your rule of the slow change of mollusca applies to aquatic species only. The land-shells, I presume, change much more rapidly ; or why are almost every species in Madeira and in each of the West Indian islands peculiar ? Being terrestrial, they are affected as insects are by physical changes, and more still by organic changes. Such changes are certainly much slower in the sea."

Later on, in May, after reading my article on "The Races of Man and Natural Selection," which Darwin thought so highly of, though at the same time he was quite distressed at my conclusion that natural selection could not have done it all, Sir Charles objected (May 22, 1864)—very naturally for a geologist, and for one who had so recently become a convert to Darwin's views—that my suggestion of man's possible origination, so early as the Miocene, was due to my "want of appreciation of the immensity of time at our disposal, without going back beyond the Newer Pliocene."

To this objection I replied (May 24) as follows : " With regard to the probable antiquity of man, I will say a few words. First, you will see, I argue for the *possibility* rather than for the *necessity* of man having existed in Miocene times, and I still maintain this possibility, and even probability, for the following reasons. The question of time cannot be judged of positively, but only comparatively. We cannot say *à priori* that ten millions or a thousand millions of years would be required for any given modification in man. We must judge only by analogy, and by a comparison with the rate of change of other highly organized animals. Now, several existing *genera* lived in the Miocene age, and also anthropoid apes allied to Hylobates. But man is classed, even by Huxley, as a distinct *family*. The origin of that family—that is, its common origin with other *families* of the Primates—must therefore date back from an earlier period than the Miocene. Now, the greater part of the *family* difference is manifested in the head and cranium. A being almost exactly like man in the rest of the skeleton, but with a cranium as little developed as that of a chimpanzee, would certainly not form a distinct *family*, only a distinct *genus* of Primates. My argument, therefore, is, that this great cranial difference has been slowly developing, while the rest of the skeleton has remained nearly stationary ; and while the Miocene Dryopithecus has been modified into the existing gorilla, speechless and ape-brained man (but yet *man*) has been developed into great-brained, speech-forming man.

" The majority of Pliocene mammals, on the other hand, are, I believe, of existing genera, and as my whole argument is to show how man has undergone a more than generic change in brain and cranium, while the rest of his body has hardly changed *specifically*, I cannot consistently admit that all this change has been brought about in a less period than has sufficed to change most other mammals *generically*, except by assuming that in his case the change has been more

rapid, which may, indeed, have been so, but which we have
no evidence yet to prove. I conceive, therefore, that the
immensity of *time*, measured in years, does not affect the
argument. My paper was written too hastily and too briefly
to explain the subject fully and clearly, but I hope these
few remarks may give my ideas on the point you have
especially referred to."

In 1867, when a new edition of the "Principles of Geology"
was in progress, I had much correspondence and many talks
with Sir Charles, chiefly on questions relating to distribution
and dispersal, in which he, like myself, was greatly interested.
He was by nature so exceedingly cautious and conservative,
and always gave such great weight to difficulties that occurred
to himself or that were put forth by others, that it was not
easy to satisfy him on any novel view upon which two opinions
existed or were possible. We used often to discuss these
various points, but in any case that seemed to him important
he usually preferred to write to me, stating his objections,
sometimes at great length, and asking me to give my views.
In reply to some such inquiries I sent him my paper on the
birds of the Lombok to Timor groups, and wrote to him at
the same time more fully explaining its bearing, as afterwards
given in my "Malay Archipelago." I also wrote him on the
curious facts as to the distribution of pigs in the whole
archipelago, as illustrated by facts he had himself given
showing the remarkable power of swimming possessed by
these animals. Another fact he wanted explained was the
presence of a few non-marsupial mammals in Australia, and
why there were not more of them, and why none were
found in the caves. On these points I wrote to him as
follows :—

"MY DEAR SIR CHARLES,

"I think the fact that the only placental land
mammals in Australia (truly indigenous) are the *smallest* of
all mammals is a very suggestive fact as to how they got
there. Mice would not only be carried by canoes, but they

would also be transported occasionally by floating trees carried down by floods. I think myself, however, that it is most likely they were carried by the earliest canoes of prehistoric man, and that they afford an example of rapid change of specific form, owing to the ancestral species having been subjected to a great change of conditions, both as regards climate and food, and having had an immense area of new country to roam over and multiply in, in every part of which they would be subjected to different conditions. These considerations, I think, fully meet the facts, and there ought to be no large rodents found in the caves of Australia, and no other rodents of very distinct type from those now living. When any such are found it will be time enough to consider how to account for them. It is, as you say, a most important fact that, in three such distinct localities as New Zealand, Australia, and Mauritius, no bones of extinct carnivora or other mammalia should be found along with the wingless birds and marsupials, while abundance of remains of these groups *are* found. We may, I think, fairly claim this as a proof that such placental mammals did not exist in those countries, and the fact that the only exception in the existing Australia fauna are *mice* indicates very clearly that they are a recent introduction. When all the known facts are in our favour, I do not think we need trouble ourselves to answer objections and overcome difficulties that have not yet arisen, and probably never will arise."

Some months later (November, 1867) he wrote me about the dispersal and the colours of the races of man. On the first point I replied at some length, principally to show why we should not expect the primary regions which show the great features of the distribution of birds, reptiles, and mammalia should also apply to man. On the question of colour I replied as follows: "Why the colour of man is sometimes constant over large areas while in other cases it varies, we cannot certainly tell; but we may well suppose it to be due to its being more or less correlated with constitutional characters favourable to life. By far the most *common* colour of man is a *warm brown*, not very different from that of the American Indian. White and black are alike deviations from this, and are probably correlated with mental or physical peculiarities which have been favourable to the increase and maintenance of the particular race. I should infer, therefore, that the *brown* or *red* was the original colour of man, and that it maintains itself throughout all climates in America because accidental deviations from it have not been accompanied by any useful constitutional peculiarities. It is Bates's opinion that the Indians are recent immigrants into the tropical plains of South America, and are not yet fully acclimatized."

In the following year, when I was living at Hurstpierpoint, in a letter I wrote to Sir Charles, thanking him for the trouble he had taken in regard to the Bethnal Green Museum, I added some remarks on Darwin's new theory of "Pangenesis," which I will quote, because the disproof of it, which I thought would not be given, was not long in coming, and, with the more satisfactory theory of Weismann, led me entirely to change my opinion. I wrote (February 20, 1868): "I am reading Darwin's book ('Animals and Plants under Domestication'), and have read the 'Pangenesis' chapter first, for I could not wait. The hypothesis is *sublime* in its simplicity and the wonderful manner in which it explains the most mysterious of the phenomena of life. To me it is *satisfying* in the extreme. I feel I can never give it up, unless it be *positively* disproved, which is impossible, or replaced by one which better explains the facts, which is highly improbable. Darwin has here decidedly gone ahead of Spencer in generalization. I consider it the most wonderful thing he has given us, but it will not be generally appreciated."

This was written when I was fresh from the spell of this most ingenious hypothesis. Galton's experiments on blood transfusion with rabbits first staggered me, as it seemed to me to be the very disproof I had thought impossible. And later on, when Weismann adduced his views on the continuity of the germ-plasm, and the consequent non-heredity of acquired characters; and further, when he showed that the supposed transmission of such characters, which Darwin had accepted and which the hypothesis of pangenesis was constructed to account for, was not really proved by any evidence whatever;—I was compelled to discard Darwin's view in favour of that of Weismann, which is now almost everywhere accepted as being the most probable, as well as being the most in accordance with all the facts and phenomena of heredity.

Towards the end of the year Sir Charles sent me a number of interesting papers to read, and among them was a criticism of Darwin by G. H. Lewes. When writing to thank him for them I replied to this criticism as follows :—

"I have just been looking through Lewes. I think that in his great argument about the *luminous* and *electric* animals he completely fails to see their true bearing. He admits the fact that the organs producing *light* or *electricity* differ in *position* and *form* whenever the animals that bear them differ in general structure, while in their essential minute structure the (corresponding) organs closely resemble each other, however widely the animals may differ. But this is a necessary consequence of such organs being modifications of muscular tissue, which is almost identical in structure throughout the animal kingdom. If electrical and luminous organs were always identical in *form* and *position* as well as in structure, it would be a powerful argument in his favour; but as it is, I do not see that it proves anything but that the required special variation of an (almost) identical tissue occurs very rarely, and has still more rarely occurred at a time and under conditions which rendered its accumulation *useful* to the animal, in which case alone it would be selected and specialized so as to form a perfect electric or luminous organ.

"Again, to suppose that because one single organ of a simple kind may be produced independently of common descent, therefore a combination of hundreds of organs, many of them consisting of hundreds of parts, should all be brought by the action of similar causes to an identity of form, position, and function (in different animals), appears to me absolutely inconceivable. For instance, I cannot conceive any two species of vertebrata developed independently from distinct primal specks of jelly (protoplasm) through the millions of forms that must have intervened ; but I can conceive vertebrata and mollusca so developed *ab initio*. If this is all Lewes claims, Darwin will, I am sure, admit it. If he maintains a distinct origin for mammals, birds, and fishes, how does he deal with the identical forms of the embryos up to a certain stage, which is still that of a vertebrate animal ? But he never tells us what he does believe in detail, and it seems to me that his views are utterly groundless if he goes beyond the four or five primitive forms, which is all that Darwin claims as essential to his system.

"His notion of the mammals of Australia having possibly developed *ab initio* is too wild to be seriously refuted, and I think he gives it up in his last part, which you have not sent me. What of the fossil marsupials in Europe? The identity of embryos? The identity of bone, tooth, hair, and nail structure ? The identical general arrangement of vertebræ, limbs, muscles, cranium, brains, lungs, tongue, stomach, and intestines—all to have been developed independently through, or out of, forms as low as medusæ and actiniæ by *general similarity of conditions!* It is too absurd!"

The subject on which Sir Charles Lyell and myself had the longest discussions was that of the effects of the glacial period on the distribution of plants and animals, and on the origin of lake basins. On the former question he was disposed to accept my views in opposition to those of Darwin, as shown by the following letter of February 2, 1869:—

"Dear Wallace,
      "The more I think over what you said yesterday about the geographical distribution of tropical animals and plants in the glacial period, the more I am convinced that Darwin's difficulty may be removed by duly attending to the effects of the absence of cold. The intensity of heat, whether in the sea or in the air, is not so important, as you remarked, as uniformity of temperature."

He then goes on to give illustrations of this, and urges that there are no recent deposits in or near the tropics containing fossil remains proving any change of fauna and flora such as Darwin had advocated. He then continues—

"I know of no evidence of this kind, and I don't think that Darwin has given any time or thought to Croll's eccentricity theory, or to my chapters upon it, and I wish much that he could see your review[1] before he came out with this new edition (the fifth) of 'The Origin;' for I am afraid that he will make too much of the supposed corroboration afforded by the imaginary warmth of the southern hemisphere, and of the equally hypothetical expulsion of tropical forms from the equatorial zone north of the line."

In the sixth edition of "The Origin," published three years later, Darwin still held to his views of the extreme severity of the glacial epoch influencing even the equatorial zone, and explaining the transmission of so many northern types of plants and insects to the southern hemisphere, as shown by the following passage :—" From the foregoing facts, namely, the presence of temperate forms on the highlands across the whole of equatorial Africa, and along the peninsula of India, to Ceylon and the Malay Archipelago, and in a less marked manner across the wide expanse of tropical South America, it appears almost certain that at some former period, no doubt during the most severe part of a glacial period, the lowlands of these continents were everywhere tenanted under the equator by a considerable number of temperate forms. At this period the equatorial climate at the level of the sea was probably about the same with that now experienced at the height of from five to six thousand feet under the same latitude, or perhaps even rather cooler " (p. 338).

In my " Island Life " I have discussed at some length all these facts, and many others which Darwin did not take into consideration, and have explained them on the theory

[1] My *Quarterly Review* article on "Geological Climates and the Origin of Species," a proof of which Sir Charles had seen.

that the glacial epoch had no effect whatever in lowering the temperature of equatorial plains, while it might easily lower the snow-line on even equatorial mountains. Those interested in this question, after reading Darwin's exposition of his views, should read the twenty-third chapter of my "Island Life," the facts and arguments in which, so far as I am aware, have never been controverted. Darwin himself, however, never accepted them.

On the question of the ice-origin of Alpine lakes I had much correspondence with Sir Charles, but I could never get him to accept my extreme views. In March, 1869, I received from him a letter of thirteen pages, and another of thirty pages, on this and allied questions, setting forth the reasons why he rejected ice action as having ground out the larger lakes, much as he states them in the fourth edition of "The Antiquity of Man." At page 361 he says that "the gravest objection to the hypothesis of glacial erosion on a stupendous scale is afforded by the entire absence of lakes of the first magnitude in several areas where they ought to exist, if the enormous glaciers which once occupied those spaces had possessed the deep excavating power ascribed to them." He then goes on to adduce numerous places where he thinks there ought to have been lakes on the glacier theory, which are the same as he adduced in letters to myself, and which I answered in each case, and sometimes at great length, by similar arguments to those I have adduced in vol. i. chap. v. of my "Studies, Scientific and Social." If any one who is interested in these questions, after considering Sir Charles Lyell's difficulties and objections in his "Antiquity of Man," will read the above cnapter, giving special attention to the sections headed *The Conditions that favour the Production of Lakes by Ice-erosion*, and the following section on *Objections of Modern Writers considered*, I think he will, if he had paid any attention to the phenomena in glaciated regions, admit that I show the theory of ice-erosion to be the only one that explains all the facts.

During the same year (1869) I find passages of interest

in my letters on quite different subjects, some of which I wrote upon at a much later period. On February 25, in a letter about the Bethnal Green Museum, I added, "Have you seen the curious paper in the *Atlantic Monthly* of February on 'The Birth of the Solar System'? It contains a new nebular hypothesis, quite distinct from the old one. The writer maintains that all we *know* about the formation of the planets is that they are slowly *increasing* in bulk from the falling in of meteoritic bodies. He maintains, therefore, that this is the *origin* of all planets and suns, space being full of cold meteoric dust, heat being produced by its agglomeration. Thus all small bodies in space are cold, all large ones hot; the earth is therefore getting hotter instead of colder, and early geological action was less violent than it is now. Is not that turning the tables on the convulsionists?

"Many of the author's statements are, I think, inaccurate, but the view of the formation of the solar system by the agglomeration of *cold* dust instead of *hot* vapour seems to have some show of probability."

This hypothesis was new to me, and I had quite forgotten all about it when I met with it in Sir Norman Lockyer's works while writing my "Wonderful Century," and definitely adopted it as more accordant with facts and more intelligible than Laplace's theory of the intensely heated solar nebula.

On April 28, after referring to Darwin's regret at the concluding passages of my *Quarterly Review* article on "Man," which he "would have thought written by some one else," I add the following summary of my position, perhaps more simply and forcibly stated than in any of my published works :—

"It seems to me that if we once admit the necessity of *any* action beyond 'natural selection' in developing man, we have no reason whatever for confining that agency to his brain. On the mere doctrine of chances it seems to me in the highest degree improbable that so many points of structure, all tending to favour his mental development, should concur in man alone of all animals. If the erect posture, the freedom of the anterior limbs from purposes of

locomotion, the powerful and opposable thumb, the naked
skin, the great symmetery of form, the perfect organs of
speech, and, in his mental faculties, calculation of numbers,
ideas of symmetry, of justice, of abstract reasoning, of the
infinite, of a future state, and many others, cannot be shown
to be each and all *useful* to man in the very lowest state of
civilization—how are we to explain their co-existence in him
alone of the whole series of organized beings? Years ago I
saw in London a bushman boy and girl, and the girl played
very nicely on the piano. Blind Tom, the half-idiot negro
slave, had a 'musical ear' or brain, superior, perhaps, to that
of the best living musicians. Unless Darwin can show me
how this latent musical faculty in the lowest races can have
been developed through *survival* of the fittest, can have been
of *use* to the individual or the race, so as to cause those who
possessed it in a fractionally greater degree than others to
win in the struggle for life, I must believe that some other
power (than natural selection) caused that development. It
seems to me that the *onus probandi* will lie with those who
maintain that man, body and mind, could have been developed
from a quadrumanous animal by 'natural selection.'"

In a letter to Darwin, written a week later and printed in
the "Life, Letters, and Journals," Sir Charles quotes the preceding argument entire, and goes on to express his general
agreement with it.

He then refers to the glacial-lake theory as follows:—
"As to the scooping out of lake basins by glaciers, I have
had a long, amicable, but controversial correspondence with
Wallace on that subject, and I cannot get over (as, indeed, I
have admitted in print) an intimate connection between the
number of lakes of modern date and the glaciation of the
regions containing them. But as we do not know how ice
can scoop out Lago Maggiore to a depth of 2600 feet, of
which all but 600 is below the level of the sea, getting rid of
the rock supposed to be worn away as if it was salt that had
melted, I feel that it is a dangerous causation to admit in
explanation of every cavity which we have to account for,
including Lake Superior."

This passage shows, I think, that he was somewhat staggered by my arguments, but could not take so great a step without further consideration and examination of the evidence. I feel sure, therefore, that if he had had before him the numerous facts since made known, of erratic blocks carried by the ice to heights far above their place of origin in North America, and even in our own islands, as described at p. 75 and p. 90 of my "Studies" (vol. i.), with evidence of such action now occurring in Greenland (p. 91), of the Moel Tryfan beds having been forced up by the glacier that filled the Irish sea, he would have seen, I feel sure, that his objections were all answered by actual phenomena, and that the gradual erosion of Lago Maggiore was far within the powers of such enormous accumulations of ice as must have existed over its site.

The following letter I quote entire, because it calls attention to a very original but much neglected book which, though probably not wholly sound in its theoretical basis, contains suggestions which may help towards the solution of a still unsolved problem :—

                                              " May 3, 1871.

"DEAR SIR CHARLES,
        " I have just been reading a book which has struck me amazingly, but which has been somewhat pooh-pooh'd by the critics, and which therefore you may not have thought worth looking at. It is W. Mattieu Williams' 'Fuel of the Sun.' Whether the theory is true or false, the book is the work of a man of original genius. Its originality is so startling that I have found it to require reading twice to take it in thoroughly ; and it is so different from all modern theories of the sun that I can quite see why such a work by an outsider should not have received due attention.   If sound, it completely solves the problem of the perpetuity of the sun's heat, and gives geologists and Darwinians any amount of time they require. It seems to be reasonable, it is beautifully worked out, it is quite intelligible, and till

shown to be a fallacy I hold by its main doctrine. I hope
you will read it, and, if you see no fallacy in it, get Sir John
Herschell to read it and tell us if there is a positive fallacy
which destroys its whole value or no."

Some weeks later Sir Charles thanked me for recommend-
ing the book to him, which he had read with great pleasure,
adding, " It is as interesting as any novel I ever read." The
fundamental idea of the book is that the sun in its motion
through space comes into contact with an excessively diffused
space-atmosphere, which it collects and condenses by its
gravitative force, thus forming the sun's photosphere. Then,
on cooling, the outer portion of this gaseous envelope is
left behind or expelled, so that the mass of the sun does
not increase. The value of the explanation will of course
depend upon whether this latter part of the theory, which
the author explains at considerable length, is dynamically
possible. In view of modern discoveries as to the nature
of matter, it might be well for some competent physicist
to re-examine this work, which is largely founded on the
author's own observations and experiments as a metallurgical
chemist.

In the latter part of 1872 I was assisting Sir Charles by
reading over the completed MSS. and afterwards the proofs
of Part III. of " The Antiquity of Man," dealing with " The
Origin of Species as bearing on Man's Place in Nature." In
one of the letters I wrote I made a suggestion (which he did
not adopt, nor did I expect him to do so), but which I will
here give as it is a subject on which I wrote afterwards, and
which I still consider to be of very great importance.
Readers of the " Antiquity " will see that part of his own
MSS. has been omitted.

" November 10, 1872.

" DEAR SIR CHARLES,

"I have read the MSS. with very great interest.
Two points of importance are, Milton's advocacy of scientific
as against classical education (which I should think would be

new to most persons), and freedom of thought as essential to intellectual progress. The latter point (occupying pp. 13–23 of your MSS.) is of such immense importance, and your opinion on it, clearly expressed, would have so much weight, that I should much wish it to be developed in a little more detail, though I cannot see how it can possibly be got into ' The Antiquity of Man.' The points that may be more fully treated seem to me to be—1st, to show in a little more detail that there *was* such practical freedom of thought in Greek schools and academies; 2nd, to put forward strongly, the fact that, ever since the establishment of Christianity, the education of Europe has been wholly in the hands of men bound down by penalties to fixed dogmas, that philosophy and science have been taught largely under the same in-fluences, and that, even at the present day and among the most civilized nations, it causes the greater part of the intel-lectual strength of the world to be wasted in endeavours to reconcile old dogmas with modern thought, while no step in advance can be made without the fiercest opposition by those whose *vested interests* are bound up in these dogmas.

"3rd. I should like to see (though, perhaps, you are not prepared to do it) a strong passage following up your con-cluding words, pointing out that it is a disgrace to civilization and a crime against posterity, that the great mass of the instructors of our youth should still be those who are fettered by creeds and dogmas which they are under a penalty to teach, and urging that it is the very first duty of the Govern-ment of a free people to take away all such restraints from the national church, and so allow the national teachers to represent the most advanced thought, the highest intellect, and the purest morality the age can produce. It is equally the duty of the State to disqualify as teachers, in all schools and colleges under its control, those whose interests are in any way bound up with the promulgation of fixed creeds or dogmas of whatever nature.

"I should be exceedingly glad if you could do something of this kind, because I look with great alarm on the move-ment for the disestablishment of the Church of England, a step which I fear would retard freedom of thought for centuries. This would inevitably be its effect if any similar proportion of its revenues, as in the case of the Irish Church, was handed over to the disestablished Church of England, which would then still retain much of its prestige and respectability, would have enormous wealth, which might be indefinitely increased by further private endowments, and might have a ruling episcopacy with absolute power, who would keep up creeds and dogmas, and repress all freedom of thought and action, and thus do irreparable injury to the nation. Besides this, we should lose a grand organization for education and a splendid endowment which might confer incalculable benefits on society if only its recipients were rendered absolutely free. What might have been the result if, during the last hundred years, the twenty thousand sermons which are preached every Sunday in Great Britain, instead of being rigidly confined to one monotonous subject, had been true lessons in civilization, morality, the laws of health, and other useful (or elevating) knowledge, and if the teachers had been the high class of men who, if unfettered, would have gladly entered this the noblest of professions?

"I so much fear that Miall's premature agitation may force some future Government to (carry) disestablishment on any terms, that I think it of the greatest importance to point out what may be lost by such a step."

The passages referred to in the beginning of the above letter were both omitted by Sir Charles, being thought, apparently, rather out of place. The book did not appear till the following summer, and from that time till his death he undertook no more literary work. My remarks on the question of disestablishment, however, seemed to me so important that I elaborated my ideas into an article, which appeared in *Macmillan's Magazine* (April, 1873), and is reprinted in the second volume of my "Studies," under the title, "Disestablishment and Disendowment : with a proposal for a really National Church of England." In putting this suggestion before the country I have done what was in my power to indicate a method by which, when the time for legislation comes, the present institution may be replaced by one that will be a great educational and moral power in every part of our land.

I do not remember when I first saw Sir Charles Lyell, but I probably met him at some of the evening meetings of the scientific societies. I first lunched with him in the summer of 1863, and then met, for the first time, Lady Lyell and Miss Arabella B. Buckley. Miss Buckley had become Sir Charles's private secretary early in that year, and she informs me that she remembers this visit because Lady Lyell gave her impressions of me afterwards—I am afraid not very favourable ones, as I was shy, awkward, and quite unused to good society. With Sir Charles I soon felt at home, owing to his refined and gentle manners, his fund of quiet humour, and his intense love and extensive knowledge of natural science. His great liberality of thought and wide general interests were also attractive to me ; and although when he had once arrived at a definite conclusion he held by it very tenaciously until a considerable body of well-ascertained facts could be adduced against it, yet he was always willing to listen to the arguments of his opponents, and to give them careful and repeated consideration. This was well shown in the time and trouble he gave to the discussion with myself as to the glacial origin of the larger alpine lake basins, writing me one letter of thirty pages on the subject. Considering his position as the greatest living authority on physical geology, it certainly showed remarkable open-mindedness that he should condescend to discuss the subject with such a mere amateur and tyro as I then was. The theory was, however, too new and too revolutionary for him to make up his mind at once, but he certainly was somewhat influenced by the facts and arguments I set before him, as shown by the expressions in his correspondence with Darwin, which I have quoted.

In the much vaster and more important problem of the development of man from the lower animals, though convinced of the general truth of Darwin's views, with which he had been generally acquainted for twenty years, he was yet loth to express himself definitely ; and Darwin himself was as much disappointed with his pronouncement in the recently published "Antiquity of Man," as he was with my rejection of the sufficiency of natural selection to explain the origin of man's mental and moral nature. Sir Charles Lyell's character is well exhibited in what he wrote Darwin soon after its publication (March 11, 1863). "I find myself, after reasoning through a whole chapter, in favour of man's coming from the animals, relapsing to my old views whenever I read again a few pages of the 'Principles,' or yearn for fossil types of intermediate grade. Truly, I ought to be charitable to Sedgwick and others. Hundreds who have bought my book in the hope that I should demolish heresy will be awfully confounded and disappointed. . . . What I am anxious to effect is to avoid positive inconsistencies in different parts of my book, owing probably to the old trains of thought, the old ruts, interfering with the new course. But you ought to be satisfied, as I shall bring hundreds towards you, who, if I treated the matter more dogmatically, would have rebelled. I have spoken out to the utmost extent of my tether, so far as my reason goes, and further than my imagination and sentiment can follow, which, I suppose, has caused occasional incongruities" ("Life of Sir Charles Lyell," vol. ii. p. 363). These passages well exhibit the difficulties with which the writer had to contend, and serve to explain that careful setting forth of opposing facts and arguments without stating any definite conclusion, which is felt to be unsatisfactory in some portions of his great works.

During the ten years 1863-72, I saw a good deal of Sir Charles. If he had any special subject on which he wished for information, he would sometimes walk across the park to St. Mark's Crescent for an hour's conversation ; at other times he would ask me to lunch with him, either to meet some interesting visitor or for friendly talk. After my marriage we occasionally dined with him or went to his evening receptions. These latter were very interesting, both because they were not overcrowded and on account of the number of scientific and other men of eminence to be met there. Among these were Professor Tyndall, Sir Charles Wheatstone, Sir Charles Bunbury, Mr. Lecky, and a great many others. The Duke of Argyll was frequently there, and although we criticized each other's theories rather strongly, he was always very friendly, and we generally had some minutes' conversation whenever I met him. Miss Buckley (now Mrs. Fisher) was a very constant guest, and would point out to me the various celebrities who happened to be present, and thus began a cordial friendship which has continued unbroken, and has been a mutual pleasure and advantage. I therefore look back upon my friendship with Sir Charles Lyell with unalloyed satisfaction as one of the most instructive and enjoyable episodes in my life-experience.

END OF VOL. I

Printed in the United States
By Bookmasters